Dead Sea

Crosscurrents

NEW STUDIES ON THE MIDDLE EAST

R. Kevin Lacey and Sari Nusseibeh
General Editors

Vol. 2

This book is a volume in a Peter Lang monograph series.
Every volume is peer reviewed and meets
the highest quality standards for content and production.

PETER LANG
New York • Bern • Frankfurt • Berlin
Brussels • Vienna • Oxford • Warsaw

Dead Sea

New Discoveries in the Cave of Letters

EDITED BY
Carl E. Savage, Philip Reeder,
Richard A. Freund, AND Harry M. Jol

PETER LANG
New York • Bern • Frankfurt • Berlin
Brussels • Vienna • Oxford • Warsaw

Library of Congress Cataloging-in-Publication Data

Names: Savage, Carl E., editor. | Reeder, Philip, editor.
Freund, Richard A., editor. | Jol, H. M. (Harry M.), editor.
Title: Dead Sea: new discoveries in the Cave of Letters /
edited by Carl E. Savage, Phil Reeder, Richard A. Freund, Harry M. Jol.
Description: New York: Peter Lang, 2018.
Series: Crosscurrents: new studies on the Middle East; vol. 2
ISSN 2381-2443 (print) | ISSN 2381-2451 (online)
Includes bibliographical references and index.
Identifiers: LCCN 2016015058 | ISBN 978-1-4331-3427-2 (hardcover: alk. paper)
ISBN 978-1-4539-1884-5 (ebook pdf) | ISBN 978-1-4331-4208-6 (epub)
ISBN 978-1-4331-4212-3 (mobi)
Subjects: LCSH: Letters, Cave of the (Israel)
Excavations (Archaeology)—Judaea, Wilderness of.
Classification: LCC DS110.L37 D43 | DDC 296.1/55—dc23
LC record available at https://lccn.loc.gov/2016015058
DOI 10.3726/978-1-4539-1884-5

Bibliographic information published by **Die Deutsche Nationalbibliothek**.
Die Deutsche Nationalbibliothek lists this publication in the "Deutsche Nationalbibliografie"; detailed bibliographic data are available on the Internet at http://dnb.d-nb.de/.

The paper in this book meets the guidelines for permanence and durability of the Committee on Production Guidelines for Book Longevity of the Council of Library Resources.

29 Broadway, 18th floor, New York, NY 10006
www.peterlang.com

Printed in Germany

Table of Contents

Figures

Abbreviations

1D	One-Dimensional
2D	Two-Dimensional
3D	Three-Dimensional
AGC	Automatic Gain Control
AMS	Accelerator Mass Spectrometry
Ar/Ar	Argon/Argon Dating
BCE	Before Common Era
BP	Before Present
C-14	Carbon 14
CE	Common Era
CM (cm)	Centimeters
CMP	Common Mid-Point
COL	Cave of Letters
DC	Direct Current
EM	Electromagnetic
ERT	Electrical Resistivity Tomography
GPR	Ground Penetrating Radar
HZ	Hertz
ICP	Inductively Coupled Plasma
INAA	Instrumental Neutron Activation Analysis
K/Ar	Potassium/Argon Dating
KG (kg)	Kilograms

KM (km)	Kilometers
KM^2	Square Kilometers
LSC	Liquid Scintillation Counter
M (m)	Meters
Ma	Million Years Ago
MBGS	Millimeters Below Ground Surface
MHz	Megahertz
MM (mm)	Milimeters
MPa	Megapascal—a unit for measuring pressure
MSL	Mean Sea Level
Na/Ca	Sodium to Calcium Ratio
NS (ns)	Nanosecond
OHM-M	Ohm-Meters
PS (ps)	Per second
PSI	Pounds per Square Inch
RPM	Revolutions per Minute
Sr/Ca	Strontium-to-Calcium Ratio
TM	Trademark
USA	United States of America
U-Th	Uranium Thorium Dating
UXO	Unexploded Ordinance
V	Volts
vPDB	Vienna Pee Dee Belemnite Standard for Marine Carbonate Sediments
XRD	X-Ray Diffraction

Acknowledgements

This volume is based on the hard work and dedication of numerous researchers who share a common interest in Israel, and have discipline-specific interests in the areas of geology, geography, religion, history and archaeology. We were inspired by the pioneering research of Yigael Yadin and Yohanan Aharoni who were the first to explore and research the Cave of Letters. The chapters contained in this volume represent, in a sense, a continuation of their research. Space cannot express the extent of heartfelt thanks to the many individuals and organizations for their contributions to this multifaceted, interdisciplinary project, but we especially thank John and Carol Merrill for the funding and support for 2000 and 2001 excavations that was kindly provided by through Hershel Shanks [Biblical Archaeology Society]. We also wish to thank the President of the University of Hartford, Walter Harrison, and the Chair of the Regents, Arnold C. Greenberg, who also provided funding and administrative support. Our participating universities, University of Hartford, University of Nebraska at Omaha, University of Wisconsin-Eau Claire, University of South Florida, and Drew University, are also to be thanked for their support.

Not all of the scholars and students who participated in the research at the Cave of Letters published chapters or addendum in this volume, but their contributions to this project were enormous. None has been more important than Dr. Rami Arav. We wish to thank Rami for his years of service to archaeology in Israel and to the Cave of Letters Project. His vision of archaeology has benefited our project, and Israeli archaeology in ways that are too numerous to name. We wish to thank those other members of our excavation team as well: Dr. Gordon Brubacher from Doane

College, and his students Laura Kraus, Denise Kraus and Karen Weatherman, classical historian Dr. Mark Smith of Albertson College, environmental scientist and cave explorer Jim Webster from the University of Georgia, biblical scholar John Greene from Michigan State University, physician Dr. Gordon Moshman from Omaha, Nebraska, biblical scholar Dr. Larry Schiffman from New York University, Matt Schuett, who was our geophysics technical assistant, and Chris Morton, University of Wisconsin-Eau Claire for website development and field support recording. There were other staff who, though not involved in the excavations, aided the project in many ways: Dreanna Hadash, artist, Charleen Green, restoration specialist, Ophir Palmon, Andrew Liakos, and Christine Dalenta, photographers. We thank the Weizmann Institute's Professor Steve Weiner and his student Michal Kaufman, for research on the ancient DNA of the bones discovered during our excavations in the Niche of Skulls.

We also need to thank the support team of Israeli climbers who ensured that we could safely enter and exit the cave: Chaggai Rosenbluh, and his supporting staff of Noam Metzner, and Oded Danon.

We also acknowledge the role of Wendi Chiarbos, Coordinator of the Bethsaida Excavations Project, who helped coordinate the preparations for the excavations in the Cave of Letters and the UNO TV television documentary on the Cave of Letters. Additionally, we thank Ms. Susan Gottlieb, the Administrative Assistant at the Maurice Greenberg Center for Judaic Studies, whose assistance in coordinating other parts of the project was invaluable.

Many companies were instrumental in helping us with our equipment needs. The Onset Computer Corporation provided the temperature and relative humidity data loggers that were used in the micro-climate studies that were completed in the cave. We also greatly appreciate the assistance provided by Robert Randhare, from Olympus Industrial who provided us with state-of-the-art industrial fiberscope equipment.

The remaining acknowledgements are for the people that do not receive mention in the course of the book. People like Amir Drori, former Director General of the Antiquities Authority of Israel who encouraged us to return to the Cave and Director General of the Israel Antiquities Authority, Shuka Dorfman, who encouraged this work because he has the vision to realize that the technological advances that we used in the Cave of Letters could be used in many other archaeological contexts. We also acknowledge the input of Uzi Dahari, Hava Katz, Gidon Avni, Baruch Brandl, David Amit, Peninah Shor, and Donald Ariel, all of the Israel Antiquity Authority for a variety of research, permits, insights and consultations that they provided in the course of the excavations. Likewise, we thank the different officials of the Parks and Nature Reserves Authority of Israel and its Director of Science in 2000, Dr. Avi Pervolotsky, as well as the head of Archaeology, Dr. Zvika

Zuk, and all of the Parks and Nature Reserves officials who helped us throughout the project.

A few final words must be added to the acknowledgements about those people who worked with us on aspects of the research, including Professor David Ussishkin, Menachem Mor, Hannah Cotton, Baruch Safrai and Pinchas Porat. Baruch Safrai is the person who really encouraged Rami Arav and Richard Freund to do the excavations that we completed between 1999 and 2001. Pinchas Porat, who worked on the Bethsaida Excavations Project for years, was also one of the great motivators toward the re-excavations in the Cave of Letters.

Finally, we thank our families. Often times our research takes us far from home, and without their support, we would not have accomplished what we set out to do. To our wives, Keen, Eliane, Carleen and Charleen; thank you.

Philip Reeder, Richard A. Freund, Harry M. Jol and Carl E. Savage

Foreword

BARUCH A. LEVINE
Emeritus, New York University

In 1963 Yigael Yadin simultaneously published English and Hebrew versions of a volume entitled *The Finds from the Bar-Kokhba Period in the Cave of Letters.* (Jerusalem: Israel Exploration Society). It presented a detailed review of the expeditions at Naḥal Ḥever, the site of the Cave of Letters (as Yadin named it), conducted in 1960–61, including a mapping of the interior of the Cave, with its burials, as then known, and of the nearby Roman camp. Yadin and his staff registered all of the material finds, including objects made of metal, glass, ceramics, stone and leather; also jewelry, firewood and objects of wood, and leather. No organized archeological activity occurred at the Cave of Letters between 1963–1999, except for occasional testing and surveying. Presenting a collection of diverse studies, the present volume reports on, and discusses the ramifications of the recent expeditions at this site in 1999–2001, conducted by the four editors, Philip Reeder, Richard Freund, Harry Jol and Carl E. Savage, and their staff of experts, who applied the most advanced technologies in the process.

"BOTH SIDES OF THE AISLE"

I was pleased by the invitation to write the Foreword to the present volume, particularly because I come from across the aisle! Ada Yardeni and I collaborated in editing the Semitic papyri found in the Cave of Letters for *The Documents from the Bar Kokhba Period in the Cave of Letters: Hebrew, Aramaic, and Nabataean Aramaic Papyri* (Judean Desert Studies III, 2002). Jonas C. Greenfield, collaborated with

Ada Yardeni on the Naḥal Ḥever papyri before his untimely death in 1995, which yielded editions of two major papyri, and he wrote numerous studies relating to the Naḥal Ḥever papyri. That volume included the Bar Kokhba letters, with new editions of two Greek letters by Hannah Cotton. Over the decades, and spanning regrettable delays, many scholars have investigated the rich textual evidence, including numerous Greek papyri, from the Cave of Letters and from other sites in the Judean Desert, which produced hordes of quotidian, mostly legal documents. In 1989 Naphtali Lewis edited the Greek papyri from Naḥal Ḥever, in a volume entitled *The Documents from the Bar Kokhba Period in the Cave of Letters: Greek Papyri* (Judaean Desert Studies II; Jerusalem: Israel Exploration Society). In 2000, Ada Yardeni published a two-volume *Textbook of Aramaic, Hebrew, and Nabataean Texts from the Judaean Desert,* in Hebrew, with English summaries and translations of texts (Jerusalem: The Hebrew University). In that publication, she lists the volume of Naḥal Ḥever papyri, then forthcoming, as editio princeps. In *Aramaic, Hebrew and Greek Documentary Texts from Naḥal Ḥever and Other Sites* (Discoveries in the Judaean Desert XXVII; Oxford, 1997), Hannah Cotton edited (in many cases re-edited), numerous Greek Papyri from Naḥal Ḥever, and in the same volume, Ada Yardeni edited Semitic papyri from the Seiyal Collection, also originally from Naḥal Ḥever, which she had previously edited in a Hebrew volume. All of these publications provide extensive bibliographies.

The first, full volume devoted to Judean Desert papyri and other textual finds appeared in 1961 under the title *Les Grottes de Murabba'ât* (Discoveries in the Judaean Desert II; Oxford). It included an archeological introduction by Roland de Vaux, with the textual finds edited by J.T. Milik and P. Benoit. From that point on, archeology and exegesis have pretty well parted company. It was certainly an achievement on Yadin's part to produce the 1963 report in such short order, but the fact that his volume lacked critical editions of the papyri may have initiated a policy of separate publication that endures to this day. Now that the texts have been edited, and renewed archeological exploration has been accomplished and reported, it may be possible to chart a new course for Dead Sea studies, generally, and specifically for assessing the significance of Naḥal Ḥever. The objective would be to arrive at a synthesis of text and artifact, of physical environment and human history. A sizable historical literature on the Bar Kokhba revolt and the Dead Sea region has arisen over the years, which needs to be pursued further, and integrated with the textual and archeological evidence.

In the past half century, specialization has been the dominant trend in almost every branch of research, in response to greatly expanded discovery. The problem posed by specialization is that it inevitably narrows the line of vision while at the same time deepening it, and consequently has the effect of reducing scholarly capabilities in related fields. One wonders whether the multi-tasking scholars of yesteryear, the likes of W.F. Albright, Benjamin Mazar, Roland de Vaux, and Yigael

Yadin, himself, could now master, or even control all of the requisite disciplines. The only way to stay afloat is through team effort, which is highly developed in archeological circles, as the present publication demonstrates, but is still lagging in the fields of textual and historical research.

The Cave

It is reassuring that many of the archeologists who contributed to the present volume have expressed serious interest in historical questions, even though they were reporting primarily on advanced geophysical technologies. As explained by Richard Freund, and several other contributors, the recent excavations bore an internal agenda. Among other things, they were intended to probe Freund's theory that there was 1st century C.E. habitation of the Cave. Freund concluded that Jews hid inside the Cave from the Romans at the time of the First Revolt, when the Jerusalem Temple was destroyed (66–70 C.E.), and during the period between these events and the Second Revolt (132–135 C.E.).

Two overviews of the recent excavations set the stage for the volume as a whole, one by Richard Freund and the other by John F. Shroder. Both summarize the wide variety of finds in the Cave, as well as the exploration of the cave complex as a whole, and address the overall significance of the expeditions of 1999–2001. Freund ("Acknowledgements and Summaries, 1999–2001") approaches the material from the perspective of his finding that there were four habitation levels, not two, and argumentation to that effect accordingly informs his comprehensive survey. Shroder ("Overview of New Geoarchaeological Discoveries in the Cave of Letters, Israel") is more detached, allowing for the possibility that there was a 1st century B.C.E. habitation level, but treating the evidence more independently of that question.

In effect, the debate over this historical issue threads through many of the studies. Thus, Carl E. Savage ("A First Century Occupation in the Cave of Letters? Results from the B/C Passage") contributes a careful assessment of habitation layers, with reference to coins and ceramics, and takes into account the fact that stratification within a cave complex poses different problems from those encountered in excavating a tell. Most interesting is Savage's comment in this regard: "If the coins and pottery had been found together outside of the Cave of Letters context, one might be able to speculate that a late first century date for the collection of finds would be plausible given the individual temporal markers for most of the pottery finds."

Making no claim to archeological competence, it is my sense that Freund's periodization makes sense in historical and sociological terms. I hasten to emphasize that Freund's position is proactive, and serves to deepen the discussion in significant ways. What we require now is a detailed history of Jewish habitation in the general area of Naḥal Ḥever and the Dead Sea, going back to the Maccabean period, and further. The likelihood that Jews took advantage of this cave complex for refuge at the time of the First Revolt depends to a considerable degree on patterns of area settlement at various periods, not only on evidence provided by the Cave, itself. It is logical to propose that, under similar stress, earlier Jews would have availed themselves of such an extensive cave complex as refuge just as the Jews of the Bar Kokhba period had done. Thus, Harry Jol, in his methodic, and clearly formulated study "Ground Penetrating Radar Subsurface Imaging within the Cave of Letters" states the following:

> Although GPR (= Ground Penetrating Radar) could locate the Bar Kokhba floor, the question remained whether the cave's lower stratigraphy might indicate other habitation levels beneath this layer. Since other caves in the area have had artifacts associated with the Jewish Revolt of the first century CE to as far back as the Chalcolithic period, the ability to identify these different archaeological levels (horizons) is an important consideration for archaeologists and historians.

In my view, this is the most succinct formulation of the challenge posed by Freund's hypothesis. Similar, technologically speaking, is Paul Bauman's contribution "Electrical Imaging and Metal Detection in the Cave of Letters." This project utilized a technique known as ERT (= Electrical Resistivity Tomography), which affords an additional method of control alongside GPR, in Harry Jol's report. Once again, the objective was to identify occupation levels and to distinguish them from rubble that fell from roof collapses. This can be accomplished, as Bauman explains, by geophysical techniques, but these same techniques cannot assign dates to such features. Nonetheless, Bauman fully accepts the conclusion that the so-called "Bar Kokhba floor" was populated when the documents found in the Cave tell us it was.

Generally speaking, all of the contributors seemed to recognize that as exacting as technological instruments and mechanisms may be, interpretation of the data they provide must be analyzed with the application of discrete judgment. I call attention to several cases in point. Amos Frumkin contributes a chapter on the geology and formation of the cave 10–7 million years ago. His report is highly technical, and deals with environmental processes utterly remote from literate culture. And yet, in summarizing the importance of the cave, he calls attention to the fact that "it contained the most important finds of the Bar Kokhba revolt." He further notes that hundreds of other caves have been located in the region, some of which yielded documents and artifacts traceable to the Bar Kokhba period. Shachar Argov contributed a chapter entitled: "Rock Breaking at the Cave of Letters" He notes that subsequent to the 2nd century C.E. "earthquakes caused ceiling rock-falls to bury artifacts hidden by the cave's dwellers." He proceeds to describe in detail the tools used to split large rocks so that they could be removed without upsetting the interior formation of the cave.

Philip Reeder's contribution: "Microclimatic Variations and Patterns of Cave Use at Cave of Letters, Israel" testifies to the cultural importance of climate (including "micro-climate") as a factor in the preservation of ancient documents. As Reeder explains, it is the exceptional aridity of the Dead Sea region that accounts for its importance as a new venue of papyrological research outside ancient Egypt. Reeder's measurements deal primarily with the question of "how human activities affect cave climate," and he concludes that "the air outside the cave is usually less moist than the air in the cave." Accordingly, six research questions relevant to air-flow and human presence in the cave were posed, and all of them answered after various projects involving the installation of monitors had been completed. Question number 6 is particularly suggestive: "will airflow patterns in the cave indicate patterns of human utilization/?" This was answered by utilizing smoke tests, especially in the "hearth area," as we would expect. Issues pertaining to climate hold a particular attraction for me. In 2003, a volume was published by E.J. Brill bearing the title: *Semitic Papyrology in Context: A 'Climate' of Creativity*. It brought together papers presented at an in-house conference held at New York University in 2000. The title was purposely allusive, referring to the hordes of ancient papyri recently discovered in the Dead Sea region.

Our discussion of "the Cave" concludes with two studies. The former is Philip Reeder's other contribution, "Mapping the Cave of Letters." In a detailed review, Reeder shows how the new mapping of the cave not only enlarged the length of the Cave to 525 meters, 275 meters longer than previously mapped, but also provided new information on passageways and other structural features of the Cave. Digital technology was employed, which allows for feeding-in new data on base maps. Here is how Reeder introduces his overall report:

> The COL (=Cave of Letters) is important because it constitutes one of the few other sites concerning the Bar Kokhba revolt. Because so little is known about this war in which both sides are suspected to have lost heavily, archaeology is one of the ways to increase the historical record.

THE ARTIFACTS

Having discussed studies of the Cave, itself, let us move to the artifacts discovered within it. These can be outlined briefly: Richard Freund presents further argumentation for the existence of a 1st century habitation level based on the evidence of an incense shovel, comparable to one unearthed at Bethsaida. His discussion, which is a model of artefactual analysis, proceeding from the specific to the general, makes the case that it would be misguided to rule out an earlier habitation level at Naḥal Ḥever, although the evidence for it is limited. Freund reviews the inventory of incense shovels previously discovered by Yadin, as well as those found elsewhere, and discusses their afterlife in synagogue iconography. Clearly, this was an important artifact, in terms of both utility and symbolism, and Yadin's treatment of the issue of its historical significance had been open to criticism. Jan Gunneweg reviews the application of "Instrumental Neutron Activation Analysis" to selected pottery sherds and other clay products, once again concluding that an earlier level is possible, but that this particular technique could not be expected to confirm it. Here is what he has to say:

> We are, therefore, inclined to conclude that more research and analyses have to be performed on local clays from the environment of the cave, itself, as well as of pottery unearthed in Engedi and/or Masada before one can claim with some certainty that the Cave of Letters' ceramics are from before 70 AD.

Nili Liphschitz presents her findings from subjecting wood, mostly branches found in the cave to radiocarbon testing. The results indicated that some of the branches were datable to the 1st century C.E., while others were later than the Bar Kokhba revolt. Whereas it can be determined that the wood was brought in from nearby, and that some of it was "old wood", it remains uncertain as to whether the evidence is sufficient to confirm an earlier habitation level. The contribution by Fred Strickert ("The Meaning of Bar Kokhba Coins and their Distribution") has direct historical relevance, as we would only expect. Four of the coins date from before the Second Revolt, and of special interest is a small coin dated to the second year of the First Revolt (68), which, as Strickert notes, "may well signal occupation at that period." A major issue raised by the assemblage of coins is the geographical extent of the Bar Kokhba revolt, including the question of Jerusalem's status. Strickert takes us through the evidence from sites all over the map as he summarizes the debate over whether or not the Bar Kokhba rebels ever seized Jerusalem or parts of the city. Ultimately, he seems to favor a more limited spread for the revolt than traditionally supposed, and at the very least, calls for revising accepted historical conclusions on that score.

Orit Shamir's report ("Textiles, Threads, and Cordage from the Cave of Letters") epitomizes the blessings of aridity! Shamir includes unusual photographs of textile fragments, and concludes that the fabrics used were probably produced at Ein-Gedi, comparing them to other similar finds in the area. Among other things, the textiles give evidence of the traditional Jewish aversion to the inter-weaving of wool and linen prohibited in the Torah (Lev 19: 19, Deut 22: 11), reminding us once again that the communities of Bar Kosiba's network were observant of Jewish religious practice. Shamir follows accepted models of reporting on textile finds, and does so meticulously, showing that with little exception, the materials were of local origin. Finally Nicolae Roddy ("the Niche of Skulls in the Cave of Letters") sheds light on burial customs practiced by the inhabitants of the Cave, namely, the secondary burial of skulls and bones, involving the use of baskets and mats. Textiles were often used as wrapping for skeletal remains. As an internal agenda, Roddy enters into a broad discussion of Jewish burial customs and their theological ramifications during the Roman period, fully utilizing Rabbinic sources in the process. Addressing the question of proper burial under critical conditions, he notes that Rabbinic law showed no objection to burial on the battlefield. The unusual arrangements of burial in the cave may reflect the practice of bone collection as temporary inhumation, although that is questionable.

This leaves two contributions of interest, which differ from the major thrust of the volume. Walter Bouzard ("The Date of the Psalm Scroll from the cave of Letters") discusses the thorny documentary history of canonical scrolls, particularly the Psalm fragments found by Yadin at Naḥal Ḥever, including the Seiyal collection, which belongs to the same scroll. He proceeds to review the views of eminent scholars on the ductus of the fragments, over which there has been unending debate over the past half-century, or so. As has been observed so many times, the evidence is equivocal. Even if the script points to a 1st century C.E. date, which is far from certain, the possibility remains that later occupants of the Bar Kokhba era brought older texts into the Cave. Finally, we have the charming essay by Alberto Roitman ("Archaeology, Museology, and Identity"), who is Director of the Shrine of the Book, part of the Israel Museum. Roitman focuses on the sociology of display; fascinating fields of study that seeks to interpret attitudes reflected in the manner that antique objects are shown to the public. In the present case, attitudes toward Bar Kokhba and his revolt have shifted, as doubts over his virtue and leadership have intensified in the Israeli ethos.

Preface

PHILIP REEDER
(Co-Editor), Duquesne University

RICHARD A. FREUND
(Co-Editor), University of Hartford

HARRY M. JOL
(Co-Editor), University of Wisconsin-Eau Claire

CARL E. SAVAGE
(Lead-Editor), Drew University

The editors of this book feel, as we are sure many of the other authors do, that the research that resulted in the 14 chapters and three addenda presented in this book was some of the most rewarding and exciting research we have conducted in our careers. While riding across the Judean desert to the cave that first morning in the "desert buggy," we are sure that more than one person thought "what have I gotten myself into." Traversing the recesses of the cave by day, fighting the fierce desert winds, and sleeping in the desert by night during the 1999 research, we were by no means comfortable. But we felt lucky to be afforded the privilege to be part of a rare project, where researchers from many disciplines came together for a common cause. That cause was to try to uncover information about the mysterious and mystical figure Shimon Bar Kokhba, and what some of his followers were doing in a cave in the Judean Desert that later became known as the Cave of Letters. From the perilous decent, hundreds vertical meters down the side of the Naḥal Ḥever, to the horizontal traverse along the narrow ledges, to the ledge below the entrance, danger seemed at just an arm's length. The anticipation we felt as we climbed the 15-meter long series of aluminum ladders that were set up below entrance #1 for the first time was exhilarating. Would all the planning and logistics mesh? Would the hundreds of kilograms of equipment we brought to the cave from around the world function as intended? Would John and Carol Merrill, who funded a large portion of the project, feel that our findings were worthy of their support? Would we indeed uncover new information about the mystical figure Shimon Bar Kokhba, and would this information allow us to address some of the nagging questions about Bar Kokhba, his followers, and this period in history? Would the United States Public Broadcasting Systems investment in the production of a documentary based upon this research for its popular Nova series, titled "Ancient Refuge in the Holyland," filmed during the second expedition to the cave in 2000, be a success? With all of these questions circulating through our conscious and subconscious thoughts, we spent 26 days in the Cave of Letters over a three-year period (1999–2001).

Richard Freund's book, "Secrets of the Cave of Letters" provides an account of his perspective on what we found in the Cave of Letters and what it means. This volume, *Dead Sea: New Discoveries in the Cave of Letters*, broadens, yet focuses this body of research, by providing detailed accounts of the multi-faceted research design by the researchers most linked to the various segments of the project. In all, 18 different authors have contributed their unique perspectives to this volume.

The Cave of Letters Project actually had its origins in another project. The University of Nebraska at Omaha (UNO) formed the Bethsaida Excavations Project in 1991. Bethsaida is an ancient city located on a hill near the north shore of the Sea of Galilee. Nine of the authors associated with this volume also have ties to Bethsaida. In 1998, Richard A. Freund, then the Chair of the Department of Philosophy and Religion at UNO, began to lay the groundwork for our first expedition to the Cave of Letters in July 1999. This was followed by subsequent expeditions in July 2000 and 2001. These would be the first Israeli government sanctioned research efforts in the cave since those of famed Israeli archaeologist Yigael Yadin in the early 1960's.

Of the 18 different authors affiliated with *Dead Sea: New Discoveries in the Cave of Letters,* twelve traversed the ledges and ladders to enter the cave and collect their data. Additionally, samples were collected in the cave and then transported to various laboratories for radio carbon, instrumental neutron activation, dendrochronological and detailed textile analysis by researchers who have also contributed chapters to this volume. Also, part of the research team was students that contributed to both the field and laboratory work associated with this project. The Cave of Letters proved to be an outstanding natural laboratory, wherein students were provided with unparalleled experiential learning opportunities.

Our discussions regarding the creation of this edited volume began in 2004 while at a Biblical Archaeology and Literature conference (the Batchelder Conference) at the University of Nebraska at Omaha. We revisited the topic for two more years at this recurring conference in Omaha, and in 2006 we worked out a framework for what we wanted the volume to include. It was decided that it would be a cross-disciplinary undertaking that would include relevant works by geographers, geologists, archaeologists, historians, biblical scholars, geophysicists and scientists, who had in some way contributed to the project. Once the framework for the creation of *Dead Sea: New Discoveries in the Cave of Letters* was established, we vowed to finish up other projects we were working on, and to begin to solicit chapters and addenda for the volume starting in early 2007. Schedules and commitments slowed the process of preparing the initial, and what has eventually evolved into what we feel is now the "final" draft of the book. It has been over nine years in the making, but the final draft of the volume was submitted to Peter Lang Publishing in 2016.

This book should be of interest to anyone with an interest in the physical, cultural and historical aspects of the Dead Sea Region and ancient Israel. The chapters contained in this volume span a breadth of research topics, yet they are linked under the banner of adding all types of new information to the existing base of knowledge about the second Jewish Revolt, and the mystical figure Shimon Bar Kokhba. Chapter topics range from a general summary of the research conducted in the cave (Chapter 1); to the physical nature of the cave, surrounding landscape and some of the important discoveries (Chapters 2 and 3); to mapping the cave (Chapter 4); using geophysics in the cave to gather data (Chapters 5 and 6); C and the microclimate of the cave (Chapter 7). The archaeology of the cave is also the focus of five chapters (8 to 12). Chapter 8 re-examines the bronze artifacts found in Hall A by Yadin; while Chapter 9 discusses the discoveries from the Niche of Skulls; Chapter 10 discusses the artifacts discovered in the BC passage in the cave; Chapter 11 the textiles, thread and cordage found in the cave; and Chapter 12, the coins. Chapter 13 discusses the date of Psalms Scroll that Yadin discovered in the cave in 1960, and Chapter 14 discusses the materials from the Cave of Letters that are displayed in the Shrine of the Book at the Israel Museum. Additionally, there are three shorter chapters in the book, which we chose to call addenda, that deal with (1) the dendrochronology of wood found in Hall B of the cave near what was surmised to be a cooking hearth (Addendum 1), (2) the Instrumental Neutron Activation Analysis of pottery shards from the cave (Addendum 2), and (3) the techniques used to break rocks in the cave, so that potential artifacts under the rocks could be accessed (Addendum 3).

We are excited that all our hard work in the cave, laboratory, and our offices has come to fruition and that *Dead Sea: New Discoveries in the Cave of Letters* is completed and in print. We feel that we have been able to compile an impressive assemblage of researchers, and writings that tell the story of our endeavors, and more importantly, shed a bit more light on Shimon Bar Kokhba, and what some of his followers were doing in the Cave of Letters. We hope you enjoy our book.

CHAPTER ONE

THE 1999–2001 Acknowledgements AND Summary OF Research AT THE Cave OF Letters

RICHARD A. FREUND
The Maurice Greenburg Center for Judaic Studies,
University of Hartford, West Hartford, CT 06117, USA

The three seasons of excavations at the Cave of Letters revealed four levels of occupation in the Cave of Letters. Yigael Yadin's[1] expedition held that there were really only two levels of occupation of the Cave of Letters—Bar Kokhba era and Chalcolithic period. I have added Level 1, the modern period, since all three of our expeditions together with all of the (illicit) visits to the cave have changed the cave's environment. The four levels present from the understanding of the evidence from our excavations are:

Level 1—contains all modern material discovered in the Cave but which is from the period after Bar Kokhba to the present (135–2002)
Level 2—the Second Revolt period (132–135 CE)
Level 3—the First Revolt period and aftermath (66 CE–132 CE)
Level 4—Chalcolithic period (4,500–3,200 BCE)

Level 1: Although Level 4 is clearly represented in the Cave of Letters, we will not comment on any developments in our understanding of this level of occupation except to confirm that both Yadin's excavations and our own had remains from this period. It is clear to all us who worked in the cave starting in 1999 that illegal/illicit (unlicensed) excavations have been going on in the Cave of Letters since 1961. Bedouin visits and pilfering have been going on for the past 50 years. We have found that some of these illegal excavations must have used generators and lighting since new wiring for some lighting was found throughout the three halls. In addition, cigarette boxes and newspapers from different periods, coins and personal items show telltale signs of investigations. It is clear from the recovery of the largely biblical textual finds earlier assigned to other caves and now reassigned to the Cave of Letters that we include ourselves, the excavations of Yadin and Yohanan Aharoni in level #1 since we too have changed the cave. Most of the illegal excavations include individuals between the excavations of Aharoni and Yadin who entered the cave in search of scrolls. The Department of Antiquities of the fledgling State of Israel began looking for scrolls after the discovery of the first Dead Sea Scrolls in caves around Qumran (then in Jordan) in 1947. There were some biblical texts that were brought in by Bedouin in the 1950s from what they said were caves in *Nahal Tzeelim* that actually came from the Cave of Letters. These texts were finally retraced back to the Cave of Letters after the painstaking efforts of some remarkable researchers. We know that the list of visitors included local amateur climbers, some cave researchers and geologists, Bedouins, treasure hunters and thrill seekers who saw the challenge of the cave access and the probability that it still contained materials important enough to risk going there without the benefit of the specialized equipment and proper licensing we were required to use. We have spoken to some of these people. We also spoke to individuals who participated in the 1950s Aharoni excavations (Baruch Safrai and others) and the 1960s Yadin Excavations, (Pinchas Porat and others). Their oral histories are preserved on film and were invaluable for reassessing some of the work done by earlier excavations.

Level 2 (The Second Revolt: 132–135 CE) finds are dealt with extensively in three volumes by the Israel Exploration Society. We added to this body of knowledge with coins (Figure 1.1) and our own analysis of elements that were discovered and published by others. Carl E. Savage's chapters in this volume on the stoneware and the pottery from the BC passage also add to this question (see Figure 1.4).

Level 3 (The First Revolt: 66–132 CE) finds were the most problematic of the different archaeological strata that we analyzed. The coins are an important part of the delineation of this stratum but are now accompanied by elements that can be re-dated back to the first and early second century, prior to the Second Revolt.

Nine coins have been found in the three excavations of the Cave of Letters. Yohanan Aharoni's expeditions in the 1950s found no coins. Yadin excavations found only two coins from the Bar Kokhba period. One coin was discovered outside the cave on the ledge at the base of his rope ladder in 1960 and the other coin was found at locus 64y (see Figure 1.4) in Hall C in 1961. Although the coins will be the subject of Fred Strickert's chapter, the information on the coins is presented here to demonstrate the importance of this stratum.

The nine coins were painstakingly found during three different excavation seasons using a variety of techniques in 1999, 2000, and the 2001 excavations. Four coins from before the Bar Kokhba Revolt were found and five coins from the period of the Bar Kokhba Revolt. If this evidence was the only evidence discovered in the cave it would be sufficient to argue that the cave was used before the Revolt and during the Revolt.

Coins are an important part of an argument since they are specific indicators of what people carried often only in specific periods. Coins were and are important propaganda media since they are handled by everyone and can carry symbols and inscription that often speak louder and more often than any other vehicle of advertising that is disseminated. Not all coins were legal tender or meaningful after the historical or political events they were meant to commemorate had passed. The symbols and inscriptions, in fact, often rendered them unfit for use in a later period. Of course, in antiquity bronze, silver, and gold coins could be melted down or re-struck and used again; they also can be historical benchmarks and kept as collectors' items of a past history that had come and gone; but for the most part coins were slides of life as they were lived in a certain time and place and tell us about the people who used them. Not infrequently coins were re-struck over earlier coins for utilitarian reasons but more often they were re-struck over other coins to make a political point. For the Jews to strike their coins over the image of the Emperor or the Roman Eagle made a statement that often swords and battles could not. It is this history that was played out in our clues from the coins. The coins in the Cave of Letters along with other information found in the cave add up to a cumulative argument that tell us a story about the life of Jews in this period and these nine coins, together with other finds, tell us about the life of the refugees who used the cave.

Yigael Yadin's excavations discovered only one coin (Bar Kokhba) inside of the cave. Our excavations discovered one Bar Kokhba period coin in the cave in 1999, two Bar Kokhba coins in 2000, and two in 2001 for a total of five Bar Kokhba coins from inside of the cave. The coins from before the Bar Kokhba Revolt are equally spread from 68 CE to 117 CE and feature an array of major figures from the first and second century CE including:

a. The Emperor Trajan (Emperor of Rome: 98–117 CE)
b. King Rabbel II (King of the Nabateans: 70–106 CE)
c. The Emperor Vespasian (Emperor of Rome: 69–79 CE)
d. A Jewish-Zealot coin from the First Revolt (68 CE).

One could make the argument solely from the coins that the Cave of Letters was either continuously in use (or known) from 68–135 CE or that most of the coins could have been in use through the time of the Bar Kokhba Revolt. This is true for all of the coins but the Zealot coin. It would have had no value in the period after the First Revolt and its poor bronze would not have been useful for re-striking in the Bar Kokhba era. Some of my colleagues have suggested that perhaps the Zealot coin was an historical artifact that had been kept by a Bar Kokhba refugee.

On July 11, 2000, we came upon a coin, which connects the use of the cave from the First Revolt with the Second Revolt. In Locus 18 (see Figure 1.4), a Zealot coin from the second year (68/69 CE) of the First Jewish Revolt against Rome was found. This coin in the Cave of Letters is similar to the other second year Zealot coins found at Qumran and Masada, two places where the Zealots found themselves during and after the First Revolt. The Zealot coin found in the Cave of Letters would not have been in use in the time of the Second Revolt. It is unusual since these coins were legal tender only for a short time period of the first uprising and have little intrinsic meaning (except as an antiquity) after the Revolt. The effort to melt down Zealot coins would not have been a worthwhile effort nor were they good candidates for re-striking. This Zealot coin is small (less than 3 grams) and made of poor bronze with an obverse symbol of an *amphora* (a large jar that became a symbol of freedom for Jews) with two handles around an inscription starting on the left which reads "Year Two." The coins are inscribed in paleo-Hebrew script. The use of the paleo-Hebrew script is an attempt to hearken back to the grand old days of ancient Israelite glory, (when paleo-Hebrew script was still being used) from the First Temple period. By the first century CE, the paleo-Hebrew script had not been in regular use for a few hundred years. The Zealot coin is so poorly made that traces of the *Tof* (X in paleo-Hebrew is a *Tof*) from the other side of the coin are visible on the obverse of the coin. On the reverse of the Zealot coin is a vine leaf with a small branch and tendril with the inscription: "*Freedom of Zion*" (*Herut Tzion*) with the "*Tof*" of *Herut* being the letter which is on the obverse.[2] This means that this coin was hastily struck on a poor quality bronze with poor equipment. Professor Kindler has reasoned that it was probably struck late in year two of the First Revolt (August, 68 CE). It was late in 68 CE that, according to Josephus Flavius, the Jewish historian who accompanied the Romans in the war to chronicle the events, those events started to go badly for the Jews in Jerusalem, although it was still possible to leave Jerusalem. By this time, poorly made coins were used. This coin is significant for a variety of reasons. While it is true that many other Roman bronze and silver coins were used by the Bar Kokhba rebels both as legal tender and for possible re-striking, small antique coins such as those from Rabbel II (70–106 CE) and especially this First Revolt (68 CE) coin would not have been good candidates for re-strikes and would have had limited value by the time of the Second Revolt. The small bronze coin of King Rabbel II and the First Revolt coin of the Jews would not have held the symbolic significance that re-striking over a Roman coin would have held for the Bar Kokhba rebels. In addition, the Rabbel and Zealot coins are so small as to make it nearly impossible to re-strike any of the denominations commonly in use in the Bar Kokhba currency. In short these two coins out of the nine signal a presence in the cave before the Bar Kokhba Revolt. Together with the correct dating of the Herodian lamps, the Roman cooking pots and the stoneware suggest that there might have been a presence in the cave from the period of the First Revolt in addition to the use of the cave in the Second Revolt.

The opposite argument can be drawn from the presence of a silver tetradrachma coin we discovered in the cave. On July 17, 2000, in Locus 27, a silver tetradrachma Vespasian coin from the year 70/71 CE, struck at Tyre, was found. The coin's weight is 14.60 grams with an obverse head of Vespasian, right laureate with the inscription of "*Vespasian the King*." The reverse has an eagle standing on a club head to the left holding a wreath in its beak with a palm branch on the right. It has a border of dots around the outside of the coin. The legend on the reverse is: "*Year 2*" or 70/71 CE. This coin was in use in the time of the First Revolt, but its value, size, and symbolic importance would have made it an excellent type of coin for re-striking a Bar Kokhba coin during the Second Revolt. This could have been one of the coins that Bar Kokhba might have wanted to purposely re-strike with his own images or the image of the Temple simply because it would satisfy the movement's need to show that they felt they were capable of defeating the Roman Eagle and Emperor. Silver coins are a rarity in Bar Kokhba coinage, but silver coins do provide an opportunity for symbolic re-striking and have intrinsic value. The only problem with this theory is that this coin was in a refuge cave which was not a place for re-striking. The coin is a fine specimen and could have been legal tender in 132 CE, despite its age. It could have been, however, a remnant of the First Revolt presence in the Cave in and around the year 70 CE. This is a coin that was certainly legal tender in 70 CE when the Zealots might have used the cave.

The other two coins found in our excavations from before the Bar Kokhba Revolt, the Nabatean kingdom coin and the Emperor Trajan coin might still be in use in this region in 132 CE despite being minted twenty to thirty years before the Second Revolt. The Trajan coin is a beautiful silver coin that would be worthy of being re-struck, but the Nabatean coin was made of bronze and a less likely candidate for re-striking. In addition, there is no indication that the people who were in this cave were hoarding coins for re-striking. They seem to have been refugees who arrived only with their possessions and perhaps some coinage.

The Bar Kokhba era coins found in the Cave of Letters indicate that the refugees possessed Bar Kokhba coin issues from the start of the Revolt until the end. The set of coins trace the three years of the Revolt and tell us that the refugees that were in the cave at the end of the Revolt.

Coins were found in almost every area of the Cave both under debris and not under debris. On July 17, 2000, in Locus 25, a large Bar Kokhba coin issued in 132/33 with a weight of 10.60 grams was found. It had the traditional wreath decoration tied at the bottom with thc palm branch. The legend reads in paleo-Hebrew: ***Shimon Prince/President of Israel***: *Shimon Nasi Yisrael* on the obverse. On the reverse it reads: "First Year of the Redemption of Israel", *Shnat echat LeGeulat Yisrael.* The obverse has the distinctive lyre with four strings. It is clear on both sides that there is evidence of over-striking of a Roman bronze coin, and the bronze coin was in excellent condition because it was under an entire debris layer. Similarly, on the same day, in Locus 16, in the AB passage, a Bar Kokhba coin dated from the end of the Revolt was found. It was small with a weight of 3.30 grams in bronze. It had the palm tree with two bundles of dates and seven branches with a border of dates hanging heavy from its branches and on the obverse a bunch of grapes. The legend on the obverse is *Shin, Mem, Ayin, Vav, (final) Nun*–"Shimon" and the legend on the reverse around from the right above is: "*For the Freedom of Jerusalem*" (*LaHerut Yerush[alayim]*).[3] The last Bar Kokhba coin we discovered, without a date, was ascribed by Professor Arieh Kindler to the last year of the Revolt. The name of "Shimon" [Bar Kokhba] is found, but without the full legend, "Prince/President of Israel".

In this cave, like bookends, were two parts of the same epic tale that had begun in 66 CE with the First Revolt of the Jews against the Romans and the destruction of the Temple in Jerusalem in 70 CE and now ended with the failure of the Second Revolt of Bar Kokhba in 135 CE. Dr. Fred Strickert chapter in this volume adds to our knowledge of the question and builds upon the work of Professor Arieh Kindler (published in 2001) who published the coins from 1999 and 2000. Professor Kindler saw and read all nine of the coins that are mentioned in this volume.

OTHER FINDS

In entrance #2 of the cave, an ancient child's sandal was found (Figure 1.3). It was recovered from entrance #2 even though the area had been carefully examined and excavated by scores of people since the time of Yadin's excavations. C-14 tests were run on the sandal.

A large 20 centimeters of ancient palm frond mat was uncovered in the team dining area in the front of Hall A near entrance 2. Additionally, a wooden comb was revealed in the AB passageway even though this area was substantially worked on by Yadin's expeditions and was the daily conduit for members' passage from Hall A to Hall B where the main work of the expedition was ongoing.

On the final day of the 1999 excavations, July 16, just before closing up Locus #4 (see Figure 1.4), a large leg of furniture piece (table, desk, bed, etc.) was found by Dr. Fred Strickert and Dr. Walter Bouzard working in the most southern eastern part of Hall B (Figure 1.2). The carpentry items found by Yadin are similar but the larger piece found by our excavations indicate more about living circumstances.[4] The ends of the several pieces are charred. The upper part of the complete leg is rounded and it tapers down to a blunt point at the bottom and had been joined to the horizontal members of the table by the tongue and socket technique. Length of Yadin's complete leg was 70 cm.[5] The leg found by our group was 110 cm and it was considerably larger in diameter than the Yadin piece. Its socket was slightly cruder but still rectangular with two openings on both sides. Pegging, dovetailing or joining thicker pieces by means of tongues and sockets could have been used for joining purposes. The piece weighed in excess of two pounds and had a blunt point at the bottom for placing in between rocks or a dirt floor. Unlike Yadin's leg, which he determined to be from a table, the larger leg we discovered may have been from a bed.[6] Its proximity to the hearth may have meant it was going to be used for firewood.

THE 1999–2001 GEOPHYSICAL SURVEY AND CLIMATE STUDIES

Part of our work was devoted to a reevaluation of the life of the cave. We conducted geophysical surveys that are in four different chapters in the book. Geoscientists, Bauman, Shroder, Reeder and Jol supervised all of these studies and this type of study has few parallels in the scientific volumes on the Cave of Letters. The climate studies that were also performed in the Cave of Letters by Professor Reeder are also a unique evaluation of the elements that affect the archaeological work in the cave. The archaeological stratigraphy of "mound" sites has been well-explored in many volumes. The idea that there can be a correlation between archaeological and geological stratigraphy inside of a cave was one of the objectives of these studies. In the cave, although there is no systematic building from one stratum to another, there are systematic roof debris layers that closely parallel archaeological strata. The roof fall after every geological event provided a consistent layering effect in the cave. We were able to distinguish where archaeological layers and artifacts found at one layer were below other artifacts that rested on another geological layer. This geological stratification of debris layers throughout the Cave of Letters is a phenomenon which needs to be considered in other cave research, especially in the Dead Sea region. Many of our finds were found below the top layer of debris. The bronze artifacts found in Hall A (locus 57y) were only discovered because of the use of a metal detector. Most of Yadin's finds were found almost at the top layer of debris because the debris could not be easily moved. In Hall B, below the top layer of debris we found the hearth and the wood finds as well as the bed leg.

THE POTTERY AND LIMESTONE WARE FINDS OF THE CAVE OF LETTERS 1960–2000

Carl E. Savage will be reviewing the pottery finds from the Cave of Letter BC passage and the limestone ware finds. Yadin dated almost all of the cave finds to one archaeological stratum. He observed:

> As no evidence of occupation earlier than the Bar Kokhba period was found (with the exception, of course, of the few Chalcolithic finds), the several lamp fragments of Roman date—from both the cave and the camp above—prove to be of importance, for they may be precisely dated.[7]

In fact, Yadin dated the whole cave from the major second century finds he discovered in Hall C. He dated Herodian lamp fragments (which were clearly dated at Masada and Qumran to the first century and second century CE to make sure even these small finds matched his conclusions).[8] Stoneware pieces found in Hall A, for example, that resembled stoneware finds from first century Jerusalem, Qumran and Masada, in type, style and size, were re-dated to the second century to match his conclusions.[9] Roman cooking pots which would normally be linked in a non-cave environment to the style and type of first century cooking pots from Qumran (by Yadin's own admission), are instead re-dated to the second century to match the major finds of the Bar Kokhba letters and the Babatha archive from Hall C.[10] Yadin concluded that the entire cave was basically one single archaeological stratum, an unusual conclusion at almost any archaeological site, and, based upon the variety of buried materials we discovered and the geological analysis that was emerging this conclusion is even more problematic.

The Herodian oil lamps from the Cave of Letters are the same type as those from Qumran and Masada.[11] The so-called "Herodian" oil lamps have been the subject of much new information. More importantly, it is unlikely that the Herodian lamps are distinctive enough to have persisted into the second century CE. The other pottery issue relates to the limestone vessels found in the Cave of Letters. The limestone vessels of the Cave of Letters are very similar to the limestone vessels found two years later by Yadin at Masada. These limestone vessels were very fragile and linked to the ritual purities laws that were well known and integrated into the lives of those living in the Land of Israel. Following the destruction of the Temple in Jerusalem a handful of the long list of ritual purities mentioned by the Rabbis continued to be in effect. Yadin's designation of the limestone vessels from the Cave of Letters as second century artifacts had an enormous ripple effect in archaeology since others began to date limestone ware all over the country to the second century CE.[12] Carl E. Savage has carefully identified in his chapter the issue of the limestone vessels thanks to many other discoveries in the past forty years. The crucial point for us was that in the 1960s, finding a limestone cup in the Cave of Letters that was similar to the ones from Qumran (and then two years later at Masada) could have led Yadin to connect the Cave of Letters with the time period of Qumran and Masada in his writings; i.e. the First Revolt. It did not. Yadin's decision to group the carved limestone vessel from the Cave of Letters Hall A with the other Second Revolt finds in Hall B and Hall C profoundly changed the scope of his work.

BIBLE MANUSCRIPTS IN THE CAVE OF LETTERS

The discovery (or better: rediscovery) of the existence of the Psalms manuscript in the Cave of Letters (Psalms 7, 8, 10, 11, 14, 15/16, 23, 24, and 29—probably from Hall A) is a crucial finding. The discovery of the Babatha Archives and the Bar Kokhba rebels are both from Hall C. The Psalms and other biblical manuscript texts probably were from Hall A. The work of Professor Bouzard[13] will be presented in this volume in order to establish the provenance of the Psalms text. It is so unlike any other written piece found in Hall C that it clearly suggests a different group was present in Hall A. The Psalms were the crux of worship in the Temple in Jerusalem and this is the only Judean cave manuscript that preserves these Psalms. It is one of only three Psalms scrolls not found at Qumran; the others having been found at Masada. This knowledge painstakingly gathered over the past forty years shows us a very different context of the Cave of Letters. The Cave of Letters Psalm Scroll is unique among Psalm Scroll exemplars. Yadin discovered only one small fragment of Psalm 15/16 of the scroll and could not assess its importance. Yadin acknowledged the uniqueness of the Psalms fragment he recovered but he was unable to see the overall importance for the rest of the finds made in Hall A. The manuscript script is clearly late Herodian and can be located in the first century BCE.[14] Yadin did not know about the fragment of a Deuteronomy manuscript or of a fuller Numbers manuscript of the Cave of Letters, but today they all have been ascribed to the Cave of Letters and give us a much fuller understanding about who the refugees were that brought them. These manuscripts all stand out as religious manuscripts as opposed to all of the "secular" (Bar Kokhba) letters and (Babatha Archive) documents found together in Hall C of the Cave of Letters that were so influential in Yadin's understanding of the Cave of Letters. As noted in the chapter in this volume by Walter Bouzard, we now know that these Psalms chapters are from a different manuscript tradition of Psalms than the other manuscript traditions of the Psalms scrolls found at Qumran and Masada. One (of the many) differences noted by Bouzard is the lack of the classic superscription "A Psalm of XXX" found on the Scrolls from Qumran and Masada. It is possible that the refugees who brought the Psalms manuscript to the Cave of Letters were from a place and time when the superscriptions were well known and additions to the manuscript. A text like this might have emerged from a group such as the priests from before the destruction of the Temple of Jerusalem, for example.

In short, the fact that the only fragment found by Yadin of the Psalms manuscript came from Hall A, which is the same place that the bronze artifacts, the limestone vessel, and the First Revolt coin is significant. The fact that all of the second century manuscripts are from Hall C (Babatha archives and Bar Kokhba letters) and Hall A seems to have had the religious manuscripts may mean that Hall A was more well known to the First Revolt religious refugees and Hall B and C more used by Second Revolt refugees. These Hall A refugees may have also been responsible for the Numbers and Deuteronomy scrolls that also came from the Cave of Letters. They represent a different group of religious individuals who came to the Cave as refugees under totally different conditions than the people that used Hall C, who clearly were Bar Kokhba refugees. Hall B, on the other hand, with the hearth and the niches of burials may have been used by refugees from the First and the Second Revolt.

THE BURIALS IN THE CAVE OF LETTERS

The chapter by Professor Nicolae Roddy in this volume will survey the complicated question of the burials in the Cave of Letters niches and the textiles chapter by Orit Shamir provides the background for the dating of some of the bones and textiles which were found together. In our 2000 season, we worked in what Yadin referred to as burial niches in Hall C. One large burial niche, the Niche of Skulls (Locus 2 and Locus 2y) (see Figure 1.4), was filled with bones and textiles in the 1960 excavation and in 2000. We encountered it on the second day of the second season of our excavations, July 4, 2000 as we surveyed the places where Yadin had excavated in 1960–61. One of our prime goals was to reinvestigate the places where Yadin had excavated to see what Yadin had left in situ. The large burial niche was first discovered by Yigael Yadin's team on Friday, March 25, 1960 by the same individual who had excavated the bronze artifacts, Pinchas Porat (Born: Pinchas Prutsky). Yadin described it thus:

> On the second day, Friday, 25 March, while most of my team was still busy with the Cave of the Vulture, a group of twelve started for the big cave and for several hours they transferred equipment: lamps, baskets, picks and ropes. Only then did they begin to work. One of the volunteers, a real maverick, Pinchas Prutzky of Kibbutz Lahavot Haviva, broke off from the rest and with a candle in his hand decided to survey the innermost hall on his own. Soon enough he rejoined his colleagues pallid and speechless. This state was caused not so much by the fact that he had crawled into a crevice and could hardly extricate himself from it, but rather from the sight which he faced when he got into that crevice. The next day, on my return from project 'Cave of the Vulture', I followed him to his crevice and understood

> exactly how he must have felt alone there. He led me to the eastern wall of hall C, a couple of meters from the spot where it begins to narrow. I saw there was a an opening to a crevice (see the plan of the cave) about half a metre across, between the wall and the adjacent blocks of rock. We crawled along it for two metres, and then the floor began to descend sharply for about a metre, forming a sort of step. Here the passage as most difficult for the height at this point was no more than fifty centimetres, and a large rock rested on the floor. Only by squeezing and pushing ourselves with difficulty did we at last manage to get in, and found ourselves in a niche, some eight metres long and about one and a half metres high. To the right of the entrance, above a ledge in the wall, a patch of soot was visible, probably from a lamp that had been placed there in antiquity. Although I had been warned what to expect, I was started at what I saw once I got used to the dim light. Near the right hand wall lay a heap of skulls, without their jaw-bones, placed in several baskets stacked on top of one another. At the far end of the niche, in the right hand corner there was a separate pile of bones and jawbones in baskets and wrapped in cloth between mats. This entire heap had been covered with a large mat. One skull rolled into a crack. In the middle of the niche on the floor were remains of a skeleton covered by a colourful rug and other textiles. Here we found a basket lined in leather, containing the skull and bones of a child wrapped in an almost complete tunic. On the rocks near the left-hand wall fragments of textiles and tufts of hair were found. The skulls and bones, with the possible exception of those of the burial in the floor, all appeared to have been deposited there at the same time, after having been gathered together, wrapped in mats, and put into baskets.[15]

According to the analysis which Yadin did with Professor H. Nathan from Hebrew University, seventeen different people were buried in the niche. Two other burials were found elsewhere in the cave (Locus 31). In total: three men, eight women, and six children. With the preponderance of bodies of women and children, the question is: Where were all of the men if this was a rebel cave? Of course, it is clear that this may be part of the story: the refugees who fled to the cave were actually fleeing in the middle or end of a war and so the men were fighting and these few people remained to hide out in the cave with a small male contingent of guards. These bones (or at least some of them-skulls and some other bones) and the bones of the Cave of Horrors, located on the other side of Naḥal Ḥever, were all re-interred together in an official state funeral on *Lag BeOmer*, the holiday dedicated to the memory of the students of the Rabbi Akiva, the main rabbinic supporter of Bar Kokhba. On May 11, 1982, these bones were re-interred in a state funeral amidst much pomp and circumstance. The question was then and was for us: Were these the bones of the Bar Kokhba fighters as Yadin thought? If they were Bar Kokhba rebels, they must have had a very strange burial practice, even given their dire situation as refugees from the Romans. For this cave to have been the burial cave of Bar Kokhba fighters, it must have been in use from the first days of the war, since it would have taken at least one year to have flesh decay from the bone and then take the skulls and place them in the palm branch baskets that were found in the cave. The possibility that the bones in the Cave of Letters were originally laid out in shrouds in the Niche of Skulls in the Cave of Letters and collected later into the palm branch baskets is obvious. The large number and variety of textiles which surround the bones in the niche raise the possibility that each body was laid out in shrouds at one time and later, without the availability of a collection box (ossuary) in the Cave, the cave dwellers collected the bones for future deposit in ossuaries. There are some important elements in the chapter on textiles, threads and cordage from the Cave of Letters that are written about in full in Orit Shamir's chapter. No mixed, wool and linen mixtures were found in the garments, (no *shaatnez*) which points to their being Jews buried in the Niche. This is another example of items that define the ethnicity of the individuals, especially in the Niche. Although we assume that the textiles that we found in the Niche are similar to the ones that Yadin found in his excavations, some of Dr. Shamir's research demonstrates that the wool textiles that she analyzed from the Niche were in fact different from those analyzed by Yadin's team and are different from the comparative sites of Masada, Muraba'at, 'En-Boqeq, and Berenike. Whether this suggests that there were different groups of people who were buried in the Niche is not clear, but it is clear that the Niche textile samples from Yadin's excavations are not from exactly the same group. In addition, some of the linen textile fragments examined were not bleached as they are at other contemporaneous sites from this period, indicating that some of the people in the Cave were wearing clothes that were from different sources other than those studied at Masada, Qumran, and from the other Yadin Cave of Letters samples.

In general, the bodies would be buried with some grave goods, such as bowls, cooking pots, lamps, etc. and, indeed, remnants of such items were found in the niche. One problem still remains regarding the bones. Unlike all other cave burials and finds from this period which places one or two bodies together in a single ossuary, the bones were found sectioned into very specific categories. The skulls were all together in baskets and in many cases, the jaw bones were removed. Yadin mentioned that in one case: "a basket was found, lined with leather, in which there were a skull and the bones of a child wrapped in an almost complete tunic."[16] In the rest of the cases: "…a concentration of skulls was found, placed in baskets; the jaw-bones were missing from all. The baskets had been stacked one on the other. The bottom-most one was torn; on this lay another, round basket containing six skulls. Above this was a flat basket with four skulls…."[17] Yadin may have left some of the bones in the niche for future work. He could have easily removed all of the bones for analysis and burial, but did not. Yadin concluded: "…all appear to have been deposited at one time, after having been gathered and put into baskets or wrapped in mats-for lack of ossuaries."[18] He even concluded that one of the bodies discovered in the Cave was Babatha's; since her archives were found across from the niche it seemed reasonable to assume that she was buried there as well.

There is a much fuller explanation of the burials in the chapter by Nicolae Roddy. The baskets of skulls, separated from the rest of the bones and the state of the burials suggests that there were two distinctive groups (from different time periods) who used the burial area. One group may not have known the other group and this facilitated their collecting the bones of the early group in baskets for removal (from the cave) at a later time. This whole process (collecting separate skulls in baskets for removal) is in such contradiction to Jewish burial practices of the period as to suggest that the people who did this did not know whose bones were being collected. In any case, the different dates for the carbon 14 dating of the bones suggests different time periods for burial.[19]

Yadin assumed that the bodies all were from the Second Revolt, making it a burial cave of sorts. Some of the bones collected may have been from those who died during the short period at the end of the Second Revolt, but more likely the bones were from an earlier period of use of the Cave. Since decomposition is a slow process in a cave like this (little or no humidity; no soil, bugs, etc. to help the process along), the ancient baskets of skulls found by Yadin suggests that the bodies had died and been laid out many years before the Second Revolt refugees arrived. As we have seen, these caves preserve organic matter in a unique fashion.

In the tiny cave located just across the Naḥal Ḥever, the Cave of Horrors, the unique climate yielded "horrific" results when they were discovered in the 1950s by Professor Aharoni. In his *In the Footsteps of Kings and Rebels*[20] Professor Aharoni and Rothenberg note that they found in what surely must have been the Cave of Letters what he believed were First Revolt refugees with *skin still intact* on the bone. It is possible that Aharoni is writing about the same burial niches (locus 2y, 3y, 8y) as Yadin.[21] Accordingly, if these were the bones of First Revolt refugees who had died in the cave a generation before the Second Revolt and were found by the Second Revolt refugees, then the placing of the skulls together in baskets for removal makes much more sense.

LOCATION DOES MATTER

The fact that so many young children and middle aged women who were found in the cave indicates that this cave must have been known for a long period of time. It is hard to imagine that Second Revolt refugees "suddenly" discovered this cave. In order to plan to bring women and children into the forbidding and difficult location it must have been well known among the refugees. Water was a factor in the choosing of this cave for the refugees. It was an unlikely place for the Romans to look. The cave's difficult access to water shows that this was for a short "holding or refuge" period only. But the problem of getting access to water must have been known beforehand to whoever directed the refugees there in the first place.

RADIO CARBON DATING IN THE CAVE OF LETTERS

Radiocarbon dating was discovered in approximately the same time period as the earliest discoveries of the Dead Sea Scrolls. In fact, the entire debate over the dating of the Dead Sea Scrolls was ultimately decided by using this technique in the 1980s and 1990s. The basic principles of carbon 14 dating are based upon the presence of the carbon 14 isotope in all organic matter such as bones, parchment and textiles made from animal and vegetable matter. Even though the isotope decays at a precisely measurable rate, the micro-conditions of any setting can affect the scientific analysis of the data. The caves and the environmental conditions of the Dead Sea are important determining factors in the analysis of the C-14 data. The Cave of Letters is an especially important micro-environment to which we paid close attention so that we could calibrate our C14 results properly. The conditions of the Cave of Letters and the Dead Sea Caves in general are so unique that the C14 analysis of items from these caves has become important sub-studies in all Dead Sea Scrolls research. Elisabetta Boaretto of the Weizmann Institute will present the items that were carbon-dated during and following the 1999–2001 excavations. She will be presenting the different period suggested by 1sigma and 2sigma dates.

The dating process has brought a new level of certitude for determining the age of certain materials. Linens from the caves were among the first items tested in the late 1940s and early 1950s with this new technique. The dispute over the dating of the Dead Sea Scrolls raged on for years and has affected our understanding of the dating of items found in the Cave of Letters. Some scholars[22] who insisted that the Scrolls were medieval (until this technique was perfected) did not accept the dating even after it was perfected. We had many organic finds from the Cave of Letters and dating was done on a statistical sample of items. They were found at different archaeological strata, which we could test and associate with dates (albeit a spread of 200 years is usually found in our work!). The dates that are used for dating the artifacts usually show the extremes of the dating and different reporting scales (1sigma and 2sigma with varying dates will be noted). Based upon an experiment performed on the Dead Sea Scrolls and the Bar Kokhba letters, the "true" date of the artifacts tested in the cave appears to be ***the older extreme*** and not the average of the age range.

The unique experiment and theory was built by Magen Broshi, curator emeritus of the Dead Sea Scrolls at the Shrine of the Book, Israel Museum, Jerusalem, and he has written about the results in a variety of different publications. Broshi needed to know how to use the radiocarbon dates he received for understanding the time period when the Dead Sea Scrolls were written. Because there was no way of knowing whether the earlier, later, or average extreme of the radiocarbon dates was correct, he performed the test on documents that had dates on them that could be correlated with the Roman calendar. None of the dated documents were from the Qumran caves, but three of the selected texts for the experiment were from Bar Kokhba Caves and two were from the Cave of Letters. Broshi's carbon dates of dated documents from the cave indicated that the ***older date*** was to be preferred when assessing both the Bar Kokhba cave artifacts and the Qumran artifacts.[23] Here are the carbon dates and the dated documents he used:

IAA-13433 1799 BP ± 57yr **130–321 CE**
5/6 Hev 21 (Pap Yad in 21-purchase of crop)-dated **September 11, 130 CE**

IAA-14987 1827 BP ± 36yr **126–234 CE**
5/6 Hev 19 (Pap Yad in 19-Deed of gift)-**dated April 16, 128 CE**

IAA-14986 1758 BP ± 36yr 144–370 CE
Kefar Bebayou (Murabba'at Cave-House sale dated to 135 CE)

<u>Dead Sea Scrolls Carbon 14 Dates</u>

IAA-13415	4Q266	1954 BP ± 38yr	5–80 CE
IAA-13419	4Q258	1823 BP ± 24yr	134–230 CE
IAA-13431	4Q344	1902 BP ± 39yr	72–127 CE

The radiocarbon dates revealed that many of the Qumran texts were produced in the first century CE and even into the second century CE. In addition, other dated materials from Qumran were tested that were similar to our materials from the Cave of Letters. For example, a linen cloth 1917 BP ± 200 yr and a date palm log 1940 BP ± 85yr[24] (dated by locus to 68 CE)[*25] In this case, and similar to our own wood finds in the Cave of Letters, the wood has a date approximating the 1st century CE Although we might wonder whether the wood was cut at an earlier time and kept around for later use (wood can be hundreds of years old), we learn from the Qumran excavations that the wood that was cut for use is usually only as old as the settlement itself. Wood was so scarce that if it was cut, it was used almost immediately for the purpose it was intended. The fact that our wood dates to the early 1st century CE, indicates that the wood was probably used in the hearth, which itself can be clearly dated to the 1st century CE (both 1sigma and 2sigma dates). The three observations that Magen Broshi wrote about were[26]:

a. The "true" dates do fall within or very near the archaeologically established calibrated age ranges. In only one case, the true age is off by ten years.
b. The age range can be quite broad. In two cases they exceed 200 years.
c. The true date is not the average of the age range. In the three papyri, the "true" date **is in fact the older extreme.** Perhaps this has to do with the preservation of the papyrus material.

It is clear that the micro-climate of the Cave of Letters is unique and is a factor in the dating process. Thus, using Broshi's observations are all the more important for the radiocarbon dates that we have in the Cave of Letters.

We found that the use of Accelerator Mass Spectrometry technique required only a small piece (0.5–1.0 mg) of carbon for dating and has overcome the disadvantages of the earlier radiocarbon dating, which required large amounts of carbon. We found that we could take already existing finds like the rope tied to one of the bronze articles in the locus 57y found by Yadin and test the rope fibers. We discovered that this particular rope had never been treated with harsh chemicals to preserve or clean it and therefore was particularly good item to test in order to date the bronze artifacts. It was never tested by Yadin because Yadin had so much dated material he did not feel it necessary to do very much carbon 14 testing and the process was not very well used in the 1960s. The dating of the rope from the bronze artifacts was determined by Elizabetta Boaretto at the Israel Museum after permission was received from the Antiquities Authority of Israel. The dating of the rope connected to the bronze hoard was 1850 BP ± 40yr[27] 60CE–260 CE (2sigma). The older extreme date of 60 CE would place the bronzes in the period of the destruction of the Temple of Jerusalem.

USE OF THE CAVE OF LETTERS IN THE FIRST AND SECOND CENTURIES

The spread of time periods found in the carbon dated artifacts suggests use of the cave in the first as well as the second century CE. In Hall A, a child's leather sandal was discovered near where the Zealot coin was discovered from 68 CE (Figure 1.3). The leather sandal was dated to between 80–340 CE (2sigma). Sandals were not simple, replaceable items purchased in a store. They were hand-crafted to a specific size and foot and this one is a child's sandal; perhaps a 3 or 4-year-old. The child's sandal, suggests together with the other first century finds of Hall A, (the coin, the fragment of the Psalms text, the fragments of the limestone cup and the analysis of the bronze artifacts in 57y) that Hall A was used in the first century.

The most important carbon dating involves the *Niche of Skulls* and the hearth area. The hearth area, which is the stratigraphically-lowest and most untouched of the areas of Hall B, revealed both wood and organic matter from the hearth. The hearth's date was determined by a piece of organic matter used in the grit of the outside pottery. It is important that it was on the outside of the hearth and not adulterated by carbonization from the inside of the hearth. It was removed and tested at the Weizmann Institute and dated to 60–260 CE (2sigma) when calibrated. According to the aforementioned method (of the older extreme as the correct calibration) this would indicate that only a short time before the destruction of Jerusalem the residents were escaping to the caves, or that this cave was already being used by the insurgents in the build up to their campaign against the Romans. This would be similar to the method of the Bar Kokhba rebels who prepared and used caves and tunnels throughout Judea to store weapons and to plan and launch battles. The rebels of the Second Revolt also used caves and tunnels as refuges in the event of losses in the larger cities. The stock-piled wood in the Cave of Letters was from local indigenous woods that were probably brought there in advance of the use of the cave. The one piece of furniture we discovered (in Figure 1.2) was dated to the early 1st century and may have been brought to the cave for use as firewood or for a low table, bed or chair). It was found at the lowest part of the hearth area at a meter below the debris level in Hall B. It was carved and its early date (1945 BP ± 30yr, 20 BCE–130 CE-2sigma) suggests it was brought there for use during the First Revolt. The wood found with it dates back even earlier. But it would have made this hearth area an important center for the cave occupants in the 1st century. Many of the personal objects, including a wooden comb, other sandal pieces and palm branch baskets fragments were all found under the collapsed roof of the of the AB passage.

If we take the principles learned from Dead Sea Scroll research seriously in comparison with our own carbon dated materials, it means that the older extremes probably indicate that the materials are from two different time periods within a 100-year period from the mid-1st century through the mid-2nd century.

THE SCIENCE OF POTTERY IN THE FUTURE

In our projects at the Cave of Letters, we have a team of nearly a dozen researchers in different parts of the world working on the artifacts and on sub-projects from our work in the Cave of Letters. Some of these reports such as an a DNA study from bones from the Cave of Letters "Niche of Skulls" (conducted with the Weizmann Institute) were not published because the study involved comparative work from other sites. Other reports appear here in summary or in an appendix.

The most ubiquitous find in our excavation were the thousands of pieces of pottery. Carl E. Savage has analyzed only a fraction of the pieces of pottery because of the lack of many diagnostic pieces. We assume that many of the diagnostic pieces were those published by Yadin. Although the pottery of the Cave of Letters is very fragmentary (thousands of pieces of different vessels mainly without diagnostic pieces) one study in particular was undertaken to see the original provenance of the pottery. We tested a variety of pieces of pottery from limestone, red ceramic ware, and pieces of the hearth as well as a fragment of what appeared to be a Chalcolithic fragment. The full report is presented in the volume.

Instrumental Neutron Activation Analysis (INAA) to learn where the pottery was made was supervised by Dr. Jan Gunneweg of Hebrew University and Marta

Balla of Budapest University of Technology and Economics. This technique was first applied to archaeological artifacts in the 1950s but has been used sparingly because of cost and the requirement for using a particle accelerator or nuclear reactor producing thermal (low-energy) neutrons. It also is a very destructive process (the samples are rendered radioactive, for example) in that it requires the researcher to take a small sample of the artifact, (in the case of pottery) reduce it to ground powder, introduce the sample to the reactor core where the nuclei of the atoms are bombarded with neutrons at a controlled rate. The final result, similar to the C14 work on organic samples, (but this time based upon the gamma rays present in radioactive isotopes as they decay) causes the different trace elements to be detectable because of their different rates of decay. This process allows us to find a "chemical signature" for the pottery and identify its origin because of the very specific trace elements present in different geographic locations. Since certain types of minerals in certain concentrations appear only in certain locations around Israel (and elsewhere) this process is able to pin-point the location of the origin of the pottery. Neutron activation is extremely sensitive and the Gunneweg database reveals over 75 different naturally occurring elements. If the pottery was brought in from Jerusalem, Bethlehem, Jericho, Masada, Ein Gedi, for example (or from any other site for which INAA data exists), it can be matched the chemical compositions of the Cave of Letters pottery. The refugees in the Cave of Letters came from specific locations to the cave and did not have the facilities to make a totally new composition. The comparison with the other locations may ultimately reveal if these are refugees from Jerusalem, Ein Gedi, Qumran or elsewhere. Since Qumran was destroyed in 68 CE, we would also learn more about the timing of the use of the Cave of Letters if the connection comes from that direction.

In fact, our INAA comparison demonstrates that the pottery and the clay probably came from an area around the Dead Sea area or specifically Qumran where a pottery industry was functioning until 68 CE.

The state of preservation of the bones of the Cave of Letters may help us to understand more about who they were and about our connection to these people. We clearly did not remove all of the artifacts we located in the Cave of Letters with the fiberscope, GPR and ERT scans. We discovered a variety of other artifacts under the rocks which, despite success with the hydraulic splitter in breaking up the rocks in the summer of 2001, we decided we could not get to them without destroying what was underneath the rocks. Shachar Argov's chapter deals with how we used the hydraulic splitter with a chemical solution to crack the boulders before splitting them. The technique was the best engineering suggestion that we had in 2001 and will await new advances. Our third and last season was devoted to one project, the breaking of massive roof debris that sat on top of artifacts that were identified underneath them using the endoscope and fiberscope in 1999 and 2000. Work at the Cave of Letters began with a set up on July 21, 2001. The whole team worked at the cave for three days, July 22nd, 23rd and 24th with two main scientific goals:

1. Break up selected boulders in the cave and excavation in these selected areas. (Three rocks have been chosen for this experimental technology.
   a. A rock in Chamber B under which bones were viewed using a videoscope in 2000.
   b. A rock in the BC passage around which fragments of palm frond basket sides were found.
   c. A rock in Chamber C under which a tibula bone was found in 2000.
2. Completion of Tomography and Magnetometry of Hall C and BB passage and High Room by Paul Bauman, Komex International, Inc.

We consulted with experts all over the world to insure that all possible methods for breaking up the rocks would be safe (for the artifacts, the operator, and our staff in the cave) as well as effective. The objective of the rock splitting method listed below provided an environmentally friendly (to the bats in the cave as well as the structural integrity of the cave-so no new debris would fall during our work) approach to removing large (1–2 tons or more) boulders under which artifacts had been located in the cave. The boulder removal team relied mostly on the use of a device called a "Hydraulic Splitter". This machine is powered by a small size 70,000 MPa electric pump, which was fed by an external generator on top of the cliff. The splitter's head was placed into a pre-drilled hole and used hydraulic pressure to crack open the boulder. The method is safe, quiet, clean (no fumes), provided little vibrations, and the direction of breaks were easily controlled. The holes were drilled with a high frequency drill and a carbide drill bit, which caused very little vibration to the surrounding rocks or roof which was a major question from the beginning. Also, dust was controlled by wetting the as the drill bit and later the splitter worked in the holes. The equipment was relatively light weight but had to be dismantled in order to get it through the AB passage. This created a major problem as it took one full day, July 22nd to get the splitter reassembled and working again. A 5,000 W generator was set up on top the cliff of the Cave of Letters (by the Roman Camp) and again wires were aligned over the cliff to operate our equipment and lights. These dates (July 22–24, 2001) were chosen because they were after the dates of nesting for the vultures (at least according to our summer experts and our experience this past summer) and fit the time that the equipment could be brought to the country.

We experimented on three rocks where artifacts were known to exist below. The fiberoptic and endoscope findings were all checked and rechecked. The artifacts that could be still seen included sections of palm branch baskets and ropes. There was a limited crew, Richard Freund, Rami Arav, Shachar Argov, and Paul Bauman with help from our climbing teams led by Chaggai Rosenbluh. Shachar Argov will present his own report in this volume on this effort. Three different methods were determined for breaking the rocks in concert to insure that if the Hydraulic Splitter did not work the other back-up systems would allow work to proceed. The three methods were:

a. Hydraulic Rock Splitter
b. Betonamit (chemical) Expansive Cement
c. Conventional and Strategic Drills and Drilling Methods

The two last Bar Kokhba coins were found during the 2001 excavation, one on each of our two days in the cave. The first one was found in the AB passage (7/23/01) (in addition, more pieces of textiles, basket, a strap from a leather sandal, and rope were found) and another Bar Kokhba coin was found in close proximity just outside of the AB passage (7/22/01) in Hall B. Unfortunately, the three methods did not yield the results that we hoped. In every case, the rock did split using all three methods together and it made for easier removal of the heavy pieces by conventional pulleys (which we used during 2000), but the Splitter was too efficient. After removal of the rock pieces we discovered at locus 23, (7/24/01), for example, the pulverized fragments of the palm branch basket. The rock fragments splintered into a number of large and sharp pieces that fell onto the identified artifacts below pulverized them. This happened each time we repeated the exercise. It was decided to not attempt any other rock splitting until we had a better method to secure the artifacts below the rocks before they were shattered. Without further equipment or time, we abandoned the work.

New technologies may yet help us to understand how to safely remove the artifacts under the rocks so they may be analyzed. Other methods and new technologies may yet be developed for identifying artifacts under the roof debris which exists in the Cave of Letters and other caves on the Dead Sea. We hope that our contributions in this volume to this ongoing discussion of cave research may prove to be helpful to future researchers.

Figure 1.1: The Coins of the Merrill Cave of Letters Excavations.

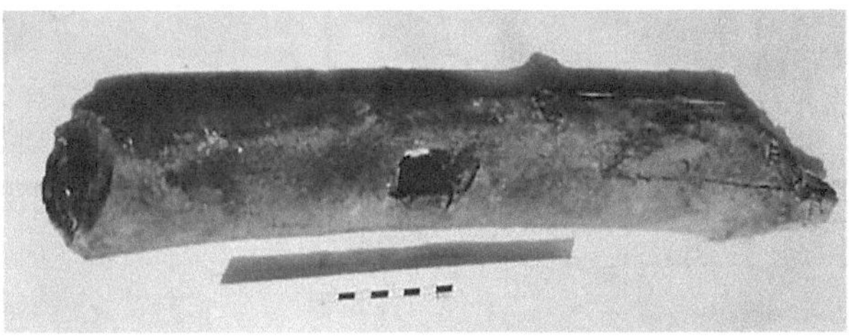

Figure 1.2: Hall B Hearth. Bed post leg (courtesy Cave of Letters Project, 1999).

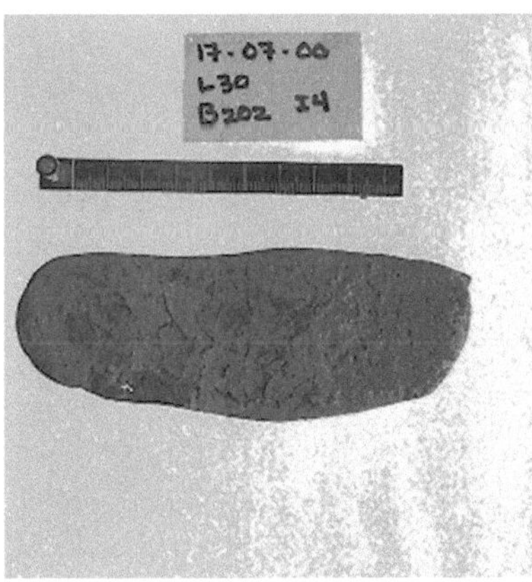

Figure 1.3: Child's sandal found in Hall A in the Cave of Letters.

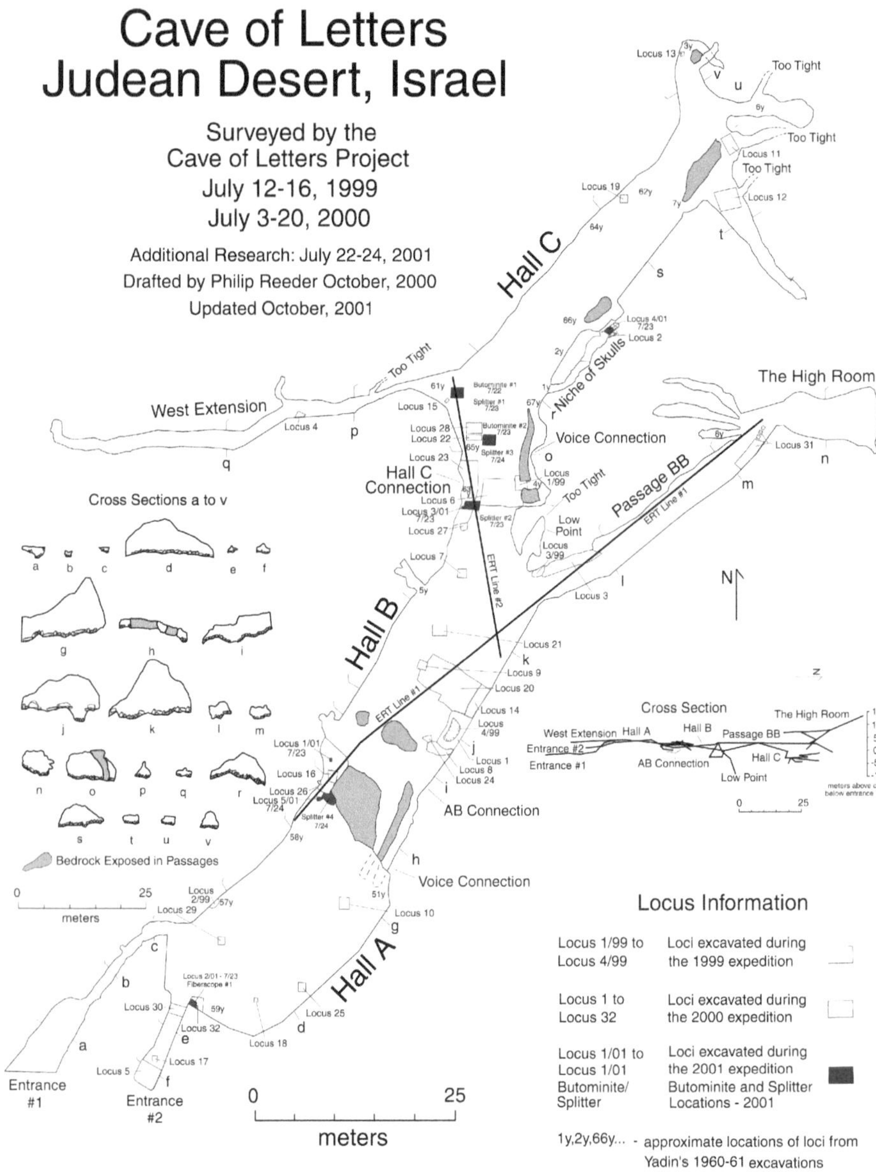

Figure 1.4: Final Survey Map of the Cave of Letters.

NOTES

1. On the Chalcolithic layer see for example: Yigael Yadin, *The Finds from the Bar-Kokhba Period in the "Cave of Letters."* (Judean Desert Studies; Jerusalem: Israel Exploration Society, 1963), 27, Figure 43, 111
2. Ya'akov Meshorer, *Ancient Jewish Coinage* (Dix Hills, N.Y.: Amphora Books, 1982), 20 and Yaakov Meshorer, "The Coins of Masada," in *Masada: The Yigael Yadin Excavations 1963–1965: Final Reports* (The Masada Reports; Jerusalem: Israel Exploration Society and The Hebrew University of Jerusalem, 1989), 102
3. Arieh Kindler, "Coins from the Cave of Letters," in *New Studies on the Bar Kokhba Revolt: Proceedings of the 21st Annual Conference on the Martin (Szusz) Department of Land of Israel Studies, March 13th 2001* (Ramat Gan: Bar Ilan University, 2001), 5–9.
4. Yadin, *Finds*, Plate 41, Figure 52. Fig. 52 lists the carpentry as a whole leg (8.3), two fragments of legs and various other fragments, of a piece of furniture, possibility a table-found at several spots in the cave.
5. Yadin, *Finds*, 134.
6. Bed and bed sockets are well known for a variety of reasons in Rabbinic literature. In *Mishnah Kelim* 19.6: "if a bed [that was made impure by the person who laid in it or came in contact with it] two longer sides were taken away and the two new ones were made for it, yet the sockets remained unchanged…." The sockets of the bed leg we discovered were clearly created with great work here and are not easily replaced since they required a large enough tree trunk to hold the pieces in place and yet not too large as to make the socket building to be extremely difficult. These beds are discussed in Rabbinic literature also in *Tosefta Kelim/Baba Metzia* 8.8.
7. Yadin, *Finds*, 112.
8. Yadin, *Finds*, 115 note 4: "On the basis of our finds, it is now definitely possible-in spite of Kahane's doubts…to continue the span of this type [Herodian type lamps] to the mid-second century A.D. "These were from the adjacent Cave to the Cave of Letters and the Roman camp-which itself raises other considerations of how long the camp was used." On why the Herodian lamps and the other lamps found in the Cave of Letters should be redated to correspond to the first century CE see Jodi Magness, "The Chronology of Qumran, Ein Feshka, and Ein El-Ghuweir," in *Mogilany 1995: Papers on the Dead Sea Scrolls Offered in Memory of Aleksy Klawek* (Qumranica Mogilanensia, 15, Krakow. Enigma Press, 1998), 55 76.
9. Found in "Hall A, two limestone vessels with knife shaved sides". Yadin, *Finds*, 112.
10. Yadin, *Finds*, 112 note 6. He writes they are: "…a direct continuation of the cooking pots found in Phase II (first century CE)"
11. Yadin, *Finds*, 114. The three from the Cave of Letters only.
12. Yadin, *Finds*, 115.
13. This is an updated version which was originally written as part of the scientific volume work of the Merrill Cave of Letters. The original article was published as Walter C. Bouzard Jr., "The Date of the Psalms Scrolls from the Cave of Letters Reconsidered," *Dead Sea Discoveries* 10, no. 3 (2003): 319–37
14. Peter Flint, "Biblical Scrolls from Naḥal Ḥever and 'Wadi Seiyâl': Introduction," in *Miscellaneous Texts from the Judaean Desert* (Discoveries in the Judaean Desert; Oxford; New York: Clarendon, 2000), 133–66, 173–82 + plates XXIV–XXVII, XXIX. (dating is on pages 143 ff.)
15. Yigael Yadin, *Bar-Kokhba; the Rediscovery of the Legendary Hero of the Last Jewish Revolt Against Imperial Rome* (London: Weidenfeld and Nicolson, 1971), 64–65.

16. Yadin, *Finds*, 31.
17. Yadin, *Finds*, 31.
18. Yadin, *Finds*, 32.
19. See Boaretto Table 1: RTT 3726–1.2, 50–340 CE, RTT 3726–1.1, 20–250 CE, RTT 3800 120–390 CE, RTT 3726, 130–350 CE. The calibrated dates were provided.
20. Yohanan Aharoni and Beno Rothenberg. *In the Foosteps of Kings and Rebels in the Judean Desert (Hebrew)* (Ramar Gan, Israel: Massada, 1960), 131.
21. Yadin, *Finds*, 34.
22. William G. Weart, "Bible Scroll 'Find' Suspected as Hoax; Dr. Zeitlin of Dropsie College Splits with Other Scholars on Dead Sea Discovery," *The New York Times* (March 4, 1949), page 19. Solomon Zeitlin's, works are remarkable in their breadth and number and his steadfast reluctance to accept the Scrolls as anything but medieval Karaite works. The first salvo began with Professor Zeitlin's 1949 Jewish Quarterly Review article entitled: "A Commentary on the Book of Habakkuk: Important Discovery or Hoax?" Solomon Zeitlin, "A Commentary on the Book of Habakkuk: Important Discovery or Hoax?" in *The Jewish Quarterly Review.* (Philadelphia: Dropsie University, 1949), 235–47. It continued unabated throughout the 1950s. See: Solomon Zeitlin, *The Dead Sea Scrolls and Modern Scholarship* (The Jewish Quarterly Review. Monograph Series.3; Philadelphia: Dropsie College for Hebrew and Cognate Learning, 1956), Solomon Zeitlin, "The Idolatry of the Dead Sea Scrolls," in *The Jewish Quarterly Review. New Series [Serial]* (Philadelphia: Dropsie University, 1958), 243–78 and Solomon Zeitlin, "More Literature on the Dead Sea Scrolls—More Pseudo-Scholarship," in *The Jewish Quarterly Review. New Series [Serial]* (Philadelphia: Dropsie University, 1959), 221–38, among the many more articles and chapters in books.
23. Magen Broshi A. J. Timothy Jull. D.J. Donohue Emanuel Tov, "Radiocarbon Dating of Scrolls and Linen Fragments from the Judean Desert," in *'Atiqot. English Series* (Jerusalem: Dept. of antiquities and Museums in the Ministry of Education and Culture: Dept. of Archaeology, Hebrew University, 1996), 64 ff.
24. M. Broshi G. Bonani I. Carmi, "Radiocarbon Dating of the Dead Sea Scrolls," in *'Atiqot. English Series* (Jerusalem: Dept. of antiquities and Museums in the Ministry of Education and Culture: Dept. of Archaeology, Hebrew University, 1991), 28.
25. F. E. Zeuner, "Notes on Qumran," in *Palestine Exploration Quarterly* (London: Palestine Exploration Fund, 1960), 27–28
26. Magen Broshi, "The New Radiocarbon Dates of the Dead Sea Scrolls and Their Significance," in *The Practical Impact of Science on Near Eastern and Aegean Archaeology* (Wiener Laboratory Publication.3; London: Archetype Publications, for the Wiener Laboratory, American School of Classical Studies, Athens, 1999), 108.
27. The dates before present (the established date is from the time that C14 testing began or 1950 CE) are corrected by a number of factors. These are the calibrated dates according to the lab.

BIBLIOGRAPHY

Aharoni, Yohanan, Beno Rothenberg. *In the Footsteps of Kings and Rebels in the Judean Desert (Hebrew).* Ramar Gan, Israel: Massada, 1960.

Bonani, G., M. Broshi, I. Carmi. "Radiocarbon Dating of the Dead Sea Scrolls." Pages 27–32 in *'Atiqot. English Series.* Jerusalem: Dept. of antiquities and Museums in the Ministry of Education and Culture: Dept. of Archaeology, Hebrew University, 1991.

Bouzard, Walter C. Jr. "The Date of the Psalms Scrolls from the Cave of Letters Reconsidered." *Dead Sea Discoveries* 10, no. 3 (2003): 319–37. *Dead Sea Discoveries: A Journal of Current Research on the Scrolls and Related Literature.* Leiden; New York: Brill.

Broshi, Magen. "The New Radiocarbon Dates of the Dead Sea Scrolls and Their Significance." In *The Practical Impact of Science on Near Eastern and Aegean Archaeology.* Wiener Laboratory Publication. 3. London: Archetype Publications, for the Wiener Laboratory, American School of Classical Studies, Athens, 1999.

Flint, Peter. "Biblical Scrolls from Naḥal Ḥever and 'Wadi Seiyâl': Introduction." Pages 133–66, 173–82 + plates XXIV–XXVII, XXIX in *Miscellaneous Texts from the Judaean Desert.* Discoveries in the Judaean Desert, Oxford; New York: Clarendon, 2000.

Jull, A. J. Timothy, D. J. Donohue, Magen Broshi Emanuel Tov. "Radiocarbon Dating of Scrolls and Linen Fragments from the Judean Desert." Page 64 in *'Atiqot. English Series.* Jerusalem: Dept. of antiquities and Museums in the Ministry of Education and Culture: Dept. of Archaeology, Hebrew University, 1996.

Kindler, Arieh. "Coins from the Cave of Letters." In *New Studies on the Bar Kokhba Revolt: Proceedings of the 21st Annual Conference on the Martin (Szusz) Department of Land of Israel Studies, March 13th 2001.* Ramat Gan: Bar Ilan University, 2001.

Magness, Jodi. "The Chronology of Qumran, Ein Feshka, and Ein El-Ghuweir." In *Mogilany 1995: Papers on the Dead Sea Scrolls Offered in Memory of Aleksy Klawek.* Qumranica Mogilanensia, 15. Krakow: Enigma Press, 1998.

Meshorer, Ya'akov. *Ancient Jewish Coinage.* Dix Hills, New York: Amphora Books, 1982.

Meshorer, Yaakov. "The Coins of Masada." In *Masada: The Yigael Yadin Excavations 1963–1965: Final Reports.* The Masada Reports. Jerusalem: Israel Exploration Society and The Hebrew University of Jerusalem, 1989.

Yadin, Yigael. *Bar-Kokhba; the Rediscovery of the Legendary Hero of the Last Jewish Revolt Against Imperial Rome.* London: Weidenfeld and Nicolson, 1971.

———. *The Finds from the Bar-Kokhba Period in the "Cave of Letters."* Judean Desert Studies. Jerusalem: Israel Exploration Society, 1963.

Zeitlin, Solomon. "A Commentary on the Book of Habakkuk: Important Discovery or Hoax?" Pages 235–47 in *The Jewish Quarterly Review.* Philadelphia: Dropsie University, 1949.

———. *The Dead Sea Scrolls and Modern Scholarship.* The Jewish Quarterly Review. Monograph Series. 3. Philadelphia: Dropsie College for Hebrew and Cognate Learning, 1956.

———. "The Idolatry of the Dead Sea Scrolls." Pages 243–78 in *The Jewish Quarterly Review. New Series [Serial].* Philadelphia: Dropsie University, 1958.

———. "More Literature on the Dead Sea Scrolls–More Pseudo-Scholarship." Pages 221–38 in *The Jewish Quarterly Review. New Series [Serial].* Philadelphia: Dropsie University, 1959.

Zeuner, F. E. "Notes on Qumran." Pages 27–28 in *Palestine Exploration Quarterly.* London: Palestine Exploration Fund, 1960.

CHAPTER TWO

Depositional Processes and Paleoenvironmental Implications of the Cave of the Letters and Other Rift-Shoulder Deposits

AMOS FRUMKIN

Institute of Earth Sciences, The Hebrew University of Jerusalem, Jerusalem, Israel

The Cave of the Letters contains a unique section of one of the highest aquatic sediments along the shoulders of the Dead Sea depression. The cave has acted as a sediment trap, preserving autogenic dolomite and detritic deposits. The dolomitic sediment may correlate with late Miocene dolomites within the rift valley. The morpho-stratigraphic setting of the sediment implies deposition within an early topographic low which existed in the Dead Sea region more than 3 million years (Ma) ago.

INTRODUCTION

The Cave of the Letters is located 3 km west of the Dead Sea (Figures 2.1 and 2.2). The Dead Sea depression, the lowest surface on Earth at −420 m MSL (in 2007 CE), is the base level for a ~43,000 km^2 endoreic catchment. The Dead Sea depression has developed as a pull-apart basin within the Dead Sea transform, containing fluvial and evaporite beds of Miocene to Holocene age. The sediments are associated with several late Cenozoic lakes which occupied the 100 km long, 20 km wide topographic depression enclosing the present Dead Sea. Lack of dating methods for most of these continental sediments and the limited morpho-stratigraphic extent of their outcrops do not allow direct determination of the age of the Dead Sea depression and its deposits. The older part of the sedimentary sequence within the basin is only partly known from limited outcrops and boreholes. Addressing the time and space dimensions of the commencement of the Dead Sea depression is approached here by using sediments in the Cave of the Letters and associated relict deposits. Understanding the depositional environments of such sediments can serve as indicators of past lake levels, rates of tectonic subsidence, and base level changes. The western Dead Sea depression escarpment is up to ~700 m high, or ~+300 m MSL within two km of the Dead Sea.

Here I discuss the early rift-related aquatic sediments at +160 m MSL of the Cave of the Letters, and combine it with aquatic sediments at higher, similar and lower elevations, to determine the initial stage of the Dead Sea depression development. The Cave of the Letters entrance is located along the northern wall of Naḥal Ḥever (Figure 2.3), an ephemeral stream draining some 180 km^2 of the Judean Desert into the Dead Sea. Much of its runoff originates from the semi-arid zone ~20 km west of the cave. It flows along most of its course in a shallow channel. Five km west of the Dead Sea it enters a deep box-shaped canyon through which Naḥal Ḥever enters the Dead Sea depression. The bottom of Naḥal Ḥever canyon below the Cave of the Letters is at −70 m MSL (below Mediterranean sea level), the cave is at +160 m (above Mediterranean sea level),and the top of the escarpment above the cave is at +230 m MSL.

Since the first encounter of aquatic sediments in the Cave of the Letters was discussed previously,[1] several outcrops of similar sediments have been found in various caves along the Dead Sea depression.[2] These two papers include references to earlier geological studies which are omitted here.

THE IMPORTANCE OF THE CAVE OF THE LETTERS IN VIEW OF REGIONAL STUDIES

The Cave of the Letters has been archaeologically excavated since 1953. The geological survey and paleoenvironmental study presented here were performed from 1981 to 1996 by the Cave Research Center of the Hebrew University of Jerusalem. Between 2000 and 2007 an interdisciplinary study of the Judean Desert caves[3,4] has been conducted by the Cave Research Center of the Hebrew University of Jerusalem and Bar Ilan University, discovering ~400 hundreds 'new' caves, some with documents[5] and treasures[6] attributed to the Bar Kokhba revolt.[7] Even after this comprehensive study, the Cave of the Letters is unique among the Dead Sea depression caves in several ways: (1) it is the largest limestone cave, by length and volume; (2) it contained the most important finds of the Bar Kokhba revolt; and (3) it contains the thickest late Cenozoic cave deposits of the Dead Sea depression.

STUDY METHODS

In order to allow human entry to the hardly accessible cave, we built a bridge over the chasm leading to the cave entrance, fixed ropes and an 11 m long ladder for climbing to the western entrance. These were used during our six expeditions to the cave (from 1981 to 1996), and later dismantled for conservation reasons.

We surveyed the Cave of the Letters by a 'Disto' laser range meter, compass and inclinometer. Our survey mainly addressed the parts of the cave not surveyed earlier by Yadin's expedition, in addition to preparing a vertical profile through the main cave passage (Figure 2.4). The geology of the cave was studied in the field using a Brunton compass and sampling equipment. An 11 m thick profile of cave sediments was described and sampled at the cave entrance (Figure 2.5).

The cave sediments were studied by the following methods and equipment:

(1) Careful visual inspection;
(2) Joel 840 scanning electron microscope with energy dispersive system;
(3) X-ray diffraction using a PW 3020 diffractometer for mineralogical analyses. Approximate proportions of the various minerals were calculated using conversion factors based on the height of the principal peaks of the XRD spectrum, calibrated with known proportions of local minerals;
(4) The chemistry of the sediments was determined by a Perkin Elmer Sciex ELAN-6000 ICP mass spectrometer;
(5) Granulometric analysis was performed by wet sieving for particles above 0.063 mm, and by Sedigraph 5000ET particle size analyzer for smaller particles,
(6) Stable isotope composition of dolomites was measured using a VG SIRA mass spectrometer with an automated carbonate analyzer.

THE GEOLOGY OF THE CAVE OF THE LETTERS

The Cave of the Letters is within the Shivta Formation limestone of Turonian age, belonging to the Judea Group. This formation contains the most carbonate caves in Israel. Within the 300 m high escarpment of Naḥal Ḥever, almost no caves are exposed in other formations, above or below the Cave of the Letters, but several other caves do exist in the Shivta Formation.

The cave is generally sub-horizontal, with local slopes mainly controlled by differential collapse piles. The main passage is divided into several large voids (A-B-C,

Figure 2.4). Some smaller side passages were also surveyed, the main ones being C-D, B-a, and bb' (Figure 2.4). The total measured length of the cave is 324 m. The main passage, from entrance I through rooms A, B, C to E is 173 m long. The cave has two entrances (Figures 2.3 and 2.4), where Naḥal Ḥever truncated two side passages.

Collapse features are common in most parts of the cave, especially in room B and the inner part of room C. Most of the cave floor is covered by collapse debris. Most of the ceiling consists of bedrock planes of weakness (fractures and bedding planes), along which the former ceiling has detached and fallen by gravitation and earthquakes. Our study of ~400 Judean Desert caves indicates that the collapse events in various caves were not coeval. Major collapse events occurred before the Chalcolithic period and between the Chalcolithic and Roman periods. The breakdown of the Cave of the Letters ceiling is enhanced compared with other caves, due to the large extent of the ceiling, rendering it relatively unstable. The collapse of the cave ceiling has commonly stabilized temporarily in a self-supporting dome-like structure.

Morphologic features indicating formation under slow-moving water conditions, such as solution pockets and smooth walls are observed in room A and in smaller passages. They indicate that the cave was formed by water which filled the cave completely, under confined conditions, with a probable hypogenic source. The confining rocks are chalk and marl of the Mount Scopus Group, which still covers most of the Judean Desert plateau today. The cave had formed in an ancient aquifer, isolated from the surface, and preceding the entrenchment Naḥal Ḥever canyon. Only in a later period was the cave breeched by the deepening of Naḥal Ḥever.

THE CAVE SEDIMENTS

The sediment fill that occurs in the Cave of the Letters is best exposed in the western entrance (I in Figure 2.4). At this point the canyon wall cutting the cave exposed a 13.5 m thick vertical section of the sediments that had filled the cave passage up to the roof (Figure 2.5). The sediments are also exposed along the walls and bottom of the entrance passage up to room A (Figure 2.6).

The lowermost ~2.5 m of the Cave of the Letters sediment fill consist mainly of thinly bedded to laminated, gray to yellowish soft dolomitic deposit (Item 2 contained within Figure 2.5). The deposit consists of ~90% dolomite crystals 0.1–0.01 mm is size, and minor deposits of detritic kaolinite and illite clays (Figure 2.7). The dolomite crystals are overgrown by ~0.001 mm quartz crystals, which comprise up to ~10% of the non-clay part of the sediment. Barite and K-feldspar crystals appear in minor quantities. Voids between the dolomite crystals are sometimes filled with gypsum. Microfossils or other fauna have not been found. Overall, the lower 2.5 m of the section indicate an estuarine—lagoonal environment, with minor detritic influx.

The dolomitic sediment has sagged after the dewatering of the cave (Figure 2.8). The sediment is also distorted by limestone boulders, up to a few m in size, embedded in the sediment fill (Figure 2.9, and item 3 in Figure 2.5). Limestone rock fragments are present within the entire section, originating from disintegration of the cave walls and ceiling while the fine-grained sediment was deposited in the cave.

The upper 11 m of the cave sediments consist (in addition to rock fragments) of fairly well-sorted silty clastics, with some clay and sand fractions (Figure 2.10, and item 4 in Figure 2.5). The clastics are moderately bedded to laminated, often with micro-ripple cross bedding structures. The individual grains are well-rounded, indicating transport by water. The dominant minerals are calcite and quartz. The clay fraction of the clastics consists of kaolinite, illite, and traces of smectite. The non-clay fraction consists of 20–60% calcite, 20–50% quartz, 10–50% plagioclase, 5–10% dolomite, and 0–5% halite.

The present cave passage from the entrance inwards along the first 25 m was created by secondary removal, mostly anthropogenic, of sediments from the upper 1.5 m clastic layers (Figure 2.6, and item 5 in Figure 2.5). The uppermost sediments are observed also in some inner parts of the cave where they are not covered by collapse debris, or where the overlying rubble has been removed by archaeological excavation.

Calcite speleothems and pedogenic clay derived from the surface, common inside sealed caves in Israel, are hardly observed in the Cave of the Letters sediments. Both speleothems and pedogenic clay are rare, like in most other Judean Desert caves, due to the present-day arid climate. Speleothems that do occur in the region are U-Th dated to glacial periods, which were considerably wetter in the Dead Sea region.

POTENTIAL SOURCES OF DOLOMITIC DEPOSITS

In order to identify the possible sources of the dolomitic intra-cave deposit, some associated dolomitic rocks are discussed briefly. Relevant dolomites can be divided into three types:

(1) Bedrock stratigraphic dolomite: Four formations crop out across the Naḥal Ḥever canyon near the Cave of the Letters, all belonging to the late Cretaceous Judea Group (Figure 2.11) Avnon, Derorim (alternatively termed Tamar2), Shivta and Nezer. Of these, only Derorim Formation (Tamar2) is composed of dolomite along the whole length of theoutcrop of Naḥal Ḥever canyon. Being just under the Shivta Formation, in which the Cave of the Letters is situated, the Derorim (Tamar2) Formation cannot be ruled out *a priori* as a potential source of the Cave

of the Letters dolomite (Figure 2.12). (2) Bedrock discordant dolomites: The Shivta and Nezer Formations are composed primarily of limestone, but they have undergone discordant dolomitization in many sites along the Dead Sea depression. The discordant dolomites are epigenetic, appearing as irregular bodies within the upper part of the late Cretaceous rocks. Such a dolomitic body cuts through the upper part of Naḥal Ḥever escarpment a few hundred m east of the Cave of the Letters, but no such dolomitization has been noticed in the limestone walls of the Cave of the Letters itself and its immediate surroundings. (3) Dolomites within the sedimentary fill of the rift valley region: These appear in two formations: Sedom Formation in the Dead Sea depression and the Bira Formation in the vicinity of the Jordan River valley. These dolomites are associated with estuarine—lagoonal—evaporitic environments within the early depressions in the vicinity of the rift valley. The evaporitic conditions in both regions were later replaced by aquatic environments, depositing the Amora Formation in the Dead Sea depression and the Gesher Formation in the Jordan River valley (Figure 2.13).

Although the Cave of the Letters dolomite is disconnected from other dolomites, the chemical composition may indicate a possible correlation across various outcrops.

The various dolomite types are distinctive from each other by their Sr/Ca ratios (Figure 2.14). The Derorim (Tamar2) dolomite and the discordant dolomites have Sr/Ca equivalent ratios of $0.1–0.2 \times 10^{-3}$ and $0.2–0.4 \times 10^{-3}$ respectively. The Cave of the Letters dolomite Sr/Ca ratios vary from 0.5 to 0.9×10^{-3}, outside the range of Derorim (Tamar2) and the discordant dolomites. Sedom and Bira Formation dolomites have Sr/Ca equivalent ratios of $0.5–0.9 \times 10^{-3}$, the same range as the Cave of the Letters dolomite. In addition, the Na/Ca ratio of the Cave of the Letters dolomite is within the range of Sedom dolomites and close to the range of the Bira Formation dolomite, but two orders of magnitude higher than in Derorim (Tamar2) and the discordant dolomites (Figure 2.15).

Isotopic composition may also help to distinguish between the various types of dolomites. The Cave of the Letters dolomite, Sedom Formation dolomite and Bira Formation dolomite have $\delta^{13}C$ values $<-1‰$ (vPDB); on the other hand, the Derorim (Tamar2) dolomite and the discordant dolomites have $\delta^{13}C>-1‰$. This is another indication that the Cave of the Letters dolomite belongs to the same group as the Sedom and Bira dolomites. However, large (~6‰) temporal fluctuations of $\delta^{13}C$ in Quaternary lacustrine deposits of the Dead Sea basin suggest that this indicator may be used only after a detailed isotopic study is performed in each sedimentary sequence.

It may be concluded that no physical or chemical process could produce the Cave of the Letters dolomite by decomposition and redeposition of the Derorim (Tamar2) Formation, or the discordant dolomites located in close proximity to the cave. Rather, a closer association with the Sedom and Bira Formations can be inferred from the Sr/Ca ratios. Other caves seem to present a more complicated picture.

DEPOSITIONAL ENVIRONMENT OF DETRITIC SEDIMENT IN THE CAVE OF THE LETTERS

The particle-size and sediment morphology of the Cave of the Letters detritic deposits are compared with recent deposits of Naḥal Ḥever fan-delta, whose deposition environment can be identified directly. As in the fan-delta, two main depositional environments may be recognized in the Cave of the Letters deposits (Figure 2.16):

(1) Shallow Fluvio-lacustrine environment, with sub-horizontal, laminar clay, silt and sand, and micro-ripple cross-bedding; (2) Fluvio-lacustrine environment, with larger percentage of silt and clay, lesser amounts of sand, and no cross-bedding. The Cave of the Letters sediments is enriched with silt and clay fractions (Figure 2.16). This indicates that a large portion of the sediment influx was injected into a low-energy water body from a nearby fan-delta under low-energy conditions. In addition, the cave itself acts as a low-energy sediment trap, allowing fine-grained particles to settle from suspension where flow velocity decreases dramatically. Overall, the characteristics of the clastic sediments of the Cave of the Letters indicate deposition in a fan-delta at the edge of an open water body with detritic influx from a nearby stream.

The high proportion of plagioclase and quartz grains within the Cave of the Letters detritus raises a question of the source of these clasts. Today there are no sources of plagioclase in the Naḥal Ḥever catchment, and fine quartz detritus is relatively rare. The nearest possible source is the Hazeva Formation which had covered much of the Negev and southern Judean Desert during the Miocene, and whose remnants are still ubiquitous today twenty km SW of the Cave of the Letters, at an elevation of +600 m. Hazeva Formation sandstones are composed of 50–80% quartz and 10 to 30% feldspar grains. Plagioclase is also common in rocks exposed on the eastern side of the Dead Sea. The topographic low of the Dead Sea depression between these rocks and the Cave of the Letters, and the eastward flow direction of Naḥal Ḥever render it improbable for detritus from the east to reach the Cave of the Letters after the cave was breeched by Naḥal Ḥever. Therefore the Hazeva sands seem to be a more likely source of the Cave of the Letters detritus. This suggests that Hazeva Formation covered much of Naḥal Ḥever catchment during the clastic deposition in the Cave of the Letters, and only later was it eroded from the surface of this part of the Judean Desert.

AQUATIC DEPOSITS IN OTHER CAVES

Following the discovery of the Cave of the Letters sediments, several other caves with aquatic sediments were found nearby, forming a cluster around 105 to 196 m MSL.[8] Secondary dolomite deposits were documented in the following caves: 301, 303, 353b, whose altitudes are +196, +130, +105 m, respectively. South of the Dead Sea, Langozky 1,2, and Langozky 3 caves, at −99, and −125 m MSL respectively, also contain aquatic dolomite. (Figures 2.1 and 2.9). The soft laminar cave dolomite significantly differs in its structure, morphology and petrography from the hard massive surrounding bedrock. Laminar aquatic marl deposits were documented in Cave 160c, +111 m MSL (Figures 2.1 and 2.17). These intra-cave sediments may indicate the spatial distribution of the aquatic bodies which deposited them. More recently, aquatic marl and silt were documented in Masada 2008 Cave, and dated by cosmogenic isotopes to ~3 Ma.[9]

BASE LEVEL EVOLUTION AND SEDIMENT TERRACES ALONG THE DEAD SEA DEPRESSION

It is commonly assumed that early to middle Miocene sedimentation within the Dead Sea transform kept pace with tectonic subsidence. Under these conditions the Hazeva Formation was deposited as a continuous cover across the rift, including the Naḥal Ḥever catchment. When the Hazeva sediment supply from the south-east was cut off, or the subsidence rate increased, the sedimentation rate no longer kept pace with the subsidence of the rift bottom. Consequently the topographic low of the Dead Sea depression was formed, accompanied by some uplift of the rift shoulders. The lowering of base level was followed by major diversion of the regional drainage system into the Dead Sea depression. This was followed by rapid down cutting of surrounding stream valleys, such as Naḥal Ḥever. In addition, the regional water table fell, adjusting to the declining base level. While base level was falling, sediments were deposited on the bottom and shoulders of the deepening tectonic trough. Long periods of deposition have left sediment terraces along the Dead Sea depression margins. As base level continued to fall, older sediments remained as elevated terraces on the depression shoulder, west of the border faults. Under such conditions of declining base level and entrenchment, high-elevation terraces are normally older than low-elevation ones on the same escarpment (Figures 2.13 and 2.17). This relative morpho-stratigraphy is sometimes the only clue to the age of sediments having neither indicative fossils, nor absolute dating control. Such sediment terraces will retain their relative position if they are all found on a specific structural block without internal deformation.

Four distinct terraces of aquatic sediments are identified along the Dead Sea depression below the presently studied uppermost cluster of cave deposits (Figure 2.17):

(1) Calcite chalks and marls at a level of 0 to +30 m, referred to as 'Mazar' sediments. The microfossil assemblage includes *Ammonia beccarii* and other species indicating brackish environment of Neogene to recent age. Similar outcrops are known further south. A hypsometric curve of the Dead Sea depression shows a prominent topographic terrace at 0 to +30 m altitude, possibly indicating that the Dead Sea depression contained a relatively stable water body at this elevation, serving as base level during a prolonged period.
(2) Silty-clay marl and oolithic limestone between −90 and −130 m MSL, reported several at sites along the Dead Sea depression (Figures 2.1 and 2.17), including 'Langozky caves' (above), and referred to as 'Samra''and 'Hamarmar' Formations.
(3) The Lisan Formation, cropping out along the Dead Sea depression and central Jordan valley, up to ~−150 m MSL. It has undergone only slight tectonic deformation and is well dated to the late Pleistocene.
(4) Holocene Dead Sea sediments. These well-preserved sediments indicate lake-level fluctuations from ~−375 m to ~−430 MSL during the mid-late Holocene.

The five levels of aquatic sediments described above, including the Cave of the Letters deposits and its associated cluster of caves, provide a basic framework for the Dead Sea depression base level development, although the higher levels are fragmentary. Naturally, tectonics and erosion have modified the upper levels more than the lower ones, because of the longer time span involved. In particular, the cluster of aquatic deposits at and near the Cave of the Letters may have been uplifted compared to other parts of the Dead Sea escarpment. The youngest deposits within the Cave of the Letters have been recently dated by cosmogenic isotopes to ~3 Ma.[9] This demonstrates the late Neogene chronology of the sediments and the associated development of the rift valley.

DISCUSSION

The dissolution morphology of the Cave of the Letters, mentioned above, indicates that the cave was completely filled with groundwater while being formed. The regional water table must have been above the cave during that period. Naḥal Ḥever could not have been entrenched down to level of the Cave of the Letters at this stage, because otherwise the cave would have been drained into the downcutting canyon. At this stage, the Dead Sea depression base level could not have been much lower than the cave, and probably the topographic depression did not yet exist. If the cave sediments are of late Neogene age, as suggested below, then the cave void was probably formed between Oligocene and middle Miocene times, prior to the formation of the Dead Sea depression.

The cave deposits show that the Cave of the Letters has acted as a trap for sediments originating outside the cave, rather than sediments of local intra-formational origin. Similar sediments have been preserved in other caves, indicating that such deposits had probably been deposited also outside the caves, but these must have been removed by subsequent erosion.

The following succession of events is inferred: The cave was formed under the water table, during a prolonged stable period. The falling groundwater levels associated with the Dead Sea depression subsidence left the cave high and dry above base level. During this period, or later, the cave was cut by the paleo-Naḥal Ḥever, whose entrenchment was also associated with the early stages of the Dead Sea depression development (Figure 2.11). Soon afterwards the cave was filled by aquatic sediments. The main cave passage, perpendicular to the canyon wall, was probably much longer before it was originally cut by the paleo-Naḥal Ḥever (Figure 2.11). Since then, the canyon has entrenched 230 m deeper, and its northern wall retreated laterally, consuming the southern part of the Cave of the Letters. Of the original cave, only the northern part has survived. The present two entrances of the cave (I and II on Figure 2.4), are located where the erosional scarp, in its present location, has cut two branches of the cave.

The high elevation of the Cave of the Letters deposit indicates that it is one of the earliest aquatic sediments on the Dead Sea depression shoulder. Only one higher outcrop has been found so far, in Cave 301 (Figure 2.17). On the eastern side of the escarpment, within the subsiding basin, the earliest sediment confined to the Dead Sea depression is the Sedom Formation, covered by a series of sediments up to the recent Dead Sea deposits. The Sedom Formation is not only the oldest in this nearby intra-rift sequence, but it is also the only intra-Dead Sea depression deposit containing dolomitic beds. These dolomites show Sr/Ca, and $\delta^{13}C$ values similar to the Cave of the Letters dolomite. A tentative correlation between the Cave of the Letters dolomite and Sedom Formation may therefore be suggested, supported also by the relative old age of both sediments. The limited dating evidence of these sediments suggest that such correlation may refer only to the upper part of Sedom Formation.

The Sedom Formation is composed of halite, interbedded with anhydrite, shales, and quartz sand, as well as dolomite. The formation crops out only at the Mount Sedom diapir, but it extends northward, giving rise to sub-lacustrine diapirs, such as the En Gedi diapir east of the Cave of the Letters. The formation was encountered in boreholes at a depth of several km within the fill of the southern part of the Dead Sea depression. The formation appears also beneath the Ami'az plateau between two border faults of the Dead Sea depression, displaying a thinner facies compared with the central basin. Layers of soft clay-like dolomite crop out in this transitional region.

The Jordan River valley—Dead Sea depression sedimentary fill includes <1% dolomite, but in each part of the region there is a lagoonal—estuarine rock unit that includes dolomitic beds (Figure 2.13): These are the Sedom Formation in the Dead Sea depression fill, the Bira Formation in the Jordan River valley, and the Cave of the Letters and associated cluster of caves in the Dead Sea depression shoulders. In each site the dolomite depositional stage was followed by lacustrine calcitic deposition which is illustrated by the following: Amora chalk and marl overlies Sedom Formation, Gesher chalk overlies Bira Formation, and Mazar chalk follows the Cave of the Letters dolomite according to its morpho-stratigraphic setting (Figure 2.13). It is possible that the separate dolomitic units belong to a single event, taking into account also their similar Sr/Ca ratios. Such an event could be associated with marine transgression into the rift depression, intensive evaporation and deposition of the evaporites in the Sedom and Bira Formations. Deposition of Ca-sulfate resulted in Mg-rich brines which deposited dolomites or caused dolomitization of calcium carbonates.

CONCLUSIONS

The 13.5 m thick fill of the Cave of the Letters is probably a relic of one of the earliest known sedimentary units at the Dead Sea depression shoulders, associated with the earliest topographic depression in this part of the Dead Sea transform. The development of the early Naḥal Ḥever canyon, which cut through the Cave of the Letters, indicates a major increase of subsidence/sedimentation ratio and the development of a deep land-locked topographic depression within the rift.

The dolomitic sediment of the Cave of the Letters indicates deposition in an estuarine-lagoonal water body, probably associated with the Sedom depositional environments.

ACKNOWLEDGMENTS

The expertise of Michael Bareket of the Israel Cave Research Center helped us in constructing the bridge to the Cave of the Letters during our second expedition (in 1982). Menashe Zaqen performed granulometric analysis at the geomorphology laboratory of the Hebrew University. Helpful discussions were held with Eli Raz of En Gedi, Prof. Hagai Ron, Prof. Amotz Agnon, Prof. Amitai Katz, Prof. Eitan Sass and Prof. Israel Zak of the Hebrew University, Prof. Emanuel Mazor of the Weizmann Institute of Science, Dr. Yoav Avni and Dr. Menahem Raab of the Geological Survey of Israel, and Prof. Amos Nur of Stanford University. The Nature Conservation Authority has kindly permitted the geological study within the Judean Desert Nature Reserve. The Dead Sea Research and Development and Israel's Ministry of Science assisted by providing research grants.

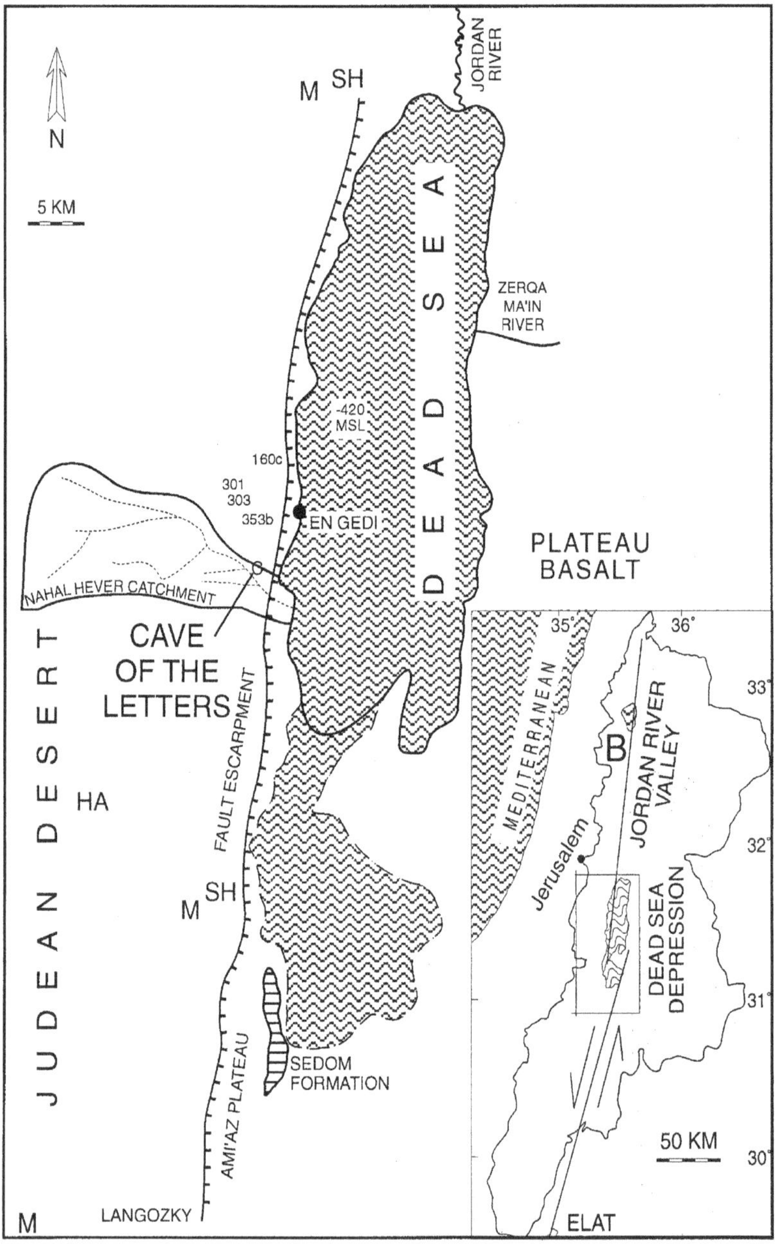

Figure 2.1: Cave of Letters location map showing additional sites of rift-related sediments on the Dead Sea escarpment.

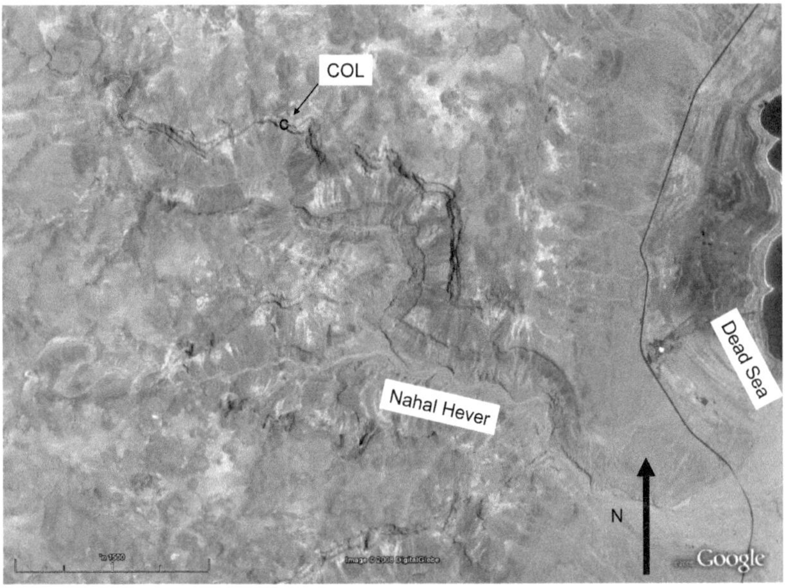

Figure 2.2: Location of the Cave of the Letters (COL) in Naḥal Ḥever: satellite view.

Figure 2.3: Location of the two entrances of the Cave of the Letters in the northern wall of Naḥal Ḥever: view from the south side of the canyon. Photo: Vladimir Buslov.

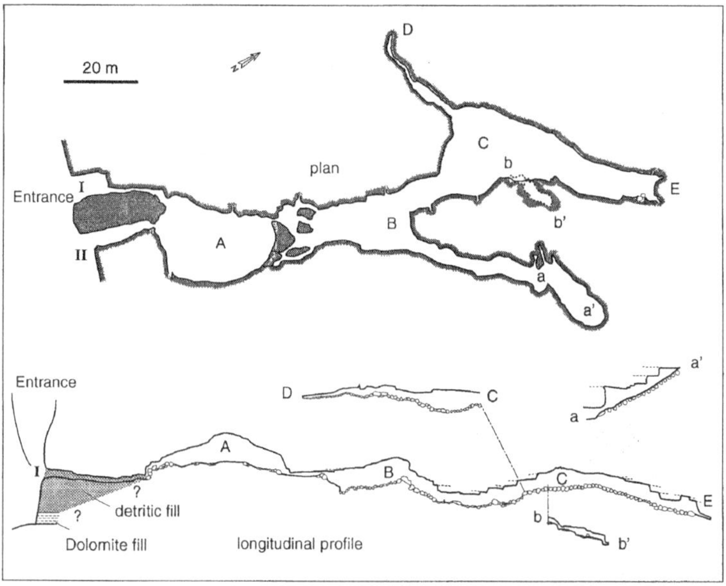

Figure 2.4: Profile and plan of the Cave of the Letters. Entrance 1 sediments are shown in the profile.

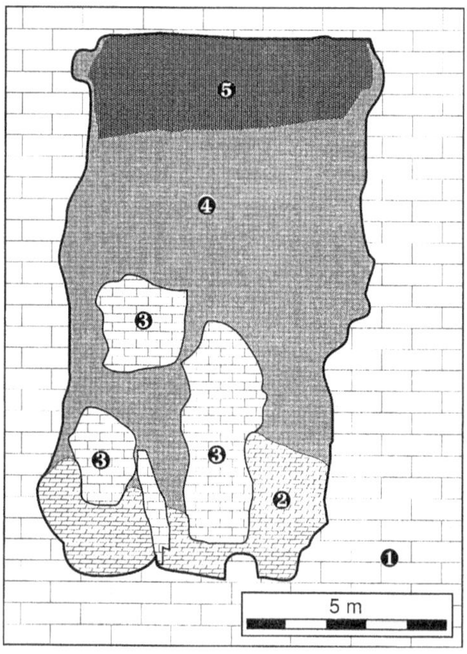

Figure 2.5: The cross-section of Entrance 1 sedimentary materials. 1 = bedrock limestone; 2 = laminar dolomitic cave sediment; 3 = collapsed limestone boulders; 4 = detrital silt deposit; 5 = cave void.

Figure 2.6: The limestone ceiling near entrance 1 and the detritic sediment fill below the ceiling.

Figure 2.7: Laminar dolomitic aquatic sediments at the bottom- left of entrance 1 section.

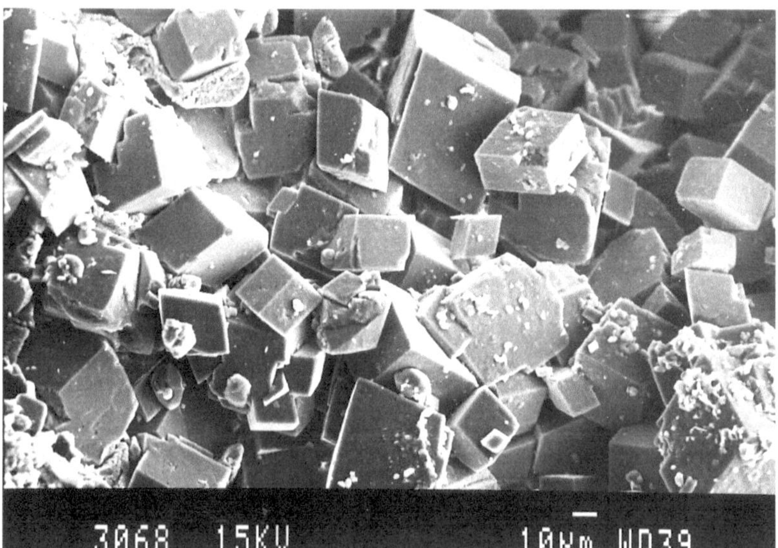

Figure 2.8: Dolomite crystals overgrown by smaller autogenic quartz crystals in a scanning electron microscope photograph.

Figure 2.9: View to entrance 1 of the Cave of the Letters (top) and its sediments.

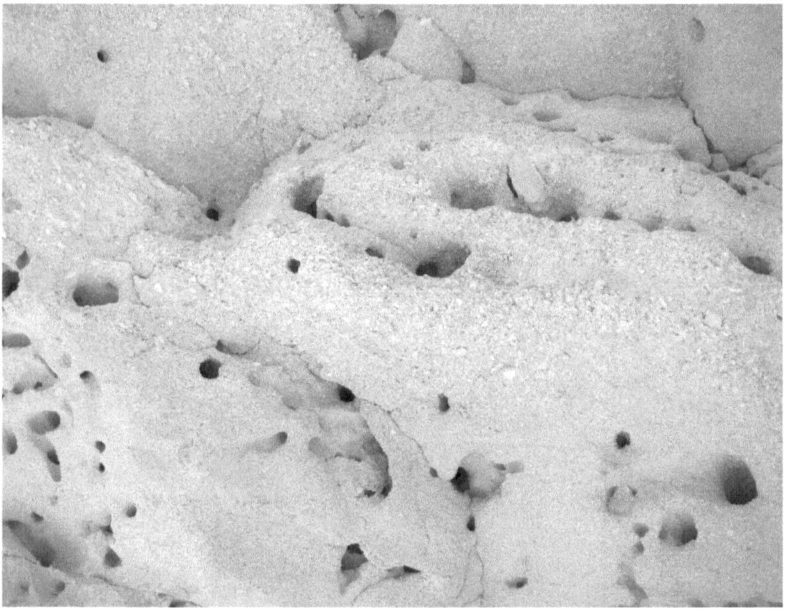

Figure 2.10: Sediments at the upper part of entrance 1.

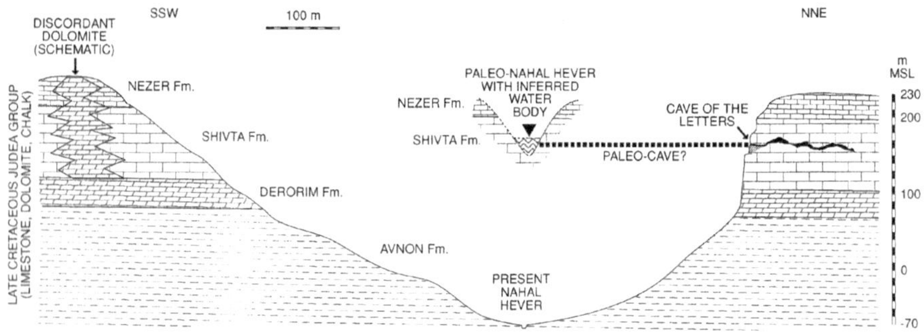

Figure 2.11: Geological cross section of Naḥal Ḥever canyon with the Cave of the Letters, showing also the probable morphology when the cave was cut by the canyon.

Figure 2.12: The Cave of the Letters sediments at entrance 1.

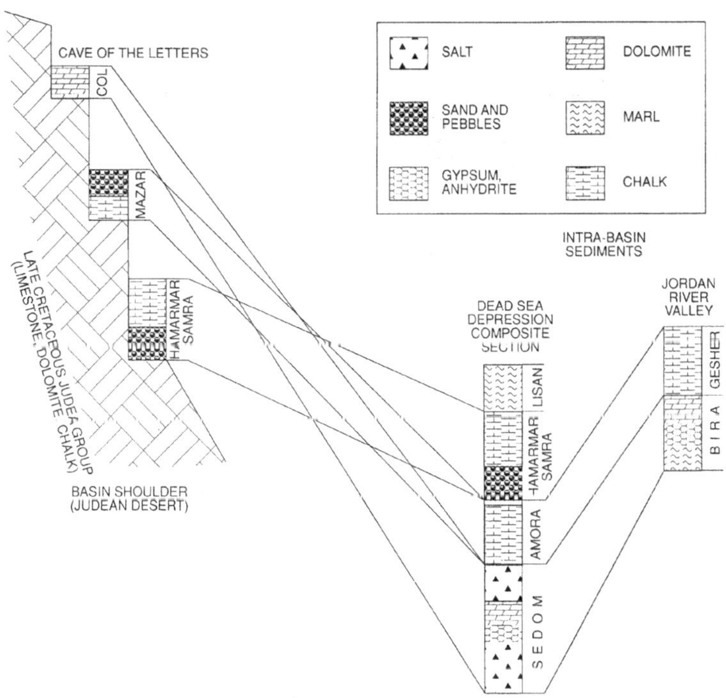

Figure 2.13: Schematic diagram showing possible late Cenozoic formations in the Dead Sea depression. 'Sedom' and 'Bira' refer to the uppermost parts of these formations.

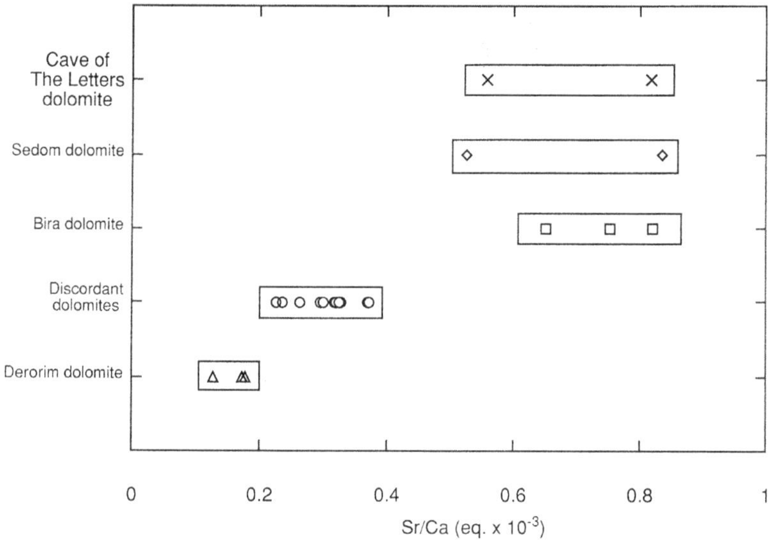

Figure 2.14: Equivalent ratios of Sr/Ca of the Cave of the Letters dolomitic sediment, compared with Judean Desert discordant dolomites and Derorim (Tamar2) dolomite and intra-rift dolomites of the Bira and Sedom Formations.

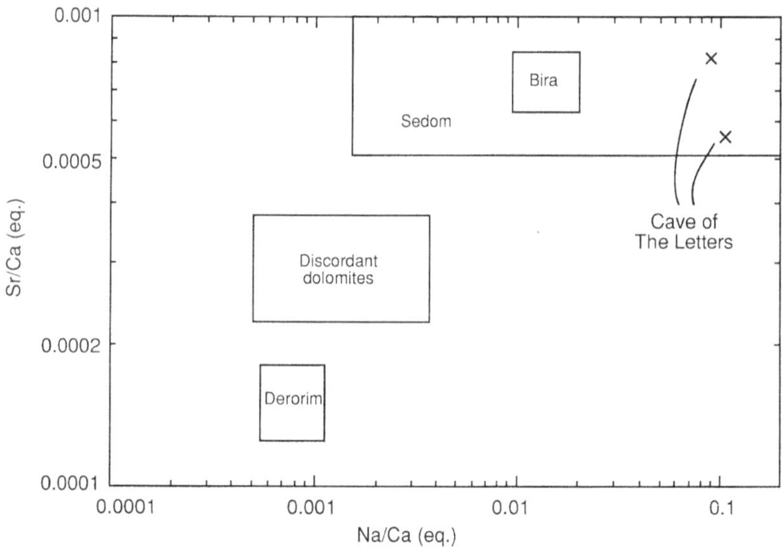

Figure 2.15: Equivalent ratios of Sr/Ca and Na/Ca of the Cave of the Letters dolomitic sediment, compared with other dolomites of the rift valley.

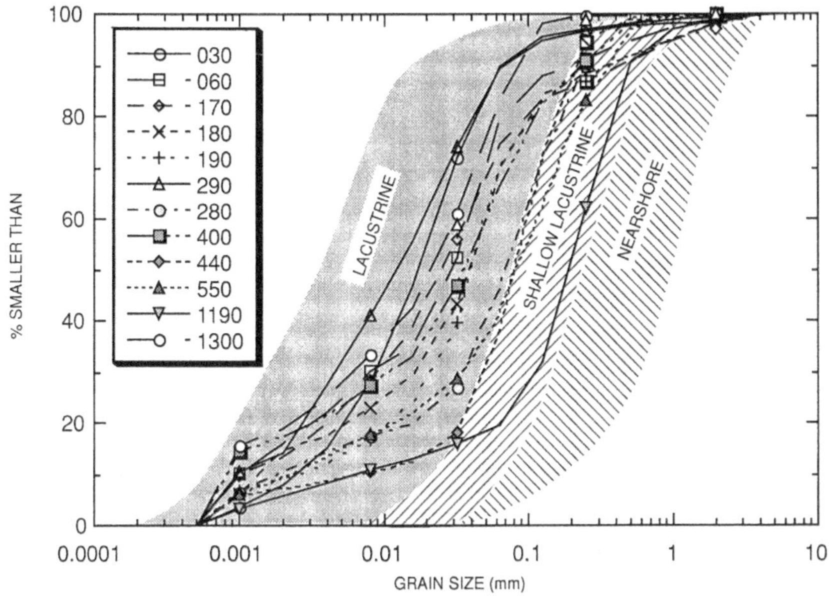

Figure 2.16: Particle size distribution of the Cave of the Letters deposits compared with size distribution of typical depositional environments.

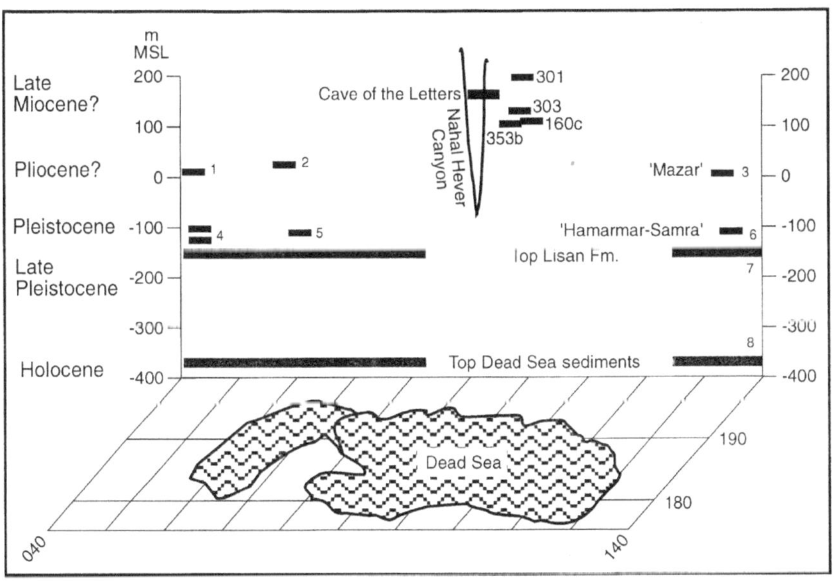

Figure 2.17: Five levels of aquatic deposits along the western shoulder of the Dead Sea, west of the major rift-border faults.

NOTES

1. Amos Frumkin, "The Cave of the Letters Sediments—Indication of an Early Phase of the Dead Sea Depression?" *Journal of Geology* 109, no. 1 (2001): 79–90.
2. Sorin Lisker, Roi Porat, and Amos Frumkin, "Late Neogene Rift Valley Fill Sediments Preserved in Caves of the Dead Sea Fault Escarpment (Israel): Palaeogeographic and Morphotectonic Implications," *Sedimentology* 57 (2010): 429–45.
3. Sorin Lisker, et al., "Late Quaternary Environmental and Human Events at En Gedi, Reflected by the Geology and Archaeology of the Moringa Cave (Dead Sea Area, Israel)," *Quaternary Research* 68, no. 2 (September 2007): 203–12.
4. Sorin Lisker, et al., "Stromatolites in Caves of the Dead Sea Fault Escarpment: Implications to Latest Pleistocene Lake Levels and Tectonic Subsidence," *Quaternary Science Reviews* 28 (2009): 80–92.
5. Nahum Cohen, "New Greek Papyri from a Cave in the Vicinity of Ein Gedi," *Scripta Classica Israelica* 25 (2006): 87–95.
6. Roi Porat, Hanan Eshel, and Amos Frumkin, "Two Groups of Coins from Ein-Gedi," *Israel Numismatic Journal* 15 (2006): 76–86.
7. Roi Porat, Hanan Eshel, and Amos Frumkin, "Finds from the Bar-Kochba Revolt from Two Caves in Ein Gedi," *Palestine Exploration Quarterly* 139, no. 1 (2007): 35–53.
8. Lisker et al., "Late Neogene rift valley fill sediments preserved in caves of the Dead Sea Fault Escarpment" (Israel).
9. Matmon, et al., 2014. Unraveling rift margin evolution and escarpment development ages along the Dead Sea fault using cosmogenic burial ages. *Quaternary Research* 82, 281–295.

BIBLIOGRAPHY

Cohen, Nahum. "New Greek Papyri from a Cave in the Vicinity of Ein Gedi." *Scripta Classica Israelica* 25 (2006): 87–95.

Frumkin, Amos. "The Cave of the Letters Sediments—Indication of an Early Phase of the Dead Sea Depression?" *Journal of Geology* 109, no. 1 (2001): 79–90.

Lisker, S., R. Porat, and A. Frumkin. "Late Neogene Rift Valley Fill Sediments Preserved in Caves of the Dead Sea Fault Escarpment (Israel): Palaeogeographic and Morphotectonic Implications." *Sedimentology* 57 (2010): 429–45.

Lisker, Soren, et al. "Late Quaternary Environmental and Human Events at En Gedi, Reflected by the Geology and Archaeology of the Moringa Cave (Dead Sea Area, Israel)." *Quaternary Research* 68, no. 2 (September 2007): 203–12.

Lisker, Soren, et al. "Stromatolites in Caves of the Dead Sea Fault Escarpment: Implications to Latest Pleistocene Lake Levels and Tectonic Subsidence." *Quaternary Science Reviews* 28 (2009): 80–92.

Porat, Roi, Hanan Eshel, and Amos Frumkin. "Finds from the Bar-Kochba Revolt from Two Caves in Ein Gedi." *Palestine Exploration Quarterly* 139, no. 1 (2007): 35–53.

———. "Two Groups of Coins from Ein-Gedi." *Israel Numismatic Journal* 15 (2006): 76–86.

CHAPTER THREE

Overview of New Geoarchaeological Discoveries in the Cave of Letters

JOHN (JACK) F. SHRODER
Department of Geography and Geology, University of Nebraska at Omaha, Omaha, NE 68182, USA

The Cave of Letters in the Naḥal Ḥever area of the Judean desert, close to the Dead Sea of Israel, was long presumed to have been exhaustively investigated over 40 years ago. Problems with the prior literature in terms of the timings and sources of certain cave-filling sediments were identified by us. Our work with geological and geomorphological survey and radiometric age dating, geophysical techniques (ground penetrating radar, electrical resistivity tomography, fiber-optic endoscopic viewing beneath roof-fall colluvium, metal detectors), cave mapping and dating, and excavation of large (1–2 m) and pervasive roof-fall boulders, showed that some additional archaeological material existed from before and during the Bar Kokhba revolt (132–135 C.E.[1]) against Roman rule. Roof fall colluvium likely produced by nearby high-magnitude seismicity, coupled with rock disruptive, apparently expansive halite speleothem growth, buried additional archaeological material. Our techniques for the analysis of the largely open-matrix, rock-fall clastics inside the cave enabled discovery of new artifacts and also offer some potential for further finds in other nearby caves, already well-known for prior discoveries of documents and other materials from antiquity.

INTRODUCTION

Background and Expedition Rationale

For the first time since the long presumed exhaustive archaeological investigations in 1953 by Yigdal Aharoni and in 1960–61 by Yigdal Yadin (1963, 1971) in the Cave of Letters (COL), Naḥal Ḥever, of the Judean Desert in Israel (Figure 3.1), our two officially sanctioned expeditions to the area in summer 1999 and 2000 have led to new discoveries, as well as new interpretations. The COL is important because it constitutes one of few other sites extant, other than a few coin hoards and the sketchiest of legendary materials, concerning the Bar Kokhba revolt (132–135 C.E.[1]) that was the final Jewish uprising against Roman rule. Because so little is known about this war in which both sides are suspected to have lost heavily, archaeology is one of the few ways to increase the historical record. An initial feasibility investigation by this author to the COL in winter 1998 led to discovery of the high likelihood of additional archaeological material preserved beneath extensive roof-fall colluvium in the cave, with the result that two officially sanctioned, full-scale expeditions (1999, 2000) were mounted by our research team thereafter. This paper is a summary of some of our overall objectives, techniques, and discoveries.

Since the Bar Kokhba period, several intrusions occurred thereafter into the fairly inaccessible cave. The first was probably Roman soldiers who may have looted the cave after all the inhabitants were starved to death. Hermits in the 4th–7th centuries C.E., which were typical in other caves of the area might have caused disorder or contamination to the remains, and Bedouin looters in the 1950s probably removed and redeposited rocks and some of the fine-sized clastic material. The Yadin expeditions continued for up to a month's time, with an average of 15 people jack-hammering, digging and moving roof-fall debris and finer clastic materials, especially in Hall A. The cumulative result has been considerable stratigraphic disorder and the all-too-common disturbance to superposition and associated temporal context.

In addition, and unbeknownst to us at the time of our expeditions because of a lack of prior publication, a number of other expeditions to COL had been mounted between 1981 to 1996 by Amos Frumkin and others. This work, which we partly replicated herein, was eventually published by Amos Frumkin in this volume and elsewhere.[2] This work is discussed in this paper, especially where our observations or interpretations differ somewhat.

Research Objectives

Beyond the importance of the COL site to further elucidation of important events in historical antiquity, a variety of reasons specific to the unusual environs of the

COL led to the designation of six major goals by excavation directors R. Arav and R. Freund (2000, 2001) from the University of Nebraska at Omaha.

- An initial investigatory probe without digging into all parts of the cave to provide focus on unusual sites requiring additional work, as well as resolving the problems of moving or removing quantities of overburden debris in the ecologically fragile internal and external environments.
- Investigation of the utility of non-disruptive geophysical techniques to probe the roof-fall colluvium overburden with ground-penetrating radar (GPR), electrical resistivity tomography (ERT), high gain and lower end 'hobby-type' metal detection, and endoscopic fiber-optic viewing beneath open-matrix boulder piles.
- Reinvestigation of outermost Hall A that was extensively jack hammered and dug up by Yadin's group.
- Use of extensive hand-held instrumental surveying and computer mapping of COL.
- Radiocarbon dating of organic material preserved on and beneath living floors in layers of rock debris.
- Documentation of the entire investigatory process underground and above ground in the field and in laboratories with video tapes and cinematography.

METHODOLOGY

Ideas about chaotic breakdown breccias in caves,[3] with new geoarchaeological understandings of possible roof-fall preservation of normally perishable artifacts,[4] led us to investigate use of several remote sensing geophysical techniques, as well as standard geological and new geographical mapping and microclimatological techniques. We used generator-driven electrical cave lighting and outlets, comprehensive, handheld instrumentation, and real-time, computer-driven, cave-mapping techniques, multiple resolution metal detectors, fiber-optic endoscopic viewing, and we constructed cross sections and tomographic imagery from probes by ground penetrating radar (GPR) and electrical resistivity tomography (ERT) systems described in detail elsewhere in this volume.

Geomorphology was analyzed using aerial photographs of the exterior of the cave, by extensive field survey, photography, and work with standard tape measurement and Brunton compass. Cave sediments and rock samples were collected in bags, and carbon-dateable materials in aluminum foil. Standard petrographic thin sections were made of selected earthenware oven fragments to determine mineralogy and petrology, and x-ray diffraction was used to determine mineralogy of speleothems. Species of wood specimens collected were determined using botanical thin sections.

Detailed cave mapping was undertaken by crawls of small agile people into many new spaces never, or only superficially traversed by previous workers. The resulting maps and cross sections differ considerably from Yadin's[5,6] prior efforts, as well as from Frumkin's[7] slight modifications of Yadin's work. As explained by Reeder (this volume) survey was with a hand-held optical compass, optical inclinometer and nylon survey tape measurements. The CAVEPLOT™ computer program enabled construction of line plots that allowed higher precision than previous estimates. Cross sections and longitudinal sections were first constructed of the cave and then exported into Adobe Illustrator™ to draw details of cave-roof falls and other morphological variation. This map was used as a basis for in-cave construction of geographic information system (GIS) layers of archaeological loci, numbers of archaeological finds, and attached real-time data bases for artifact locations and descriptions that were entered on a touch-screen laptop computer within the cave. This work added detail to the understanding of the development of the geomorphology of the cave.

A variety of 'hobby-type' metal detectors were used in this study (see Bauman, this volume). The whole cave was swept for possible deeply buried large metal objects using high-end metal detection instrumentation, but nothing was found. This was followed by the smaller hobby units that located a variety of small metal objects such as nails and coins. The higher sensitivity varieties of hobby-type or treasure hunter metal detectors were more effective at finding shallowly buried coins and small metal objects because of their small coils and shorter 'ramp' times.

GPR and ERT techniques were used together (Jol, this volume; Bauman, this volume) to reinforce detection of Bar Kokhba living floors because GPR provides reflection sections of unparallel resolution, but not always clearly as to what is constituting the reflection.[8]

The goal of the ERT survey was to image the contact between recent roof collapse, which would have a large percentage of air-filled void spaces (and a correspondingly high resistivity), and the underlying and therefore older cave-floor material, which would have significantly greater interstitial packing of fine clastics (and correspondingly lower resistivity). Thus the large boulders with open matrices in hyper-arid cave conditions seemed likely to limit ERT use because of the difficulty of putting current into such ground. Nevertheless, the presence of plentiful halite or rock salt in the cave (unreported heretofore), together with ~200 ml of water poured around each electrode point made feasible the transmission of galvanic (DC) current into the cave substrate. These methods contributed further to an understanding of the long-term geomorphologic development of the cave system.

The GPR and ERT procedures were followed by fiberoptics viewing by research associate Gordon Moshman into the open matrices beneath boulders with a hospital-model endoscope, which is more commonly used for internal medical probes into the human body. The use of the endoscopic viewing was done where geophysical indicators and visual inspection showed its potential for use.

After geophysical and endoscopic examination, large boulders up to ~2 × 3 m were subsequently removed with hand winches, followed with standard archaeological excavation techniques in the sediment beneath, where possible. In addition, the fragile ecological environment, with its nearby rare-bird fledging and protected bats, required delicate logistical movements of personnel and environmental protection during cave access and excavation, with the result being that no debris could be removed from the cave itself.

The different types of probes enabled location of a number of different kinds of artifacts, as well as contributing substantially to understanding of the complex roof-fall boulder accumulations, and the destruction of normal speleothems because of the probable seismic disturbances. Artifacts, sediments, and speleothems were collected and catalogued.

RESULTS

Geology and Geomorphology

The COL was originally eroded as a typical cave opening into cherty, Turonian limestone of the Shivta Formation,[9] and as is typical of such caves, it formed beneath a water table that existed somewhat above the cave level in the bedrock at the time of initial formation. Subsequently the canyon of Naḥal Ḥever was eroded into, and down through, the cave system. The present-day openings into COL are now situated in an inaccessible cliff face ~70 m below the upper surface of the cliff and the Roman camp at its edge, and ~230 m above the gorge bottom. The cliff face exposes (Figure 3.2), below the larger of the two modern cave openings, an 11-m thickness largely of cave-roof colluvium, as well as infillings of travertine and rock-salt (halite) speleothems, coupled with lowermost lacustrine deposits (Figure 3.3) and middle and upper lenticular fluvial layers. On the basis of external breaks in cliff topography and on geophysical evidence presented elsewhere in this volume, this thickness of cave-roof colluvium was not thought to extend to large depths everywhere inside the COL. Instead two strong E-W and NW-SE joint sets were hypothesized to intersect at what is now the cliff face, originally allowing deeper local solution of a cave pit into which the fine-grained lacustrine sediments and boulders accumulated before the much later exposure at the cliff face, which was controlled by the local expression of the joint sets (Figure 3.4).

While the COL was forming (Figure 3.5) the Naḥal Ḥever probably did not exist above on the surface, and indeed in any case, had not yet been cut down to the level of the cave when that opening was forming. Subsequently after cave formation, in its final infilling phase (Figure 3.6) the cave was filled up successively from the bottom with: (1) layers of fine-grained, thinly layered and variably colored, clay-and silt-sized grains of lacustrine sediment that was either primary dolomite (Frumkin, this volume), or was later dolomitized[10]; (2) roof-fall boulders from an overlying cave roof; and (3) intercalated and uppermost silicate clastics washed in from outside. The close proximity of the cave 3-km west of the active Dead Sea transform fault system indicates the high probability of periodic, seismically generated roof falls[11,12,13] which over time raised the cave opening progressively upward and reduced its overall volume with open-matrix boulders and smaller clastics (Figure 3.7). The middle and upper fluvial facies of this infilling has clasts of angular to sub-rounded, locally derived carbonates and chert, with minor diverse rounded and frosted silicate clasts of more remote provenance, possibly from the east side of the Dead Sea rift or reworked from the Hazeva Formation nearby on the west side of the rift. The larger pebble-sized clasts in both cave entrances are directionally imbricated and cross stratified, sloping toward the north and into the cave, indicating flow in that direction, and originally filling the cave opening at the cliff face completely to the roof (see Figure 3.4) and as far as 20 m inside the cave.

Within the COL, as it developed in the infilling stage, polygenetic, fine-grained sediment also accumulated between the interstices of the roof-fall boulders deep inside the cave during periods of low-frequency seismicity or high human and animal activity to form living floors. Sources of infilling sediment include: (1) seismically generated roof and wall breakdown; (2) seismic jostling of clasts with consequent attrition; (3) wall- and roof-disruptive haltite speleothem expansion (salt-crystal weathering); (4) in-washing and reworking of fluvial infillings from the cave mouth; (5) fine-clastic residue from local solution of carbonates; (6) elutriation of fine clastics from the desert surface above downward through the plentiful bedrock joints; (7) bat guano, animal and human feces; and (8) human importation of materials.

Discussion of the COL Sediments

Any discussion of geologic history of this region requires an understanding of the basic tectonics and depositional stratigraphy of the region by which the geological order of events can be worked out. It would appear from the literature, wherein a number of different interpretations of the timing of events and rock formations in the Dead Sea Rift are offered,[14,15,16,17] and in a number of these papers without reference to prior opposing points of view, considerable controversy must exist. For example, most people agree that tectonic subsidence of the Dead Sea Rift began in Late Oligocene time, perhaps at least 25 million years ago. The first unit of interest is the Hazeva Formation. Red beds exist within this formation, and during this time rivers from the Transjordan flowed entirely from SE to NW across the tectonically quiescent Dead Sea Rift, and flowed into the Mediterranean Sea in Early to Late Miocene, perhaps 23–5 million years ago,[18,19] but perhaps as recent as Early Pliocene (~4 million years ago).[20] This was followed by renewed tectonism in the Dead Sea Rift, and a marine incursion into the Yizre'el and Jordan valleys that produced the great thicknesses of evaporites in the Sedom and Bira formations in Late Pliocene to early Pleistocene,[21] but this only occurred in latest Miocene to Early Pliocene according to Garfunkle.[22] This marine or salt-water transgression appears to have continued on up into the Pleistocene with the deposition of the Amora and Lisan formations, passing into the hypersaline Dead Sea of today. The poorly constrained dating of these events has generated considerable controversy regarding interpretation of related events.

Frumkin[23] (within this volume) has utilized the unusual Sr/Ca composition of the COL dolomite to correlate the cave entrance deposit of <10 m^2, with the Sedom and Bira dolomites, which are the products of the rather poorly dated marine transgressions into the Dead Sea depression, perhaps as long ago as 7.5–5.3 Ma, or in the Late Miocene to Early Pliocene,[24] or as young as a few million years, or the Late Pliocene-Early Pleistocene.[25] These divergent views occur because of the quite different dates on the units, which means that the age of the COL dolomitic sediments are equally uncertain. Unless some alternate subterranean source of such dolomite can be postulated, in opposition to Frumkin's 2001 work[26] (and his chapter contained in this volume), his correlation of the COL dolomite compositionally with the Sedom and Bira would seemingly require an intrusion of marine waters into the cave mouth to deposit the dolomite sometime between 7.5–5.3 million years ago in the Late Miocene, followed by roof-fall boulders into the dolomite, and then a fluvial infilling of silicate clastic sediment later.

All of this would seem to violate the principle of parsimony or Occam's Razor in hypothesis construction, for the following reasons:

- The Sedom evaporates that could be a source of the Sr/Ca ratios in the COL dolomites seem to have been deposited in the marine incursion in Late Pliocene to early Pleistocene, which is either only a few (~2–3) million years ago,[27] or twice as many (5–7) millions of years ago according to Garfunkle.[28]
- At the present time the COL is perhaps 160 m above mean sea level, so Frumkin's hypothesis would presumably require at least that much uplift since any marine incursion at sea level, and although the Dead Sea graben down-faulting is well known, any possible associated uplift of that magnitude above sea level is less discussed.
- The juxtaposition of three quite different environmental events within such a short time and small space, or the temporally and spatially limited environment of cave systems, seems possibly somewhat forced.

- The relative lack of cementation and unconsolidated friability of the COL sediments within the surrounding richly carbonate environment for the postulated 7.5–5.3 Ma seems contrary to what generally occurs in that considerable length of time. Such deposits elsewhere in the world in soluble carbonate terrains are commonly well cemented after much shorter time periods, even in arid sites.
- The presence of plentiful rock salt in the roof of COL, unmentioned by Frumkin[29] (in his chapter in this volume), with its implications for saline-groundwater or diagenetic brines, could be a source of the unusual Sr/Ca ratios in the dolomite.
- The fact that so many other dolomites are secondary replacements indicates that the role of the plentiful and highly varied saline and Mg-rich ground waters[30] in diagenetic alteration of the COL sediments cannot be ruled out, in which case, final judgment on the likelihood of a marine incursion producing the COL dolomitic sediments directly could be withheld pending further study.
- Alternatively, Zak and Freund[31] and Mart[32] show Early Pliocene (~5 Ma) rivers flowing northwest across this region from the Jordanian highlands to the Mediterranean Sea, which if true would seemingly preclude Frumkin's scenario for dolomite formation in COL during the time period he suggests.

A silicate-clast-bearing, west-flowing river system thus appears to have occurred in the region during middle to late Tertiary time.[33,34] As the Dead Sea rift formed transverse to the river system in Late Cenozoic (Garfunkel, 1987), drainage was apparently reversed and eroded downward over a long period of time to ultimately breach the cave roof and deposit silicate lithologies characteristic of outcrops in Jordan occurring east of the deepening rift (see Figure 3.6). Frumkin[35] (in his chapter in this volume) thinks it is most likely that these silicate-clastic deposits were reworked from the Hazeva Formation in Israel, and that may be possible. A more definitive interpretation might have to await a detailed assessment of trace mineralogy of these deposits and a more careful sedimentological study of the mineral provenance than has been done up to now.

Cave Mapping, Airflow, and Cooking in the Cave in Antiquity

Remapping of the COL by Reeder's team (this volume) showed considerable difference from the maps of Yadin,[36,37] although the overall plan is similar (see Figure 3.8). The new length of the cave is 540 m, which is –175 m longer than previously supposed (Frumkin[38] reported a 324 m length). Many new crawlways were entered and mapped. The interior of COL is comprised of three large domed halls (A, B, & C) with several heretofore unmapped subsidiary passages. Hall A has considerably more in-washed sand and silt clastics than the other chambers, both of which are characterized mostly by irregularly boulder-covered floors from profuse roof falls. Ceiling domes in halls A and B that resemble corbel vaulting were caused by roof falls derived from the medium-to thin-bedded (<1 m) limestone layers overhead.

A previously unmapped western extension between halls B and C extends >30 m, from a zone of close-spaced jointing through which plentiful daytime airflow issues into the cave interior, before passing out through the front entrances. The temperature and relative humidity data collected in summer 2000 indicated good air exchange between Hall A and the cave's two entrances. Although the exchange between the deep recesses of the cave and Hall A are more limited, it does occur. A smoke candle was detonated late in the day at a hearth site discovered on the southeast side of Hall B to observe air flow, which was seen to hug the west wall of halls A and B and issued from the main entrance of the cave. Bar Kokhba or other people may have utilized the normal cave airflow discussed above to evacuate smoke during their cooking. Nearby egg shells and probable goat/sheep dung in a hole beneath the hearth site may constitute remnants of a kitchen midden from the probable cooking site.

ARCHAEOLOGICAL FINDINGS

Entryway

Some early people expended a good deal of effort to penetrate into the interior chambers of the COL through all the fill material and rock that had filled both entrances (see Figure 3.4). Access into the cave would have required scaling the 11-m high precipice of loosely consolidated but vertical rubble before arriving at a floor-to-ceiling south-sloping cave plug of largely siliciclastic sands and gravels. This blockage was dug through with a trench about 1 m deep to gain entrance into Hall A and beyond. Similarly in the east entrance, part of the roof rock was hewed out to obtain entrance into the cave interior. Whether or not this was done first by the earlier Chalcolithic people who were known to investigate underground chambers in their time, or only by later people seeking refuge from Roman conquest, is unknown.

Sparse Chalcolithic (~6300–5300 yr B.P. or ~4300–3200 B.C.E[39]) potsherds and plentiful human bones, unusually aggregated and with no obvious affiliations through grave goods or symbols to Judaism, were assumed by Yadin[40,41] and others to be Jewish anyway. These remains suggest a possible earliest use of the cave as an ossuary or other refuge typical of times when extensive habitation of the Judean Desert and its caves was common.[42] The considerable inaccessibility of the cave rendered it useful during times of crisis as a place of refuge and storage of precious documents and artifacts.[43,44] Our finds of first century coins and potsherds indicate that possibly during the first Jewish revolt (66–73 C.E.), and certainly during the second in the Bar Kokhba war (132–135 C.E.), the cave was extensively occupied.

Our discoveries with the metal detectors included four overstruck Bar Kokhba copper coins in mint condition; a silver Trajan coin (113 C.E.) from Bostra, capital of the Roman Province of Arabia; a Nabatean coin (106 C.E.); a silver tetradrachma from Tyre; an iron ring; 2 nails; an important small bronze perutah coin from the time of the first revolt (67 C.E.), that could indicate that COL might have been used as an earlier refuge. Also discovered by our expedition: a child's sandal; a comb; a morticed furniture piece; fragments of a wooden plate and wooden bowl; ~20 sections of fabric; plentiful twisted cordage; basketry; papyrus fragments; fired earthen hearth fragments; animal dung (goat/sheep?); food remnants; plentiful human bones with sufficient collagen for dating; human hair and feces; and >100 pieces of fiber-cord-bound firewood (Figure 3.11). Over 8000 pieces of pottery were discovered, with a preponderance of pottery finds in the Hall C connection to Hall B. This may be where Yadin's team sorted their finds.

Hearth

Deep inside the COL and beneath ~2–3 m of roof-fall boulders on the southeast side of Hall B many (>20) fragments of poorly fired 2–3 cm thick, coarse potshards were found, which when aggregated together proved to be part of an earthen oven or *taboon* (Figure 3.9). This oven seems to have been constructed with raw clays imported into the cave, mixed there with additional grogs and tempers, and then bound together with internal reeds with a round cross section to help keep it intact after firing had dried and cracked the clay. The reeds have now been completely oxidized by fire, leaving only carbonate residue, perhaps from original surficial precipitation on the plants growing in alkaline waters, and silica residue from the plant itself. One large piece of the oven about 9 × 12 cm preserved enough curvature to allow measurement and calculation of the chord and its rise to establish a radius of ~1 m of the upper oven, with a ~1–2 basal diameter. Because such ovens commonly have a side chamber for wood stoking, as well as the bottom, such parts are commonly fired at low enough temperatures to preserve un-carbonized remains of plant or other organic material. A piece of hide or leather with black hair resembling that of a goat or cow was recovered from an oven fragment and was ^{14}C dated at a calibrated 120 C.E. (Figure 3.10). This date, with its statistical uncertainty of ±40 yr, or range in age from 80 C.E. to 160 C.E. puts the oven most likely in the time of the Bar Kokhba war, although it may have been built pre-war to prepare for the coming strife. In addition, one sample of hearth had a piece of clear, thin (~1 mm), fine glass in it, indicative of a Roman age (Figure 3.9).

Thin sections of the hearth were cut for petrologic assessment and showed that the earthenware was composed of clay that would have to have been imported from an outside source. This clay was extended by addition of broken-up pottery as grog, and with a temper of silicate sand grains, mostly quartz. Some of the quartz is well rounded, indicating an eolian source, which is commensurate with the location of the cave beneath the Judean Desert floor, as well as the presence of such material in the sandy infillings at the cave mouth. Numerous pieces of angular calcite and limestone also occur in the hearth fragments and probably reflect the nearby sources of the cave roof and walls. No exotic minerals were noted in the thin sections, which overall probably reflects a preponderance of local materials used to extend the clay. Even though halite speleothems are a major constituent of the cave walls and ceiling, no halite was noted in the hearth, which is probably a reflection of the fact that such an inclusion likely would have caused disruption of the hearth when being fired, so it may have been purposefully excluded. The thin sections also have numerous linear and circular (in cross section) void spaces that are where reeds were originally emplaced before they were later incinerated and destroyed by the heat of the oven.

Firewood

Numerous pieces of wood occur in COL, most of which appears to have been collected as firewood. The wood of the COL was distributed mainly at several different levels in Hall B. Probably as the roof blocks fell they caused many of the sticks and branches bounce around the inside of COL. The wood is of several different species typical of the Negev and Judean deserts and the Dead Sea rift valley (Liphshitz, this volume), including white broom (*Retema roetam*), Christ thorn (*Jojoba* sp.), and an acacia. Previously ignored as putative 'Bedouin' remnants, our ^{14}C dating (Figure 3.10) and extensive beetle infestation proved otherwise, although the younger beetle debris or other contamination from bat urine or other source is probably what produced some of the dates younger than Bar Kokhba time. Two of the three samples dated, however, have statistical uncertainties that fall within Bar Kokhba time, and the third was either brought into the cave later or could have been contaminated and made to appear younger than it actually was.

Human Bones

Plentiful human remains were discovered by prior expeditions[45,46,47,48] and on the assumption that the bones represented ancient Jewish patriots starved to death by the Romans; many were reinterred on the Judean Desert in a special ceremony. Nevertheless, considerable human material remained in Yadin's 'Niche of Skulls' when we were there, and Safrai[49] had indicated the probability of other bones buried beneath roof-fall blocks (see Roddy this volume). Fiber-optic viewing and our excavation revealed Safrai's supposition to be correct and some of the bones were sent to the Weizmann Institute in Tel Aviv, but the uncalibrated ^{14}C date from the bone collagen came back (email communication. from J. Zias, 11/14/2000, and E. Boaretto, 11/30/2000) at C.E. 210–330 ±20 yr, which is slightly younger than the Bar Kokhba time (Figure 3.10). This suggests either contamination from later carbon exchange with the external environment (bats, handling with bare hands, human perspiration), or that later people died or were interred in the COL. Bone is, in fact, notoriously hard from which to get an accurate date.

Fabric Analysis

Small pieces of fabric collected in COL were assessed by others[50] and found to be predominantly wool and some plant fibers (flax). The molecular and supermolecular structure of both keratin (wool) and cellulose (flax) were completely intact. In many of the fibers, mineral crystals were intimately connected with the fibers, which suggested the possible use of metal-containing mordants for fixation of the organic dyes (also see Shamir in this volume).

CONCLUSIONS

Now that the role of roof falls in the burial of artifacts in COL are better understood and mapped as a result of this study, Yadin's 1960–61 discovery of letters, documents, and personal effects, together with our new finds, indicate the possible archaeological potential of other nearby caves in this seismically active area. Comprehensive cave-survey techniques, sensitive metal detectors of different resolutions, and advanced GPR and ERT equipment demonstrate that such noninvasive techniques work well over rough terrain, as well as in open- and filled-matrix boulder masses inside caves. The methods are a practical and cost-effective means to obtain high-resolution profiles at depth to aid in the identification of subsurface features for archaeological probes in such difficult environments. The hundreds of caves in the Dead Sea region of Israel and Jordan, which are already famous for the prior discoveries of the Dead Sea scrolls and other ancient materials, could therefore offer considerable further archaeological potential.

The identification of problems of interpretation of timing and character of paleo-geomorphological development of the Dead Sea depression and its peripheral cave formations and their internal cave-fill stratigraphies, offer potential for new work to resolve difficulties between the works of Zak and Freund,[51] Mart,[52] Garfunkle,[53] and Frumkin.[54] Certainly the incompatibilities expressed between these works argues for future interpretive care in ironing out some of the remaining difficulties to better elucidate long-term landform and cave development in the region.

Finally, the video and cinematography taken in the two summers we were in the COL have resulted in two hour-long programs that were aired locally on Nebraska Educational Television (NETV)[55] (Milone and Schrader, 2001), and nationally on Nova[56] (Hochman et al., 2004). Such presentations are popular with the public and in the long run are helpful to education about archaeology, as well as in fund raising for other similar or popular scientific endeavors.

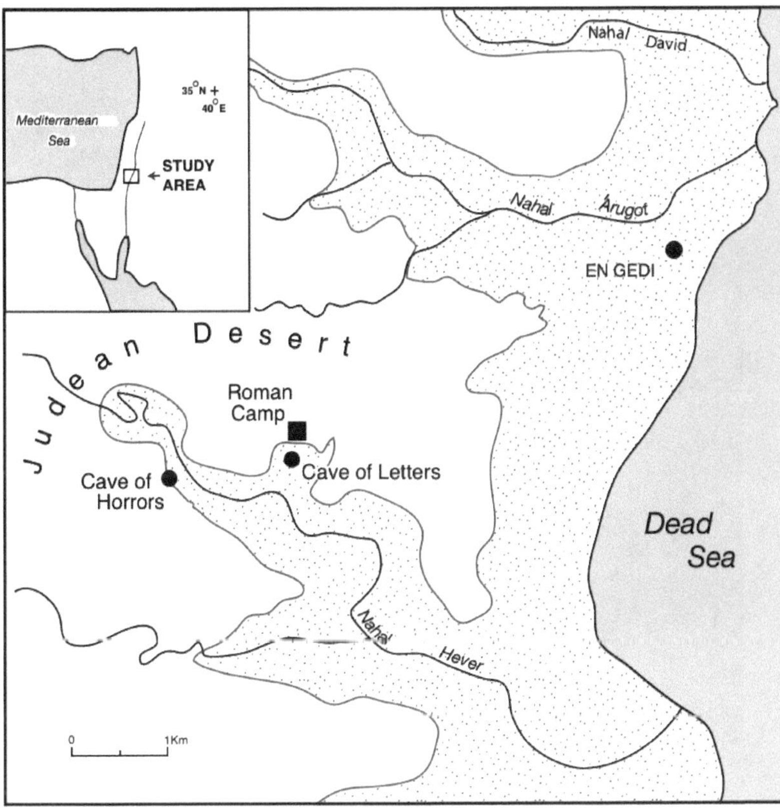

Figure 3.1: Location map of Cave of Letters in Naḥal Ḥever on the west side of the Dead Sea Rift, Israel.

Figure 3.2: Photograph of the cliff face below the Cave of Letters opening.

Figure 3.3: Close-up photograph of lowermost lacustrine dolomitic sediments at the bottom of the infilling at the mouth of the Cave of Letters.

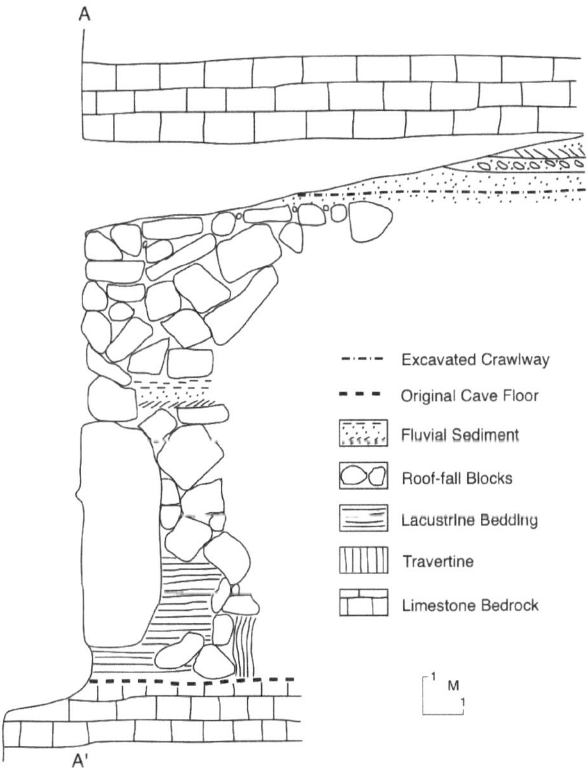

Figure 3.4: Diagram of cross section A–A' through the Cave of Letters cliff face showing rooffall infilling blocks overlain by clastic sediments.

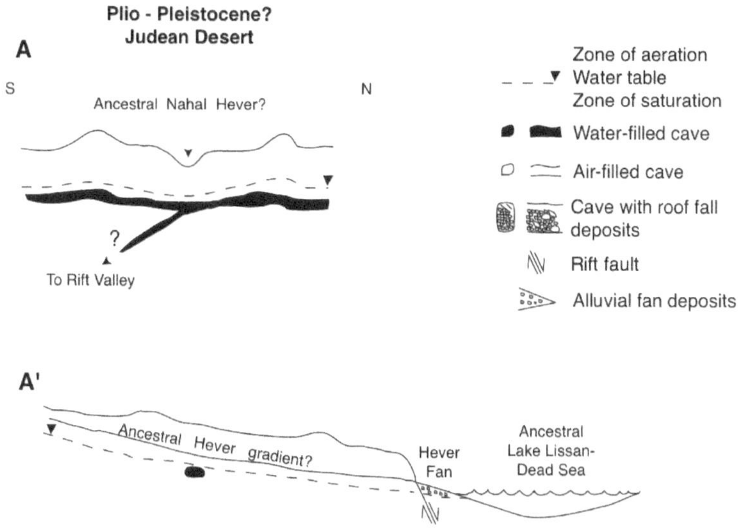

Figure 3.5: A–A': Development of a solutional cave in Plio-Pleistocene(?) time beneath a water table probably unassociated with the ancestral Naḥal Ḥever.

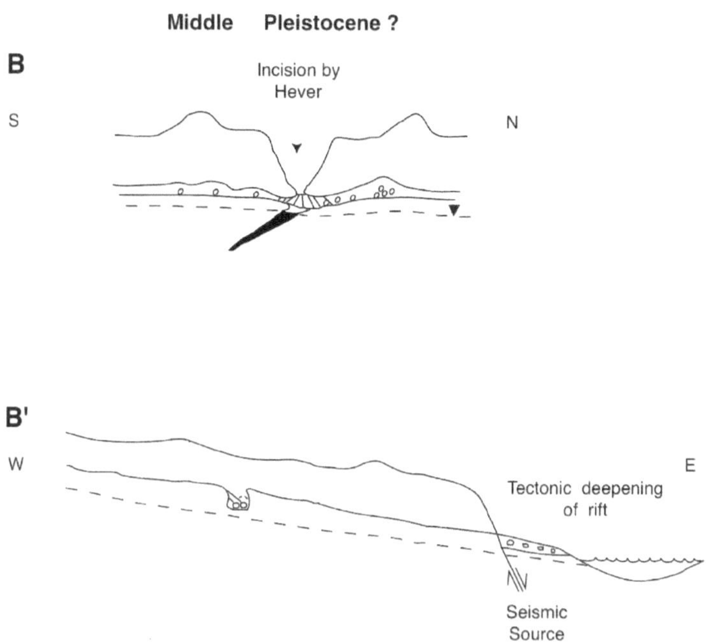

Figure 3.6: B–B': By about middle Pleistocene(?) time a low point in the cave has had deposited fine dolomitic sediments as a result of primary cave waters or secondary, diagenetic saline groundwaters.

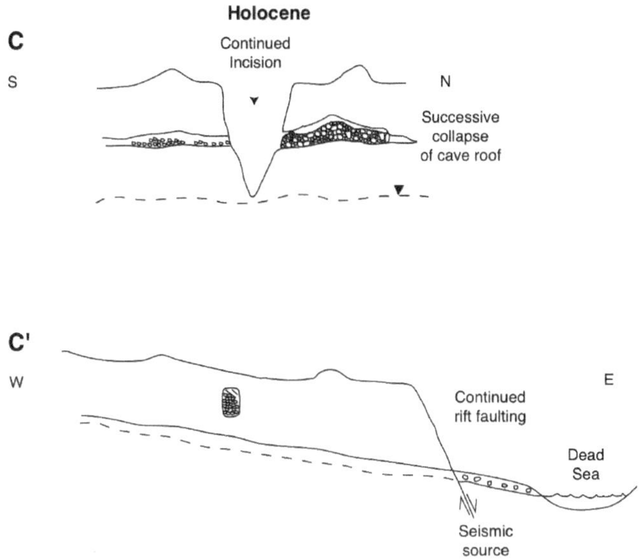

Figure 3.7: C–C': By Holocene time (10,000 yr BP) the Naḥal Ḥever has eroded well below the cave mouth and intermittent seismicity has caused recurrent roof collapse.

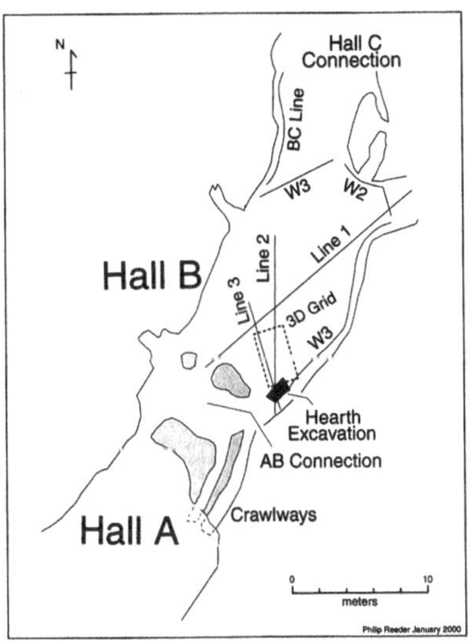

Figure 3.8: Location sketch map of GPR transect lines, the GPR #D grid, and the location of the hearth in Hall B of the Cave of Letters.

Figure 3.9: Hearth fragment #3 (hand lettered on sample) measuring ~5 × 9 × 2 cm.

Material Dated Conventional Age	Cailibrated Calendrical Mean Date	2 Sigma Calibrated Age (95% probability)
Wood		
1850±60 yr BP [BA]	140 C.E.	45–330 C.E.
1780±60 yr BP [BA]	245 C.E.	95–405 C.E.
1619±70 yr BP [BA]	430 C.E.	260–610 C.E.
Hearth Leather		
1880±40 yr BP [BA]	120 C.E.	50–230 C.E.
Human Bone Collagen		
210–330±20 yr BP [WI]		

Figure 3.10: Radiocarbon dates on materials from Cave of Letters obtained from Beta Analytic, Inc. (BA) and Weizmann Institute (WI). The later institute did not provide calibrated dates or 2 sigma data. yr BP = years before present; C.E. = Common Era.

NOTES

1. Abbreviation for Common Era, the preferred secular calendrical usage to replace to religiously motivated designation, 'A.D.' Also 'B.C.E.' for Before Common Era.
2. Frumkin, A., 2001, "The Cave of Letters sediments—Indication of an early phase of the Dead Sea depression?" *Journal of Geology*, 109: 79–90.
3. Louchs, R.G., 1999. "Paleocave carbonate reservoirs: origins, burial-depth modifications, spatial complexity, and reservoir implications." *American Association of Petroleum Geologists Bulletin*, 83: 1795–1834.
4. Safrai, B., 1993. "Recollections from 40 years ago—More scrolls lie buried." *Biblical Archaeology Review*, Jan-Feb: 51–57.
5. Yadin, Y., 1963. *The finds from the Bar Kokhba period in the Cave of Letters*. The Israel Exploration Society, Jerusalem.
6. Yadin, Y., 1971. *Bar-Kokhba*. Weidenfeld & Nicolson, London.
7. Frumkin, "The Cave of Letters sediments—Indication of an early phase of the Dead Sea depression?"
8. Jol, H.M., Shroder, J.F., Reeder, P., Freund, R.A., 2000. "Return to the Cave of Letters, Israel: A GPR archaeological expedition." *Proceedings SPIE* (International Society for Optical Engineering), eds. Noon, D.A., Stickley, G.F., Longstaff, D., Eighth International Conference on Ground Penetrating Radar, 4084: 882–886.
9. Frumkin, "The Cave of Letters sediments—Indication of an early phase of the Dead Sea depression?"
10. Frumkin, "The Cave of Letters sediments—Indication of an early phase of the Dead Sea depression?"
11. Ben Manahem, A., 1991. "Four thousand years of seismicity along the Dead Sea rift." *Journal of Geophysical Research*, 96: 20,195–20,216.
12. Ben Tor, A., 1992. *The Archaeology of Ancient Israel*. Yale University Press, New Haven, CN, USA.
13. Amiran, D.H.K., Arieh, E., Turcotte, T., 1994. *Israel Exploration Journal*, 44: 260–305.
14. Zak, I. and Freund, R., 1981. "Asymmetry and basin migration in the Dead Sea Rift." *Tectonphysics*, 80: 27–38.
15. Horowitz, A., 1987. "Palynological evidence for the age and rate of sedimentation along the Dead Sea Rift, and structural implications." *Tectonophysics*, 141: 107–115.
16. Mart, Y., 1991. "The Dead Sea Rift: from continental rift to incipient ocean." *Tectonophysics*, 197: 155–179.
17. Garfunkle, Z., 1997. "The history and formation of the Dead Sea Basin." In: Niemi, T.M., Ben-Avraham, Z., Gat, Y., eds. *The Dead Sea—the lake and its setting*. Oxford University Press, Oxford, UK: 36–55.
18. Zak, I. and Freund, R., "Asymmetry and basin migration in the Dead Sea Rift."
19. Garfunkle, Z., "The history and formation of the Dead Sea Basin."
20. Mart, Y., "The Dead Sea Rift: from continental rift to incipient ocean."
21. Zak, I. and Freund, R., "Asymmetry and basin migration in the Dead Sea Rift."
22. Garfunkle, Z., "The history and formation of the Dead Sea Basin."
23. Frumkin, "The Cave of Letters sediments—Indication of an early phase of the Dead Sea depression?"
24. Garfunkle, Z., "The history and formation of the Dead Sea Basin."
25. Zak, I. and Freund, R., "Asymmetry and basin migration in the Dead Sea Rift."

26. Frumkin, "The Cave of Letters sediments—Indication of an early phase of the Dead Sea depression?"
27. Zak, I. and Freund, R., "Asymmetry and basin migration in the Dead Sea Rift."
28. Garfunkle, Z., "The history and formation of the Dead Sea Basin."
29. Frumkin, "The Cave of Letters sediments—Indication of an early phase of the Dead Sea depression?"
30. Mazor, E., 1997. "Groundwaters along the western Dead Sea shore." In: Niemi, T.M., Ben-Avraham, Z., Gat, Y., eds. *The Dead Sea—the lake and its setting*. Oxford University Press, Oxford, UK: 265–276.
31. Zak, I. and Freund, R., "Asymmetry and basin migration in the Dead Sea Rift."
32. Mart, Y., "The Dead Sea Rift: from continental rift to incipient ocean."
33. Picard, L.Y. and Golani, U., 1992. *Survey of Israel geology map*. Reprint of 1965 edition. Survey of Israel.
34. Mart, Y., "The Dead Sea Rift: from continental rift to incipient ocean."
35. Frumkin, "The Cave of Letters sediments—Indication of an early phase of the Dead Sea depression?"
36. Yadin, Y., *The finds from the Bar Kokhba period in the Cave of Letters.*
37. Yadin, Y., *Bar-Kokhba.*
38. Frumkin, "The Cave of Letters sediments—Indication of an early phase of the Dead Sea depression?"
39. Mazor, E., "Groundwaters along the western Dead Sea shore."
40. Yadin, Y., *The finds from the Bar Kokhba period in the Cave of Letters.*
41. Yadin, Y., *Bar-Kokhba.*
42. Ben Tor, A., *The Archaeology of Ancient Israel.*
43. Yadin, Y., 1960. "Expedition D", in *Judean Desert Caves*. The Israel Exploration Society, Jerusalem.
44. Yadin, Y., 1961. "Expedition D—The Cave of the Letters." In *Judean Desert Caves*. The Israel Exploration Society, Jerusalem.
45. Yadin, Y., "Expedition D, in Judean Desert Caves."
46. Yadin, Y., "Expedition D—The Cave of the Letters."
47. Yadin, Y., *The finds from the Bar Kokhba period in the Cave of Letters.*
48. Yadin, Y., *Bar-Kokhba.*
49. Safrai, B., "Recollections from 40 years ago—More scrolls lie buried."
50. Müller, M., Murphy, B., Burghammer, M., Snigireva, I., Riekel, C., Gunneweg, J., Pantos, E., 2006. *Applied Physics A: Materials Science & Processing*, 83(2): 183–188.
51. Zak, I. and Freund, R., "Asymmetry and basin migration in the Dead Sea Rift."
52. Mart, Y., "The Dead Sea Rift: from continental rift to incipient ocean."
53. Garfunkle, Z., "The history and formation of the Dead Sea Basin."
54. Frumkin, "The Cave of Letters sediments—Indication of an early phase of the Dead Sea depression?"
55. Milone, C. and Schrader, C., 2001. *Return to the Cave of Letters*. UNO Television. One-hour documentary aired 4 November on NETV.
56. Hochman, G., Wolfinger, K., Collins, R., 2004. *Ancient refuge in the Holy Land*. Nova Television, 23 November 2004.

CHAPTER FOUR

Mapping THE Cave OF Letters

PHILIP REEDER
Bayer School of Natural and Environmental Sciences,
Duquesne University
Pittsburgh, PA 15282, USA

INTRODUCTION

As part of research conducted in the Cave of Letters in 1960 and 1961, Yigael Yadin not only discovered a priceless collection of artifacts including scrolls, pottery, coins and bronze objects, but he also collected the necessary data to produce the first maps of the cave.[1] No new archaeological research was conducted in the cave between 1961 and 1999, and it was not until research expeditions in July 1999 and July 2000, that utilized state-of-the-art technology, as well as traditional geologic, archaeologic and surveying techniques, that new information was added to the existing knowledge base about the Second Jewish Revolt, and the viability of geophysical research in caves.[2] Also, as part of the research, the necessary spatial data were collected to produce a new base map of the Cave of Letters, and subsequent maps were built-upon the base map that depicted various aspects of research in the cave.

From July 12th to July 16th 1999, and July 3rd to July 20th 2000, data was collected by research expedition members Philip Reeder (at that time affiliated with the University of Nebraska at Omaha), Ms. Carolyn Phillips from Doane College in Nebraska (1999 expedition only), and Dr. James Webster from the University of Georgia (2000 expedition only), which facilitated the production of the detailed maps of the Cave of Letters. The efforts to map the cave became a major component of the overall research design for the 1999 and 2000 expeditions because:

1. A new base map of the cave was created
2. The base map could be stored in a digital format and was thus easily updated

3. Layers of data could be added to the map in real-time
4. Mapping of the cave facilitated a thorough exploration of all passages in the cave
5. Maps could be created which highlighted the various components of the research (i.e. geophysics, geology and archaeology).

As part of Yadin's research in the cave, a series of plan view and cross-sectional maps were produced.[3] Yadin did not include the smaller passages found in the cave in his maps, hence his series of maps were incomplete. Between 1981 and 1996, six expeditions visited the cave, mostly to study the geology and sediments associated with the cave.[4] As part of this research, a new survey of the cave was completed that focused on parts of the cave not surveyed by Yadin, and on completing a vertical profile through the main passages of the cave.

As part of the research design for the 1999 and 2000 expeditions to the cave, it was decided to map the cave again, so that the map data would be in a format familiar to the researchers. This new survey mapped both the large chambers in the cave, as well as the numerous small crawlways. The data generated as part of this new survey was entered, stored and manipulated in a plotting program (CAVEPLOT),[5] which allowed the plot of the cave to be continually updated as new survey data was generated. This data was eventually exported to a graphics program (Adobe Illustrator),[6] and a base map of the cave was completed. This map then served as the starting point for other maps related to various segments of the research. Maps related to the physical make-up of the cave, the geophysical research (ground penetrating radar (GPR) and Electrical Resistivity Tomography (ERT)) that was on-going in the cave, and the archaeological research, were produced by overlaying the base map data and the specific data collected as part of the various components of the project.

DATA COLLECTION AND MAP PRODUCTION

Cave surveying is the process of measuring and drawing/sketching a cave so that an accurate map of the cave can then be produced. Without a complete and accurate survey of the cave, any resulting map will be potentially inaccurate and of limited use (http://cavingintro.net/surveymap.html).[7] The collection of survey data and the eventual production of a series of maps related to the 1999 and 2000 research completed in the cave took place as a series of steps that included (1) data collection and acquisition, (2) data entry, storage and manipulation, (3) plotting, (4) data export to a graphics program and de-construction of data plots created in the plotting program using this graphics program, and (5) map production using the graphics program. More specifically:

(1) The survey of the Cave of Letters was completed using standard cave mapping techniques that utilized hand-held survey instruments and included a) an optical compass, b) an optical inclinometer, c) a nylon survey tape, and d) a field note book.[8] A total of 66 survey stations were established within the cave during the 1999 survey. An additional 16 stations were established during the year 2000 expedition. Survey stations were established within the cave and their locations were indicated with a piece of flagging tape marked with the station identification number. The first survey station was established just inside entrance #1, and the subsequent 65 stations from 1999, and additional 16 stations from 2000, were linked to this station or to survey stations that were linked to this station. Data was collected between each station included azimuth (using the optical compass), inclination in degrees (using the optical inclinometer), and distance (using the nylon survey tape). This data was recorded in a field note book. At each survey station the distance from the survey station to the left and right cave walls, and the cave ceiling and floor were measured when possible or were estimated. At each survey station a sketch of that segment of cave passage was also produced, which included information on the location of roof fall (sometimes called breakdown) on the cave floor, holes in the cave floor, the topography of the cave floor, and features in the cave ceiling.[9] The 1999 survey took place in the most accessible parts of the cave. The year 2000 survey concentrated on very small crawlways which were not entered in 1999.

(2) The cave survey data was entered into the cave plotting program CAVEPLOT. The cave survey data that was collected, as described above in item (1), was entered into spreadsheets provided by the CAVEPLOT software. Each survey station identification number was entered along with the corresponding distance, azimuth, and inclination to the next survey station. The distance to the left and right walls, and ceiling and floor, were also entered for each station (Figure 4.1). Data for the additional 16 stations which were established during the 2000 expedition were added to the original 66 stations in the database, hence, the survey station data base contained 82 stations after the year 2000 expedition. The CAVEPLOT Program then creates a table of X, Y, and Z values for each survey station based on the survey data (Figure 4.2). The X and Y intersection marks the location of the survey station and the Z value indicates changes in elevation between stations. The Z-values are a vital component in constructing a cross-section of the cave, whereas the X and Y values are important components for constructing the plan view.

(3) A line plot of the cave can then be produced by the CAVEPLOT program. The program uses the distance and azimuth data, depicted in Figure 4.1, to create the X and Y coordinates that are contained in the table depicted in Figure 4.2. The line plot is created by linking all the survey stations together by joining the first station to the second, and the second to the third, and so on. By connecting all the survey stations together in this fashion, a plan view (view from above) line plot of the cave was produced (Figure 4.3).

The CAVEPLOT program also allows the user to utilize the distance left and right, and to the ceiling and floor at each station to produce a more complete plot of the cave. Addition of the left and right data facilitates the creation of a plot that contains the cave walls, and addition of the ceiling to floor data facilitates the creation of a cross-section of the cave. The software accomplishes these operations by creating a polygon around each survey station that takes into account the left, right, ceiling, and floor measurements as depicted in the copy of the data entry screen presented in Figure 4.1. The program then links all the polygons at each station together to provide a more complete plan view depiction of the cave (Figure 4.4). The CAVEPLOT software also allows a cross-sectional depiction of the cave to be created by rotating the plan view plot 90 degrees (Figure 4.5). Because numerous passages in the cave are at similar elevations, some of the passages are drawn on top of each other, and are thus obscured. Due to this fact, only a line plot cross-section of the cave was eventually exported into the graphics program. Also, since the survey data is stored in a digital format within the plotting program, the mapping data collected during the 2000 expedition was added to the data base from the 1999 research, and a new, updated plot of the cave was created that contained the data from the additional 16 stations surveyed in 2000.

(4) The actual production of a base map for the cave was facilitated by exporting the plot data, from the CAVEPLOT program into Adobe Illustrator software. CAVEPLOT is a program for plotting cave survey data, and although it does produce some basic graphic outputs, the quality is generally poor. In order to produce a high-quality map product based upon the cave surveys, the basic line plot and the plot that contains the left, right, ceiling and floor data were exported to the more powerful graphics program Adobe Illustrator. Once in the graphics program, the plan view plot that contains the left and right walls can essentially be de-constructed using Adobe Illustrator. When CAVEPLOT draws the polygons in a plot, numerous lines are drawn on top of each other. Using Adobe Illustrator, all the lines in the plot were deleted except for those representing the walls of the cave. Once the removal of all unnecessary lines was accomplished, an accurate plan view depiction of the cave could be produced. The first step in the production of this map, which was to serve as the base map for the project and all subsequent maps produced related to the specific areas of the project, was to create a data layer in the graphics program that depicted the walls of the cave. Another layer was created on top of this layer, and the walls were drawn in using the pen tool in Abobe Illustrator (Figure 4.6). This more stylistic layer replaced the generally angular and blocky lines drawn by the plotting program. Additionally, the data from the 2000 survey of the cave could be added to the CAVEPLOT data base, and the steps outlined above could be followed to create a new, updated base map of the cave.

(5) Using the map that depicted the walls of the cave in plan view (Figure 4.6) and the sketch that was produced during the survey of the cave, a general base

map of the Cave of Letters was produced. This map was then updated after the 2000 expedition to reflect the data associated with the 16 additional survey stations (Figure 4.7).

According to the 1999 base map, the length of the Cave of Letters, based upon the most direct route through the cave, was estimated to be 335 meters. The 474-meter surveyed length of the cave reflects the distance between the 66 survey stations which were used to produce the map. In the three large chambers in the cave (Halls A, B and C), the layout of survey stations meandered through chambers, not taking the most direct route. Hence, the 474-meter distance value is an over estimation, although it does reflect the surveyed length of the cave. The addition of the 16 survey stations from the year 2000 survey added 65 meters to the surveyed length of the cave.

The cross-section line plots, both with walls (Figure 4.5) and without, that were produced using CAVEPLOT were exported into the Adobe Illustrator program. It was determined that the cross-section plot that contained the wall and ceiling data would not produce a useful cross-sectional map of the cave because too many of the passages are at the same elevation and hence obscure each other in cross-sectional view. The line plot in cross-sectional view did not have this problem because the thinner lines crossed each other, but did not completely block the view of the intersecting lines. Details extracted from the plan view base map (see Figure 4.7) were added to the cross-sectional depiction, and the line cross-section map was added to the plan view map thus providing a depiction of the cave in both plan view and cross-section view on the same sheet (Figure 4.7).

A very detailed map of Cave of Letters was prepared after the 1999 expedition, and it was updated after the year 2000 expedition. Using the base map presented in Figure 4.7, precise details related to the location of roof fall (breakdown) on the cave floor, holes in the cave floor, the topography of the cave floor, and features in the cave ceiling, were added. The line plot of the cross-section of the cave was also added to this map, again allowing the map to depict both the plan view and cross section of the cave on the same map sheet. Additionally, detailed cross-sections which depict 22 different locations in the cave (lettered "a" through "v" on the map) were added to the map (see Figure 4.8). The map created after the 1999 expedition, and updated after the year 2000 research, represents the most detailed map ever produced of the Cave of Letters.[10]

Data related to archaeological finds in the cave were plotted on a version of the map discussed above. An archaeological finds map was produced after the 1999 expedition, and an updated map was also completed after the year 2000 expedition. Since Adobe Illustrator allows data to be added in layers, these data can be turned on and off, resulting in the production of various map depictions (Figures 4.8 and 4.9) based upon one single base map (Figure 4.7). By turning off the layers associated with roof fall (breakdown), holes in the cave floor, and topography of the cave floor, and adding layers associated with the archaeological finds, versions of the map depicting the newest archaeological finds were produced (Figure 4.9). Additionally, because the map data was in digital format, it could be stored in a hand-held touch screen computer and information related to the archaeological loci and associated baskets could be added to the map data base in real-time. This process greatly facilitated updating the archaeology research map of the cave. Also, utilizing the same base map (Figure4.7) and the detailed physical and archaeology maps (Figures 4.8 and 4.9), maps were created that depict the location of GPR or ERT survey lines, as well as the locations where rock splitting efforts were concentrated in 2001.[11] The plan view map of the cave in Figure 4.9 depicts the location of all the archaeological loci established in the cave during the 1999, 2000 and 2001 research, as well as the location of Yadin's finds, the ERT lines, and the rock splitting locations.

DESCRIPTION OF THE CAVE OF LETTERS SURVEY DATA

For the 1999 survey of Cave of Letters there were a total of 65 survey readings taken in the cave, but there are 66 survey station references because a base station was established at Entrance #1. It was from this station that the first survey shot (A1) was taken. The total surveyed distance of 474.51 meters was calculated by adding all the distances together between the 66 survey stations. This number does not take into account true horizontal distance between survey stations. The true horizontal distance for the survey, which accounts for changes in elevation between survey stations, is 460.48 meters. The average distance per survey shot is 7.3 meters, and the longest survey shot is 18.3 meters. The maximum north extent of the survey, based upon survey station A1 established at Entrance #1, is 127.44 meters. This station is located at the furthest extent of Hall C. The minimum north extent of the survey is −1.97, which occurred at station B3 (located at Entrance #2). The maximum east extent of the cave is 110.55 meters (High Room at station A22), and station A1 marks the minimum east extent of the cave. Station A22, located in the High Room, at 13.41 meters, marks the maximum vertical elevation above the datum established at station A1. Station W3, located at the junction between the BB passage and the BC connection passage, at −5.94 meters, marks the lowest point in the cave below the A1 datum.

For the year 2000 survey of Cave of Letters, there were a total of 16 new survey stations established in the cave. The total number of survey stations in the cave after the 2000 survey was 81. There are 82 survey station references because, as previously mentioned, a base station was established at Entrance #1, and it is not actually a survey station. The total surveyed distance of 540.01 meters was calculated by adding all the distances together between the 82 survey stations. This number again does not take into account true horizontal distance between survey stations. The true horizontal distance for the survey, which accounts for changes in elevation between survey stations, is 526.0 meters. The average distance per survey shot is 6.7 meters. The longest survey shot is 18.3 meters and the shortest is 1.0 meters. The maximum north extent (127.44 m), the minimum north extent (–1.97), the maximum east extent (110.55 m), and the minimum east extent (0.0) are all similar to the 1999 survey, as are the maximum vertical (13.41) and the minimum vertical extents (–5.94).

GENERAL DESCRIPTION OF THE CAVE OF LETTERS

The following description of the Cave of Letters is based upon the maps created after the year 2000 expedition (Figures 4.7, 4.8 and 4.9). Entrance #1 is located 10 meters above a 6-meter-wide ledge located on the north side of the Naḥal Ḥever. The 7-meter-wide opening leads to a 12-meter-long passage that averages 1.8 meters in height. The passage dimensions then decrease, on average, to 1.6 meters wide by 1.2 meters high. This passage can be traversed for approximately 15 meters until it merges with Hall "A". To enter Hall "A" it is necessary to crawl 2 meters up through roof fall material (breakdown blocks). Once in Hall "A" the passage size greatly increases. Along the south wall, approximately 8 meters from where Hall "A" is entered, a passage extends to the south that connects with entrance #2. This passage is approximately 12 meters long, 1.5 meters wide, and for most of its length it is less than 1.2 meters high. The ceiling height increases rapidly to 2.5 meters near the terminus of the passage. The passage ends at a 12 meter drop which leads down to the same ledge that is located below entrance #1.

Hall "A" is approximately 33 meters long, has an average width of 16 meters and an average height of 7 meters. The cave floor is completely covered with roof fall material (breakdown blocks), ranging in size from cobbles to large boulders. The room extends to the northeast and ends at a large breakdown pile that blocks the passage from ceiling to floor. On the left side (northwest) of this breakdown, a small passage, named the AB Connection, can be traversed which leads to Hall "B". The passage that enters the AB Connection is 1.3 meters wide and 1.2 meters high. After 6 meters the passage's dimensions increase to 12 meters wide and 1.6 meters high. This small room is actually an open area that exists in the breakdown that separates Hall "A" from Hall "B". Hall "B" can be entered from this small room in three ways. A small 1-meter-wide, 0.8-meter-high crawl connects from the left (northwest) side of the room. There is also a 1.5-meter-wide, 1-meter-high crawl that connects from the center of the small room. Hall "B" can also be entered from the right (southeast) side of this small room. A tight crawl over breakdown leads to a 4-meter-wide, 2.5-meter-high passage that slopes down to merge with Hall "B". From this same passage, extending in the opposite direction (southwest) from Hall "B", access to two crawlways can be gained. Both of these crawlways are approximately 12 meters long, 1-meter-wide, and have ceiling heights of 1 meter. These passages actually extend back under the breakdown pile at the northeast terminus of Hall "A". A voice connection between these crawlways and Hall "A" is possible.

Hall "B" is approximately 25 meters long, averages 15 meters wide, and has an average ceiling height of about 5 meters. As is the case with Hall "A", the floor in Hall "B" is covered with breakdown blocks. It is in this room that GPR and endoscope[12] work were concentrated during the 1999 expedition. An area in the southwest section of the room was also the site of limited excavation in 1999 and extensive excavation during the year 2000 expedition (Figure 4.9). Because this area may have been the location of a hearth, based upon the abundance of fire wood found in the area, this excavation was tentatively named the Hearth Excavation. Also during the year 2000 expedition, extensive excavation was carried out extending from the hearth excavation to the center of Hall B.

At the northeast end of Hall "B", passages extend both toward the northeast (passage BB) and almost due north (Hall "C" Connection Passage). A small crawlway also extends just slightly east of north. This passage slopes steeply down through the breakdown blocks that covers the cave floor. The passage gets too small to follow after approximately 7 meters. This point is the lowest elevation in the cave below the datum set at survey station A1.

The passage that extends northeast from Hall "B" (the BB Passage) is generally 4 meters wide and 2 meters high. It extends for approximately 35 meters until it ends at a large breakdown pile. Extensive excavation was also carried out at the base of this breakdown pile during the 2000 expedition. A one-meter-wide, 3-meter-long passage leads down through the breakdown pile to where it ends at a rubble choke. The surface of the breakdown pile can be climbed up to a room named the High Room. Along the route to this room, three small passages, which trend to the west, are encountered. The southernmost of these passages can be entered for three meters before it becomes too tight. The passage just north of this passage also trends toward the west, and it can be followed for 11 meters before ending in breakdown. The northern most of these small passages can be followed for 10 meters before ending in breakdown. Beyond these small side passages, the main passage continues upward and turns sharply to the east. This passage ends in a 5 meter × 5 meter domed room.

The Hall "C" Connection passage exits Hall "B" trending almost due north. This passage is 30 meters long, averages 7 meters wide, has an average ceiling height of 6 meters, and is inclined downward from Hall "B". The floor of the Hall "C" Connection passage is covered with ceiling breakdown. Large amounts of ceiling breakdown we removed along the west wall of the passage and an extensive excavation was carried out in 2000. Along the left (west) wall where the Hall "C" Connection passage joins Hall "C", a small crawl extends toward the west. This crawl, named the West Extension Passage, was partially explored and mapped during the 1999 expedition and was determined to be approximately 38 meters in length. It averaged 1.5 meters wide, and the ceiling height ranged from 1.5 meters to 0.3 meters. Exploration of the passage was stopped at a point where the passage was 1.3 meters wide and 0.3 meters high. At this point, air was felt blowing from the as of yet unexplored portion of the West Extension Passage, toward Hall "C". It was thought that the presence of air circulation in this passage may indicate a connection with the surface at the terminus of the passage. The passage was completely explored during the 2000 expedition. An additional 12 meters of passage was mapped in the main West Extension Passage, and a side passage was opened up that extended an additional 5 meters. The main West Extension Passage ended in a one-meter-wide, 0.3-meter-high passage that was blocked by bedrock and sediment. The airflow into the passage was noted to be coming from a 3-cm wide opening in the bedrock. Where the West Extension merges with Hall "C", a 6-meter-long passage that extended to the west-southwest was discovered. This passage basically extended under the West Extension Passage until becoming blocked by large breakdown blocks.

Hall "C" is oriented toward the northeast. It is almost 40 meters in length, it averages 11 meters wide, and the ceiling height varies from 10 meters to as little as one meter at the northeast terminus of the passage. The floor of Hall "C" is covered by ceiling breakdown (rubble of all sizes, from cobbles to very large boulders). Near the northeast terminus of Hall "C" five crawlways can be entered. Along the east wall a 2-meter-wide, 7-meter-long crawl was entered in 1999 and it ended in rubble fill. During the 2000 expedition some of this rubble-fill was removed and the length of the passage was extended by 10 meters. This extended passage ended at a bedrock constriction where the ceiling height decreased to 0.2 meters. North of this crawl a small body length crawl can also be entered which ends in rubble fill. At the same point this crawl is entered another crawl extends in the completely opposite direction. Entry to this crawl is gained by climbing up approximately two meters. This crawl extends approximately due west for 10 meters, and it actually crosses over and above the northern most extension of Hall "C". From a point a few meters north of where this passage can be entered, a larger crawl extending almost due east can also be entered. This crawl is 8 meters in length, averages 2 meters wide, and it to ends in rubble fill. A passage also extends slightly west of north from the terminus of Hall "C". This passage is 13 meters in length, averages 4 meters wide, and ends in a tight crawl that is blocked by rubble. During the year 2000 expedition some of the rubble in these passages was moved to open up additional passages, thus adding more length on to the Cave of Letters. Over 20 meters of new passage was explored in these areas which were opened up via rubble removal. Also, several archaeological finds were made in these areas.

The surveyed length of 540 meters is the final surveyed length for the Cave of Letters. All existing passages were thoroughly explored, and no known leads exist for possible future exploration. The only possibility for adding additional length on to the Cave of Letters survey is that a new passage or passages are uncovered during any future excavation in the cave.

THE LENGTH OF CAVE OF LETTERS (COMPARED TO YADIN'S SURVEY)

Because of the detailed nature of the surveys carried out in Cave of Letters in 1999 and 2000 to produce the previously discussed maps, the length values obtained for the cave are much greater than those obtained as part of Yadin's 1963 survey.[13] What follows is a summary of the survey data compiled after the 2000 expedition, and how it compares to Yadin's survey.

1. Straight line distance from entrance 1 to back of Hall "C" is approximately 150 meters based upon the year 2000 survey of the cave. Yadin refers to a similar number in his report on the cave when he states "the entire length of the cave with its three halls, from the entrance to its innermost end, is one hundred and fifty meters.[14]"
2. If all walking passages in the cave are taken into consideration and a straight-line distance is calculated, the compilation of distance data would go as follows (this does include the AB connection passage):
   - From entrance 1 to back of Hall "C" = 150 m
   - BB passage from beginning at Hall "B" to the back of High Room (not including any crawlways) = 50 m
   - Entrance 2 from lip of drop to Hall "A" = 10 m

   These lengths added together = 210 meters
3. If one were to add in all the lengths associated with all easily accessible crawlways in the cave, this would yield a total length of the cave of approximately 350 meters.
4. If the length of all of the passages associated with all of the tight crawlways which were mapped in 1999 and 2000 are included (like the extension added on to the west extension, the voice connection passages in the AB connection, a new crawlway under the west extension and one off the west extension, the behind the wall crawl voice connection passage in the BC connection, the low point, and all the really low and small crawlways in the back of Hall "C") the length of the cave is approximately 500 meters.

5. If we consider the surveyed length of the cave, which is derived by adding together the distance between the 82 survey shots used to create the map after the 2000 expedition, the uncorrected surveyed length of the cave is approximately 540 meters. This is derived by adding together the surveyed length of the cave from the 1999 map (equals 475 meters) and the surveyed length of all new passages added to the map during the 2000 expedition (equals 65 meters).
6. In the cave literature, cave length is generally reported as the horizontal length of the cave.[15] This is a measure of the length of the cave which takes into consideration that the cave passages go up and down as well as extend horizontally. You essentially calculate true horizontal distance between survey stations and then add all of these distances together to determine the length of the cave. The horizontal length for the cave after the 1999 expedition was 460 meters. We added approximately 65 meters of horizontal distance to the cave during the 2000 expedition. So, the horizontal length of the cave including all passages from both the 1999 and 2000 surveys is approximately 525 meters.

The 1999 and 2000 surveys in the Cave of Letters added 375 meters on to the length of the cave, compared to Yadin's 150 meter length, because (1) a much more thorough survey was completed, and (2) Yadin was reporting the length of the cave "with its three halls, from the entrance to its innermost end" as a distance value to walk through the cave, not as a surveyed distance.[16] The 1999 and 2000 surveys included low body tight crawls, and crawlways dug open and surveyed by the exploration and mapping team. The 1999 and 2000 surveyors went, in many cases, where Yadin's people did not go, or if they did explore these passages, they were not included on the map. Hence, the map produced by the 1999/2000 exploration in the Cave of Letters was much more detailed compared to Yadin's map. His map showed the general plan and cross-section of the main passages of the cave, and the maps produced after the year 2000 expedition are much more detailed because, as a multidisciplinary project, it was necessary to produce maps that convey the many components of the projects research design.

SUMMARY AND CONCLUSIONS

Based on the 1999 and 2000 survey of the Cave of Letters, a set of maps were produced, and the length of the cave was determined to be 525 meter, which is 375 meters longer than previous surveys indicated. The set of maps that were produced include: (1) a new base map of the cave (Figure 4.7), (2) a map depicting the physical features of the cave (Figure 4.8), (3) a map depicting the location of the archaeological finds in the cave (Figure 4.9), and (4) a set of maps depicting the locations where ERT lines were sited in the cave, as well as the locations where rock splitting occurred (Figure 4.9). Because the data used to create the maps were stored digitally, and this data base was easily updatable, new versions of the maps could be created as the volume of the data base expanded. Additionally, data could be added to the data base in real-time, which also facilitated the production of updated maps of the cave.

By utilizing a multidisciplinary approach, and working within the framework of the emerging field of geoarchaeology, unique insights were gained into the physical nature of the cave and its cultural importance. By merging the complexity of the geophysical research that occurred in the cave, with more traditional archaeological and geographical methodologies, a new and unique understanding of the Cave of Letters emerged. The surveying of the cave, and the maps that were subsequently produced, provided a mechanism for understanding the spatial complexities/inter-relationships of the research, and it also provided a means of conveying this information to anyone interested in this subject. Very few researchers, or members of the general public, have had the privilege to visit the Cave of Letters, but the set of maps produced as part of this research have contributed to establishing a better understanding of the geography of the Cave of Letters.

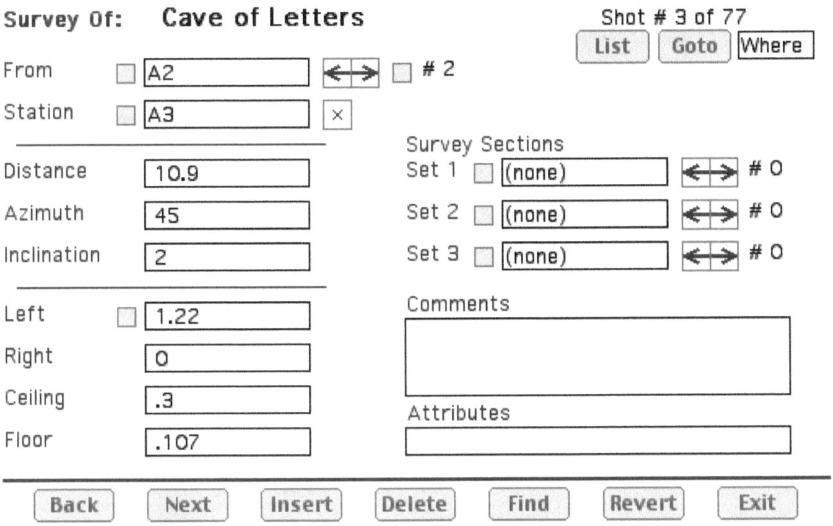

Figure 4.1: CAVEPLOT data entry screen.

XYZ Coordinants for: Cave of Letters

Num	Station	North (X)	East (Y)	Vert (Z)	Status	#	From
1	A1	+0.00	+0.00	+0.00	Plot	(@ None)	
2	A2	+0.75	+2.07	+0.72	Plot	1	A1
3	A3	+8.46	+9.78	+1.10	Plot	2	A2
4	A4	+14.75	+13.56	+0.84	Plot	3	A3
5	A5	+18.31	+18.83	+1.62	Plot	4	A4
6	A6	+16.87	+21.89	+4.78	Plot	5	A5
7	A7	+18.04	+24.93	+5.29	Plot	6	A6
8	A8	+13.59	+25.71	+4.82	Plot	7	A7
9	B1	+12.73	+22.47	+3.92	Plot	8	A8
10	B2	+9.51	+21.30	+3.56	Plot	9	B1
11	B3	1.07	+15.46	+2.80	Plot	10	B2
12	A9	+20.05	+30.24	+5.09	Plot	8	A8
13	A10	+31.19	+38.03	+5.57	Plot	12	A9
14	A11	+35.38	+40.65	+4.24	Plot	13	A10
15	A12	+38.88	+42.28	+3.77	Plot	14	A11
16	A13	+41.30	+44.71	+3.23	Plot	15	A12
17	C1	+42.91	+41.20	+3.48	Plot	15	A12

Figure 4.2: Data table created by the CAVEPLOT program.

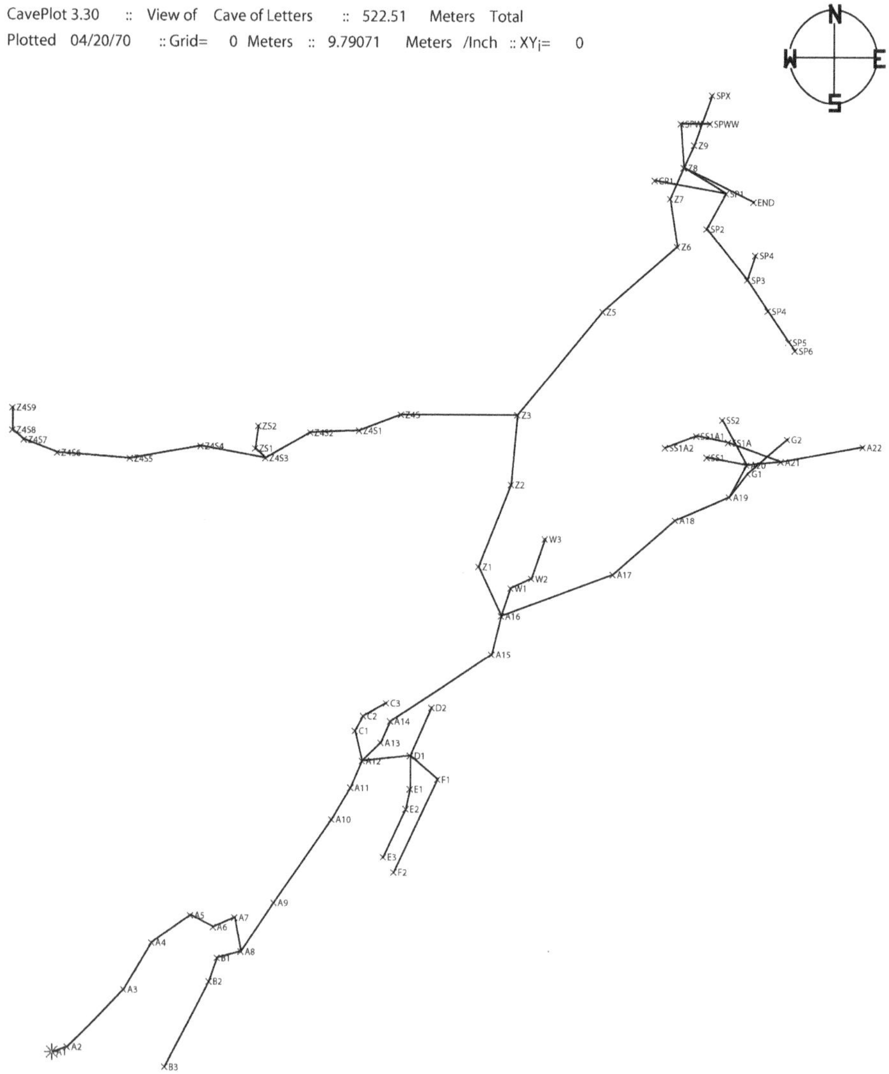

Figure 4.3: Line plot of the cave created using CAVEPLOT.

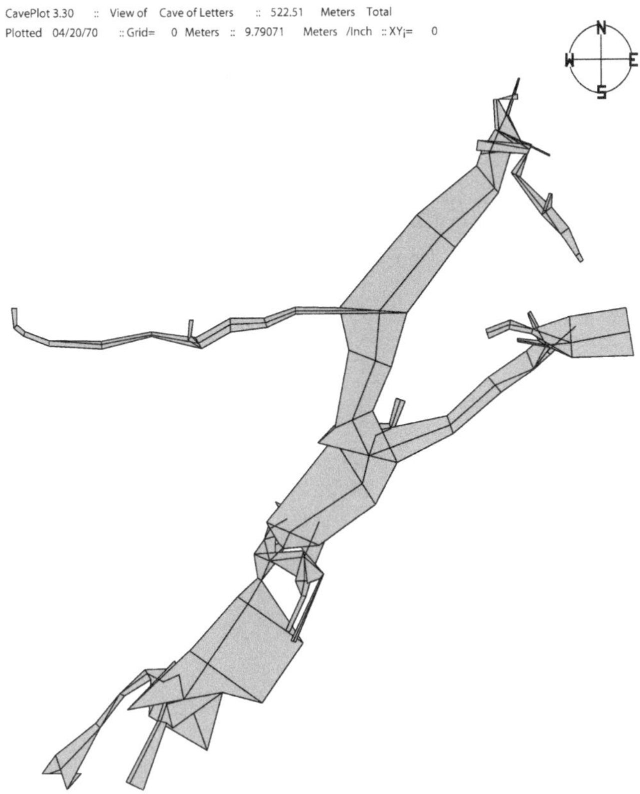

Figure 4.4: Line plot of the cave created using CAVEPLOT, with the data for distance to left and right walls included.

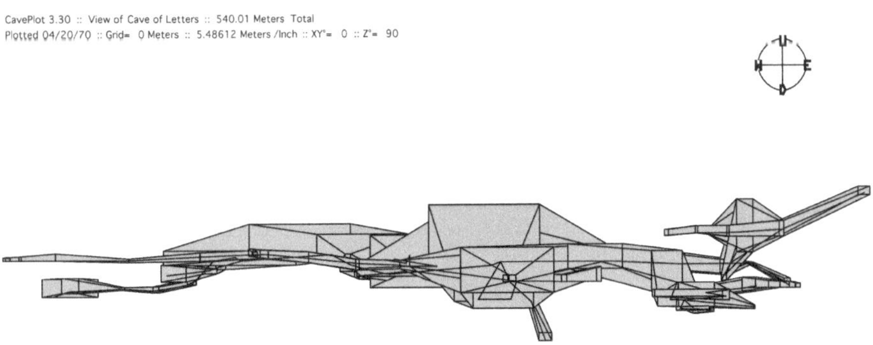

Figure 4.5: Cross-section view of the cave created using CAVEPLOT, with the data for distance to the cave walls, and the ceiling and floor included.

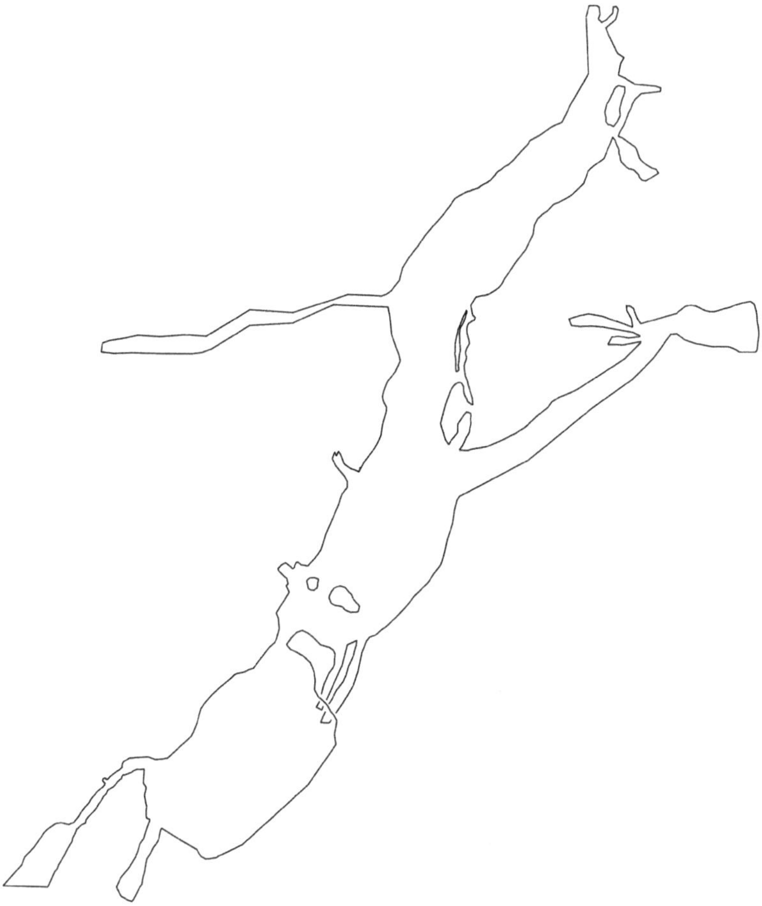

Figure 4.6: Plan view of the cave created using Adobe Illustrator.

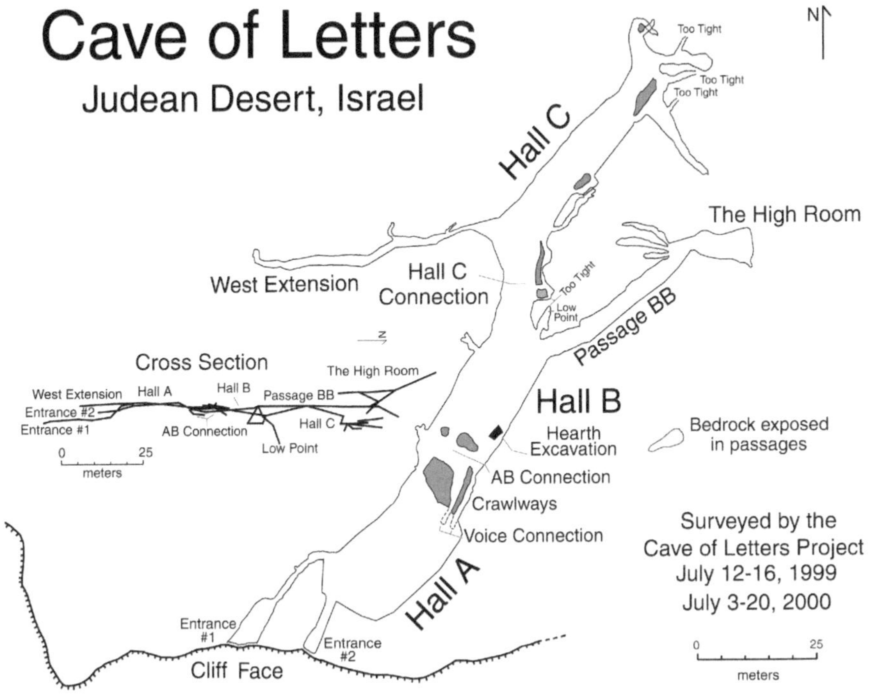

Figure 4.7: Map update from 1999 to include year 2000 data

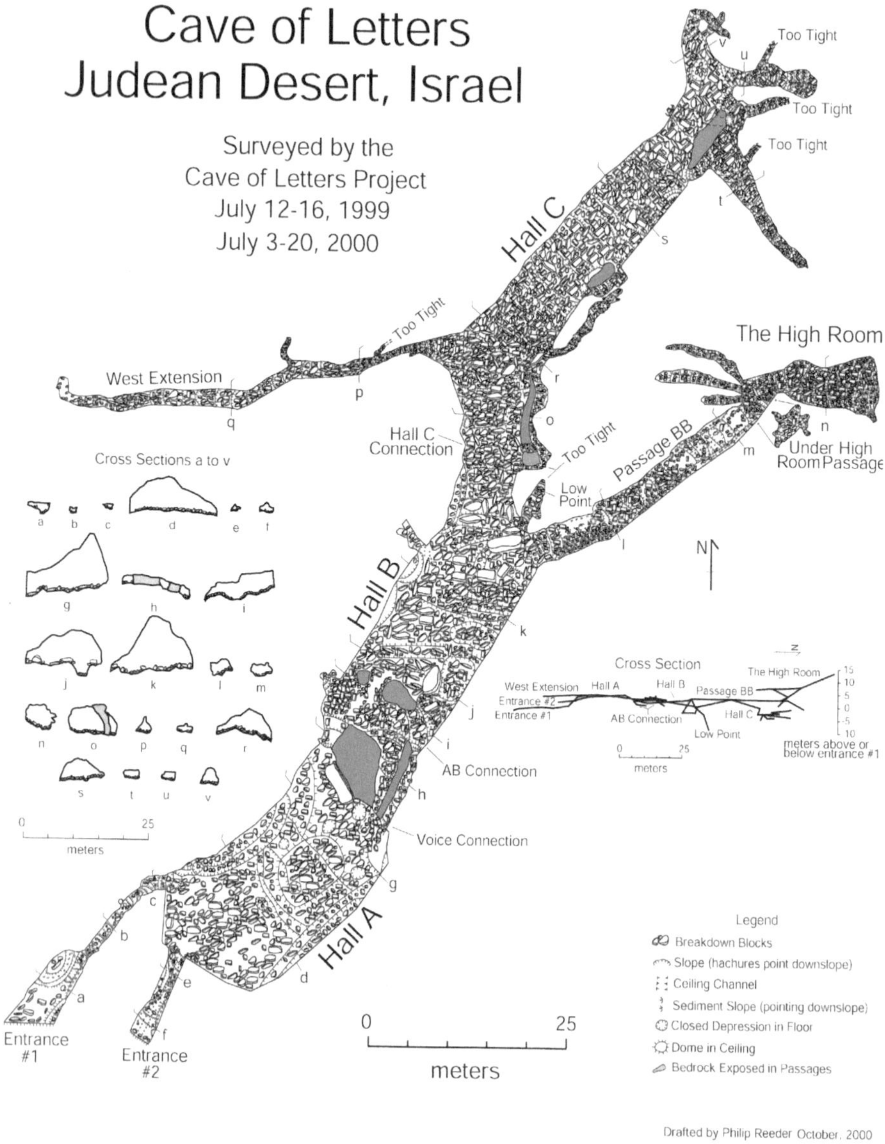

Figure 4.8: Map that includes cave cross-sections and rock debris.

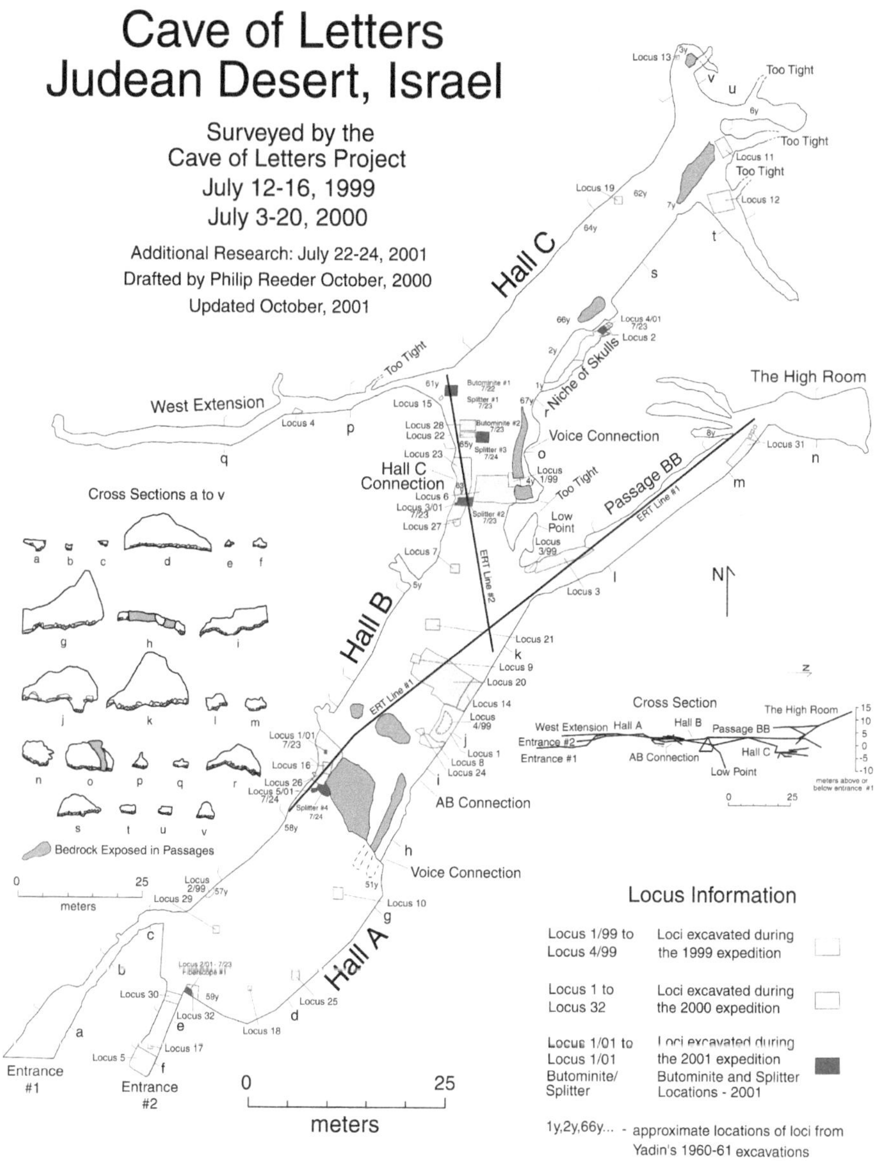

Figure 4.9: Map showing the ERT lines

NOTES

1. Yigael Yadin, *The Finds from the Bar Kokhba Period in the Cave of Letters.* (Judean Desert Studies; Jerusalem: Israel Exploration Society, 1963).
2. Philip Reeder, et al., "Multidisciplinary Research at the Cave of Letters, Israel: A Melding of Physical and Social Sciences," in *Transkart 2004 Conference Proceedings* (Hanoi, Vietnam, 2004).
3. Yadin, *The Finds from the Bar Kokhba Period in the Cave of Letters.*
4. A Frumkin, "The Cave of Letters Sediments—Indication of an Early Phase of the Dead Sea Depression?" *The Journal of Geology* 109 (2001): 79–90.
5. *Caveplot Software, 4th Edition* (Oren, Utah: Speleop Resources, 1996).
6. *Adobe Illustrator Software Wersion 7.0 for MAC* (San Jose, California: Adobe Systems, 1998).
7. Bob Robbins, "An Introduction to Caves and Caving: Cave Surveying and Mapping" (2005), http://cavingintro.net/surveymap.html.
8. Robbins, "An Introduction to Caves and Caving: Cave Surveying and Mapping."
9. Robbins, "An Introduction to Caves and Caving: Cave Surveying and Mapping."
10. Reeder, et al., "Multidisciplinary Research at the Cave of Letters, Israel: A Melding of Physical and Social Sciences."
11. Harry M Jol, et al., "Return to the Cave of Letters (Israel): A Ground Penetrating Radar Archaeological Expedition," in *Proceedings of the Eight International Conference on Ground Penetrating Radar* (GPR 2000) May 23–25, Gold Coast, Australia, D. A. Noon, et al., *SPIE* 4084 (2000): 882–86.
12. An endoscope is a medical instrument used to view inside the human body, but it was specially modified to view beneath the roof fall in the cave. By using this instrument in this fashion, portions of the Bar-Kokhba era (Second Jewish Revolt) floor could be viewed without excavating tons of roof fall debris covering the surface.
13. Yadin, *The Finds from the Bar Kokhba Period in the Cave of Letters.*
14. Yadin, *The Finds from the Bar Kokhba Period in the Cave of Letters.*
15. Robbins, "An Introduction to Caves and Caving: Cave Surveying and Mapping."
16. Yadin, *The Finds from the Bar Kokhba Period in the Cave of Letters.*

BIBLIOGRAPHY

Adobe Illustrator Software Version 7.0 for MAC. San Jose, California: Adobe Systems, 1998.

Caveplot Software, 4th Edition. Oren, Utah: Speleop Resources, 1996.

Frumkin, A. "The Cave of Letters Sediments—Indication of an Early Phase of the Dead Sea Depression?" *The Journal of Geology* 109 (2001): 79–90.

Jol, Harry M., Jack F. Shroder, Philip Reeder, and Richard A. Freund. "Return to the Cave of Letters (Israel): A Ground Penetrating Radar Archaeological Expedition." Proceedings of the Eight International Conference on Ground Penetrating Radar (GPR 2000) May 23–25, Gold Coast, Australia. D. A. Noon, G.F. Stickley, and D. Longstaff. *SPIE* 4084 (2000): 882–86.

Reeder, Philip, Paul Bauman, Richard Freund, and Harry Jol. "Multidisciplinary Research at the Cave of Letters, Israel: A Melding of Physical and Social Sciences." In *Transkart 2004 Conference Proceedings.* Hanoi, Vietnam, 2004.

Robbins, Bob. "An Introduction to Caves and Caving: Cave Surveying and Mapping," 2005. http://cavingintro.net/surveymap.html.

Yadin, Yigael. *The Finds from the Bar Kokhba Period in the Cave of Letters.* Judean Desert Studies. Jerusalem: Israel Exploration Society, 1963.

CHAPTER FIVE

Ground Penetrating Radar Subsurface Imaging WITHIN THE Cave OF Letters

HARRY M. JOL
Department of Geography and Anthropology, University of Wisconsin-Eau Claire, Eau Claire, WI 54702–4004, USA

INTRODUCTION

The Cave of Letters, a historically significant and fascinating archaeological site, is located within the northern face of a 300 m cliff within the Naḥal Ḥever canyon, Judean Desert, Israel. The floor, roof and walls of the cave are primarily composed of limestone with much of the floor covered by various-sized angular boulders of similar sedimentary composition. Due to the active seismicity of the region[1] (Dead Sea rift zone), previous earthquakes have resulted in many cave roof falls resulting in the very irregular built up floor. In 1953, Yohanan Aharoni carried out a reconnaissance archaeological survey of the Cave of Letters,[2] while in 1960 and 1961, during a renewed excavation lead by Yigael Yadin, extraordinary finds were recovered at this remote and hard to access site.[3] During Yadin's archaeological investigations it became evident that this was a location where Jewish refugees hid themselves during the Bar Kokhba revolt against Roman rule in 132–135 C.E. Due to the significance of Y. Yadin's findings and the cave's extreme inaccessibility, it has never been fully investigated and excavated using the tools of modern archaeology. One of several objectives of the John and Carol Merrill Cave of Letters archaeological expedition was to test if ground penetrating radar (GPR) could image the subsurface within the cave. If successful, the next objective was to locate and map, below the irregular cave floor surface, the floor/habitation layer upon which the Bar Kokhba people lived and/or artifacts associated with this surface. In addition, a secondary objective was to use GPR to see if there were more floors/habitation layers beneath a Bar Kokhba layer which could indicate earlier occupation(s) in the cave (e.g. Chalcolithic). The purpose of this paper is to demonstrate how the GPR results from within the Cave of Letters met these stated objectives.

ARCHAEOLOGY AND GROUND PENETRATING RADAR

The interpretation and reconstruction of subsurface environments is an important task that continually faces geoarchaeologists. In addition, as archaeological permits are becoming more difficult to attain, and excavations are becoming increasingly expensive, archaeologists are more interested in utilizing non-invasive, non-destructive geophysical methods to gain information remotely about the subsurface. Of the various geophysical techniques employed in such archaeological work, GPR has increasingly been used to image and assess sites in a cost effective and time efficient manner. Within archaeology, GPR has traditionally been successful at locating tombs, shallow graves, buried walls, tunnels and other voids, mapping historic-period fortifications, defining the outlines of buried features such as buildings, and providing an immediate image of the site stratigraphy.[4,5] In the correct sedimentary context, the GPR technique can detail the location, extent, and character of subsurface features and/or artifacts by providing near surface, high resolution, and near continuous profiles of the site being investigated.

Caves provide a very unique and challenging environment to carry out GPR investigations. The Cave of Letters provided one of the first archaeological and geomorphic sites to conduct GPR studies inside a cave environment. An earlier study showed that GPR was used in the el-Wad Cave (Mount Carmel, Israel) to confirm the existence of a series of karstic pits in several chambers of the cave that were first detected using seismic refraction, while GPR was used in Peak Cavern (Castleton, Derbyshire, United Kingdom) to show the sedimentological and archaeological patterns within the cave-fill.[6,7] Other than these two published results, few GPR studies have utilized GPR to understand the subsurface stratigraphy inside a cave system and still fewer who have used GPR to map out potential archaeological sites within a cave. Discussions with researchers who have worked in caves indicated that their earlier GPR projects in caves had not been successful for several reasons, including a) difficulty of getting the equipment into the cave, b) equipment problems due to the environmental conditions posed by working in a cave (e.g. dust, humidity), c) very shallow depths of penetration due to the conductive nature of the sedimentary deposits beneath the cave floor (which led to significant signal attenuation), and d) anomalous signals/ reflection patterns often interpreted as not originating from subsurface stratigraphy but resulting from reflections bouncing from the cave ceiling and/or wall. It should be noted that GPR studies have, however, been successfully conducted on the surface of limestone outcrops while others have mapped sinkholes (collapsed caves) in karst environments for both geomorphic and archaeological purposes.[8,9,10,11,12,13,14]

GROUND PENETRATING RADAR—METHODOLOGY

Subsurface profiling with GPR is similar to seismic reflection, except it uses electromagnetic (EM) energy, not acoustic energy. A short pulse of high frequency EM energy, usually in the 10–1000 megahertz (MHz) range, is transmitted into the ground, generating a wavefront which propagates downward. Due to a change in bulk electrical properties of different subsurface litholgies, mineralogy and/or the character of the sediment interface, some of the energy is reflected back to the surface.[15] At the surface a receiver monitors reflected energy versus delay time. The pulse delay time from the energy transmitted into the ground and reflected back to the receiver is a function of the EM propagation velocity through the sediment and the depth of subsurface reflectors. Strength of the reflected signal is approximately proportional to the difference in dielectric constants at the sediment interface.[16] These effects enable the subsurface stratigraphy and structure to be inferred from the character of the radar return signals. Changes in the dielectric constant and electrical conductivity can also affect the rate of energy attenuation. Fine-grained sediments such as silts and clays, which have high conductivities, cause significant signal attenuation and reduce the penetration to a few meters or less, while environments of low conductivity, such as sands, gravels and limestone, allow for greater depth of penetration. GPR provides a profile of horizontal survey distance (m) versus vertical two-way travel time in nanoseconds (ns, 1 ns = 10^{-9} seconds). By measuring the propagation velocity of the sediment, depth of reflections can be determined. Further theory and methodology are adequately explained elsewhere.[17]

A field portable, digital pulseEKKO™ 100 and 1000 GPR system was employed in the reflection survey mode to obtain the GPR transects in the cave during the 1999 and 2000 John and Carol Merrill Cave of Letters archaeological expeditions. Frequency is an important design consideration for any GPR survey, and therefore, four antennae frequencies, 100, 200, 225 and 450 MHz, were tested. Higher frequency antennae will provide finer resolution but will not allow penetration of the signal to depths as great as with lower frequency antennae.[18,19] To provide detailed horizontal resolution of the subsurface structures, 0.05 m to 0.25 m steps were used for the surveys, while antennae spacing (source-receiver offset) ranged depending on the frequency of antennae used. Various transmitter powers (200 V, 400 V and 1000 V) were used to optimize depth of penetration and continuity of reflections. To reduce data collection time, particularly in the rugged cave environment, a backpack transport system was employed. Each trace was vertically stacked 64 times with a minimum sampling rate of 800 ps. The digital profiles were processed and plotted (wiggle trace format) using pulseEKKO™ software with automatic gain control (AGC), signal saturation correction, trace stacking (horizontal averaging) and point stacking (down trace running average). Further processing was conducted on several of the profiles. Near surface velocity measurements were calculated from common mid-point (CMP) surveys and averaged to 0.1 m/ns. In addition, transects, where necessary, were corrected for topography. An advantage of the GPR method was that it provided real time information on depth of subsurface layers and/or location to targets. The application of radar stratigraphic analysis, an approach for interpreting sedimentary environments, on the collected data provided the framework to investigate both lateral and vertical geometry of the reflection patterns[20,21,22] (distinct signature patterns). The ability to recognize major reflection patterns aided in understanding the complex subsurface architecture of the cave's floor material, which is composed of broken up pieces of roof fall composed of limestone—an electrically resistive (low conductivity) material. In addition, after editing, processing, printing, and interpreting the GPR profiles while in the cave, probes were made at selected locations using an endoscope, metal detector, and/or by excavating to identify GPR anomalies.

For the sake of clarity, the horizontal scale of all profiles is distance in meters, whereas the vertical scale is shown as both two-way travel time (ns; left side) and depth/elevation in meters (m; right side) based on near-surface velocity of a radar pulse through the sediment investigated. The two uppermost continuous reflections in all profiles represent air wave and ground wave arrivals, respectively, and should not be considered part of the stratigraphic data.

RESULTS AND INTERPRETATION

After numerous trips through the rugged Judean Desert and much effort of carrying the GPR equipment safely into the Cave of Letters, a series of experiments along transects were conducted to test the feasibility of using a GPR system within a cave environment during the 1999 expedition. Based on the success of the initial experiments, grids of parallel transects were collected in each of the 3 chambers (A, B, C) as well as all adjoining passageways. Of the >50 transects collected, a selection of representative GPR profiles from Chamber B will be described below.

The lowest GPR antennae frequency used in the Cave of Letters was 100 MHz. An example 27 m long 100 MHz profile was collected along Line 1 with antennae separation of 1.0 m and a step size of 0.25 m (Figures 5.1 and 5.2). Throughout the profile reflections are semi-continuous, and sub-horizontal to inclined. The profile can been divided into 3 packages, which have similar internal reflection patterns, with the packages themselves divided by near to semi continuous, horizontal to sub-horizontal reflection(s). The lower package, shaded darker gray, is interpreted as the local karst bedrock (limestone). The reflections show some of the internal layering within the limestone (also visible within the cave walls) and the bedrock does come to surface a few meters off the end of the profile. Throughout the cave, based on this reflection pattern, the bedrock (original cave floor) undulates beneath the roof fall deposits that lie on top. The middle reflection package, shaded light gray, is interpreted as roof fall material that has shaken loose during earthquakes and fell onto the cave floor, slowly raising the cave chamber's floor while at the same time also raising the cave's ceiling. The upper package, not shaded, is similar in appearance as the middle package but is separated by a near continuous, undulating reflection. This undulating reflection is interpreted as the Bar Kokhba "floor", while the upper package is interpreted as the the most recent roof fall material which has accumulated on and covered up the proposed Bar Kokhba habitation layer.

The 2nd example utilizes the higher frequency 200 MHz antennae along a 19 m portion of Line 2 (Figures 5.1 and 5.3). The data was collected with an antennae separation of 0.5 m and a step size of 0.1 m. The resulting profile shows semi-continuous to continuous, sub-horizontal to inclined reflections. These reflection patterns are similar to the upper two packages interpreted on the 100 MHz line. With the higher resolution of the 200 MHz and a closer step size (0.1 m) during data collection, more details within the upper two reflection packages were visible. The nearly continuous, undulating reflection that separates the two packages is interpreted as the Bar Kokhba floor, and it is also more apparent than on the 100 MHz line (Figure 5.3). Above this reflection (package boundary), there are several semi-continuous, sub-parallel reflection patterns which are interpreted as indicating several roof fall events since the cave was abandoned during Bar Kokhba times. With the higher frequency antennae of this survey (using both a 400 V and 1000 V transmitter) the depth of penetration has been decreased and therefore the lower interpreted bedrock surface of the 100 MHz line is not visible in this mid-chamber transect. However, other 200 MHz surveys in the cave were able to image the undulating bedrock surface (original cave floor). Based on the collected GPR profiles and follow-up test probes, this bedrock surface was, on average, shallower in the passageways (at or near surface to approximately 2 m depth) than within the chambers (at or near surface to approximately 4–5 m depth).

To provide a comparison to the 200 MHz data, the 3rd example was collected with the 450 MHz antennae along the first 4 m of example 2 (Line 2, Figures 5.1 and 5.4). The data, from the highest frequency used in the Cave of Letters, was collected with an antennae separation of 0.17 m and a step size of 0.05 m. With the increase in frequency to 450 MHz and decrease in transmitter power (200 V), the profile shows a decrease in depth of penetration to approximately 3 m but an increase in the resolution. The upper 60 ns of time are plotted with a noticeable increase in the number of reflections within the upper two packages. In both packages, reflections are semi-continuous to discontinuous and sub-horizontal to inclined in nature, with the boundary between the packages less apparent. The reflections shows both, more details of the roof fall material and, surprisingly, the internal stratigraphy of the larger boulders (roof fall) that the survey crossed.

Three-dimensional (3D) datasets can greatly aid in interpreting the framework of the subsurface materials and provide a more detailed view of the geometry of individual units/packages within the Cave of Letters. Such analysis could indicate optimum zones of archaeological excavation and may also aid in the determination of volume estimates for sediment/rock removal. Therefore, within the Cave of Letter, in addition to the 2D lines surveyed, grids of parallel lines (full length of chamber) were collected in each chamber so that 3D cubes could be developed. The 4th example presented in this chapter is a grid that was collected with the 200 MHz antennae with an antennae separation of 0.5 and a step size of 0.1 m. The parallel lines were separated by 0.25 m resulting in 5.5 m by 2.5 m grid collected in Chamber B (see Figure 5.1 for location). The GPR lines were rendered as a cube (chair diagram) using 3D software (T3D) and provide a 3D perspective of roof fall material and the interpreted Bar Kokhba horizon (bottom of the cut out portion of the cube as seen in Plate 4). The reflections discussed in earlier examples are now visible in a 3D perspective and one can visualize the semi-continuous nature of the roof fall deposits that occur throughout the cube. The more continuous reflections in the cube, particularly below the Bar Kokhba horizon, could be former cave floors (horizons) upon which archaeological remains may be located.

Based on the interpretation of the datasets collected, the GPR method was able to a) image the subsurface of the rugged, Cave of Letters floor, b) record reflection signals from depths of greater than 10 m, often into the local bedrock, and c) provide profiles where three reflection packages and associated boundaries were interpreted, including the Bar Kokhba horizon.

DISCUSSION

To meet the primary objective of the project, to locate and map the Bar Kohkba horizon, the 450 MHz antennae proved difficult to interpret since it provided "too much detail" while the 100 MHz antennae provided good depth of penetration but not enough detail/resolution. As a compromise between depth of penetration, continuity of reflection patterns, and resolution, the 200 MHz antennae with a 400 V transmitter was chosen to collect the majority of transects in the Cave of Letters. Based on the interpretations of the GPR profiles, follow up investigations were undertaken at some selected sites and involved endoscopic viewing, probing and the use of metal detectors. The major features that were considered for further archaeological investigation/excavation were the interpreted habitable floor (horizon) from the Bar Kokhba time period and the associated artifacts. The results indicated that the occupation surface was undulating and covered with a thin layer of finer grained sediments. This horizon varied in depth from approximately 1.5–2.0 m below the present cave floor surface to just below or at the cave floor surface (for example where the bedrock came to surface). The archaeological finds along this occupation layer (reported in various chapters in the volume) included rope, pottery shards, linen, coins, and various sizes of wood (some carved). Although GPR could locate the Bar Kokhba floor, the question remained whether the cave's lower stratigraphy might indicate other habitation levels beneath this layer. Since other caves in the area have had artifacts associated with the Jewish revolt of the first century CE to as far back as the Chalcolithic period, the ability to identify these different archaeological levels (horizons) is an important consideration for archaeologists and historians.

The objectives of the 3D GPR surveys conducted in the Cave of Letters were to construct subsurface images and to utilize this data to better understand the internal stratigraphy and geometry of the cave deposits as well as direct locations for possible excavation. The 3D images, or cubes, constructed from the high resolution 200 MHz GPR datasets provided a unique representation of the subsurface and directed several of the archaeological investigations that led the discovery of the Bar Kokhba horizon and the recovery of associated artifacts. In addition, based on the surveys and associated cubes, there are horizons below the interpreted Bar Kokhba layer that are similar in perspective and may be a former occupied surface. Based on the experience of this archaeological project, the ability to excavate below approximately one meter in this rough terrain is a major impediment to excavating to lower depths.

CONCLUSIONS

1) The project demonstrated that GPR data can be collected in a non-invasive, non-destructive manner over very rugged roof fall blocks within a limestone cave environment (karst).

2) GPR proved to be an effective method for mapping the subsurface stratigraphic architecture within the Cave of Letters. GPR was successfully brought into the cave, set up and data was collected using several antennae frequencies (100–450 MHz).

3) The study demonstrated that GPR can assess depths up to 10 m and image the lateral extent of a cave deposits. Based on the interpretation of the GPR results endoscopic viewing, metal detectors, and initial archaeological probing were directed. Even with minimal probing the Bar-Kokhba layer was identified and many associated artifacts were located.

4) GPR was effective in imaging and mapping the roof fall deposits, previous cave floors (horizons), base of the cave (bedrock), and the internal structure of the local bedrock (limestone).

5) The results of the project demonstrate that GPR can be utilized in a cave in a time-efficient and cost effective manner to obtain shallow subsurface 3D images. The use of 3D rendering software allowed for viewing the collected datasets from any angle and for analyzing the continuity of reflections through cut-aways of the cube. For example, the imagery enhanced our ability to interpret the three-dimensional geometry of the Bar Kokhba floor. In addition, results from the 3D GPR study indicate the possibility of several previous cave floors and revealed various thicknesses of roof fall from previous earthquakes.

6) Based on the GPR interpretations and probes, it was discovered that archaeological artifacts lay buried on the proposed Bar Kokhba floor (horizon) that lied beneath the last roof rockfall(s). Many artifacts, including pottery, linen, wood fragments were found that had previously been unknown to investigators and are of Bar Kokhba in origin (or possibly earlier).

7) Based on the results and discoveries of this study, GPR surveys are warranted to image and map cave deposits within other limestone caves worldwide.

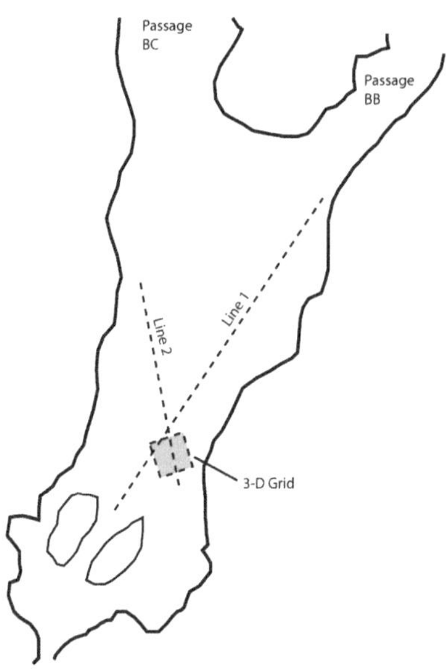

Figure 5.1: Location map of select GPR transects and three-dimensional grid.

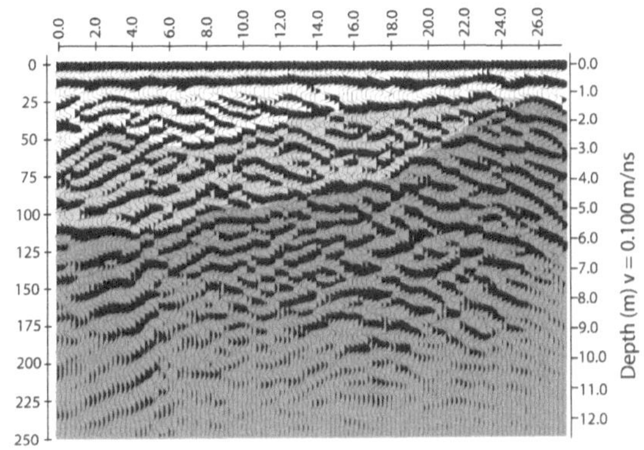

Figure 5.2: Topographically corrected 100 MHz GPR line shot along transect line 1.

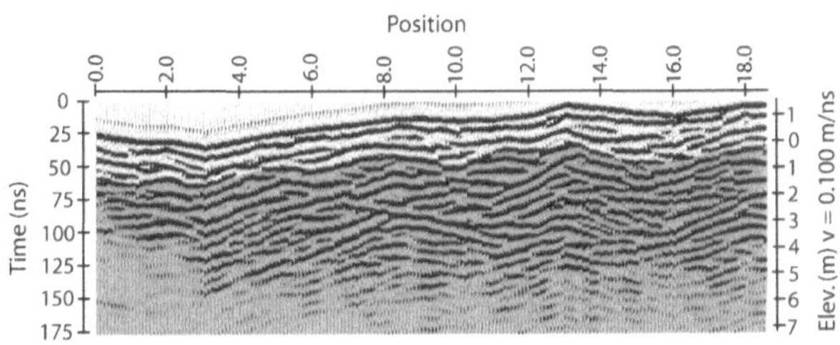

Figure 5.3: Topographically corrected 200 MHz GPR line shot along transect line 2.

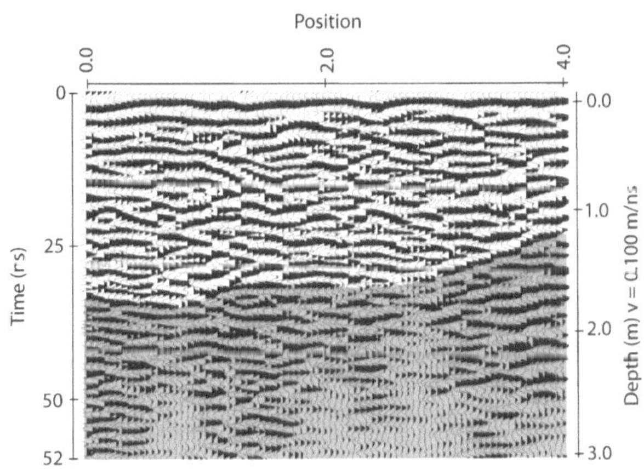

Figure 5.4: Topographically corrected, high resolution 450 MHz GPR profile collected along the first 4 m of transect line 2.

NOTES

1. A. Horowitz, *The Quaternary of Israel* (New York: Academic Press, 1979).
2. Yohanan Aharoni and Beno Rothenberg, *In the Footsteps of Kings and Rebels* (Rarnat-Gan, Israel: Massada, 1960).
3. Yigal Yadin, *The Finds from the Bar Kokhba Period in the Cave of Letters* (The Israel Exploration Society, Jerusalem, 1963), 279.
4. L.B. Conyers, *Ground-Penetrating Radar for Archaeology* (Walnut Creek, CA: AltaMira Press, 2004).
5. G. Rapp Jr. and C.L. Hill, *Geoarchaeology: The Earth-Science Approach to Archaeological Interpretation* (New Haven: Yale University Press, 1998).
6. A. Beck and M. Weinstein-Evron, "A Geophysical Survey in the El-Wad Cave, Mount Carmel, Israel," *Archaeological Prospection* 4, no. 2 (1997): 85–91.
7. J.K. Pringle, et al., "Survey, Geophysics, Geology, and Archaeology to Create a 3D Digital CAD Model Investigating Peak Cavern, Castleton, Derbyshire, UK: Integrating Cave," *Cave and Karst Science*, 29 (2002): 67–74.
8. B.R. Pratt and A.D. Miall, "Anatomy of a Bioclastic Grainstone Megashoal (Middle Silurian, Southern Ontario) Revealed by Ground-Penetrating Radar", *Geology* 21 (1993): 223–26.
9. C.L. Liner and J.L. Liner, "Ground-Penetrating Radar: A Near-Face Experience from Washington County, Arkansas", *The Leading Edge* 14 (1995): 17–21.
10. T. Overgaard, "Application of Ground Penetrating Radar for Limestone Exploration in Fakse Kalkbrud, Denmark," *Journal of Applied Geophysics* 34 (1995): 167.
11. G.A. McMechan, et al., "Ground Penetrating Radar Imaging of a Collapse Paleocave System in the Ellenburger Dolomite, Central Texas," *Journal of Applied Geophysics* 39 (1998): 1–10.
12. A.T. Chamberlain, et al., "Cave Detection in Limestone Using Ground Penetrating Radar", *Journal of Archaeological Science* 27 (2000): 957–64.
13. G. Dagallier, et al., "Ground Penetrating Radar Application in a Shallow Marine Oxfordian Limestone Sequence Located on the Easter Flank of the Paris Basin, NE France," *Sedimentary Geology* 130 (2000): 149–65.
14. M. Beres, M. Luetscher, and R. Loivier, "Integration of Ground-Penetrating Radar and Microgravimetric Methods to Map Shallow Caves," *Journal of Applied Geophysics* 46 (2001): 249–62.
15. J.L. Davis and A.P. Annan, "Ground-Penetrating Radar for High-Resolution Mapping of Soil and Rock Stratigraphy," *Geophysical Prospecting* 37 (1989): 531–51.
16. J.L. Davis and A.P. Annan, *Ground-Penetrating Radar for High-Resolution Mapping of Soil and Rock Stratigraphy*.
17. J.L. Davis and A.P. Annan, *Ground-Penetrating Radar for High-Resolution Mapping of Soil and Rock Stratigraphy*.
18. H.M. Jol, "Ground Penetrating Radar Antennae Frequencies and Transmitter Powers Compared for Penetration Depth, Resolution and Reflection Continuity," *Geophysical Prospecting*, 43 (1995): 693–709.
19. D.G. Smith and H.M. Jol, "Ground Penetrating Radar: Antennae Frequencies and Maximum Probable Depths of Penetration in Quaternary Sediments," *Journal of Applied Geophysics*, 33 (1995): 93–100.
20. M. Beres Jr. and F.P. Haeni, "Application of Ground- Penetrating-Radar Methods in Hydrogeologic Studies," *Groundwater* 29 (1991): 375–86.

21. H.M. Jol and D.G. Smith, "Ground Penetrating Radar of Northern Lacustrine Deltas," *Canadian Journal of Earth Sciences* 28 (1991): 1939–47.
22. H.M. Jol and C.S. Bristow, "GPR in Sediments: Advice on Data Collection, Basic Processing and Interpretation, a Good Practice Guide" in *Special Publication 211 of GPR in Sediments* (Geological Society of London, 2003), 9–27.

BIBLIOGRAPHY

Aharoni, Yohanan, and Beno Rothenberg. *In the Footsteps of Kings and Rebels.* Rarnat-Gan, Israel: Massada, 1960.

Beck, A., and M. Weinstein-Evron. "A Geophysical Survey in the El-Wad Cave, Mount Carmel, Israel." *Archaeological Prospection* 4, no. 2 (1997): 85–91.

Beres, M., Jr., and F.P. Haeni. "Application of Ground-Penetrating-Radar Methods in Hydrogeologic Studies." *Groundwater* 29 (1991): 375–86.

Beres, M., M. Luetscher, and R. Loivier. "Integration of Ground-Penetrating Radar and Microgravimetric Methods to Map Shallow Caves." *Journal of Applied Geophysics* 46 (2001): 249–62.

Chamberlain, A.T., et al. "Cave Detection in Limestone Using Ground Penetrating Radar." *Journal of Archaeological Science* 27 (2000): 957–64.

Conyers, L.B. *Ground-Penetrating Radar for Archaeology.* Walnut Creek, CA: AltaMira Press, 2004.

Dagallier, G., et al. "Ground Penetrating Radar Application in a Shallow Marine Oxfordian Limestone Sequence Located on the Easter Flank of the Paris Basin, NE France." *Sedimentary Geology* 130 (2000): 149–65.

Davis, J.L., and A.P. Annan. "Ground-Penetrating Radar for High-Resolution Mapping of Soil and Rock Stratigraphy." *Geophysical Prospecting* 37 (1989): 531–51.

Horowitz, A. *The Quaternary of Israel.* New York: Academic Press, 1979.

Jol, H.M. "Ground Penetrating Radar Antennae Frequencies and Transmitter Powers Compared for Penetration Depth, Resolution and Reflection Continuity." *Geophysical Prospecting*, 43 (1995): 693–709.

Jol, H.M., and Bristow, C.S. *GPR in Sediments: Advice on Data Collection, Basic Processing and Interpretation, a Good Practice Guide*. Vol. Special Publication 211 of GPR in Sediments, 9–27. Geological Society of London, 2003.

Jol, H.M., and Smith, D.G. "Ground Penetrating Radar of Northern Lacustrine Deltas." *Canadian Journal of Earth Sciences* 28 (1991): 1939–47

Liner, C.L., and Liner, J.L. "Ground-Penetrating Radar: A Near-Face Experience from Washington County, Arkansas." *The Leading Edge* 14 (1995): 17–21.

McMechan, G.A., et al. "Ground Penetrating Radar Imaging of a Collapse Paleocave System in the Ellenburger Dolomite, Central Texas." *Journal of Applied Geophysics* 39 (1998): 1–10.

Overgaard, T. "Application of Ground Penetrating Radar for Limestone Exploration in Fakse Kalkbrud, Denmark." *Journal of Applied Geophysics* 34 (1995): 167.

Pratt, B.R., and A.D. Miall. "Anatomy of a Bioclastic Grainstone Megashoal (Middle Silurian, Southern Ontario) Revealed by Ground-Penetrating Radar." *Geology* 21 (1993): 223–26.

Pringle, J.K., et al. "Survey, Geophysics, Geology, and Archaeology to Create a 3D Digital CAD Model Investigating Peak Cavern, Castleton, Derbyshire, UK: Integrating Cave." *Cave and Karst Science* 29 (2002): 67–74.

Rapp, G., Jr., and Hill, C.L. *Geoarchaeology The Earth-Science Approach to Archaeological Interpretation*. New Haven: Yale University Press, 1998.

Smith, D.G., and H.M. Jol. "Ground Penetrating Radar: Antennae Frequencies and Maximum Probable Depths of Penetration in Quaternary Sediments." *Journal of Applied Geophysics*, 33 (1995): 93–100.

Yadin, Y. *The Finds from the Bar Kokhba Period in the Cave of Letters*. The Israel Exploration Society, Jerusalem, 279 pp, 1963.

CHAPTER SIX

Electrical Imaging AND Metal Detection IN THE Cave OF Letters

PAUL D. BAUMAN
Advisian WorleyParsons, Calgary, Alberta, Canada

The Cave of Letters was a refuge for Jewish rebels and their families toward the end of a failed rebellion against the Roman occupation of Palestine in 135 of the Common Era (C.E.). The geophysical technique of two-dimensional electrical resistivity tomography (ERT) was used as an accompaniment to intrusive archaeological investigations to explore for the presence of a packed earthen floor, or occupation level, at some depth below the roof collapse rubble. Such a layer was identified in Halls 'B' and 'C' of the cave. In the Hall C connection passage, the ERT survey suggested that the floor was exposed at the surface (though hidden under large boulders). Archaeological investigations confirmed the existence of a Roman period occupation level at the surface in this location. ERT investigations also determined that the thickness of roof collapse rubble in the cave varies from essentially 0 m, to a maximum of approximately 6 m. As would be expected, variations in the thickness of the roof collapse rubble approximately mimic changes in height of the roof of the cave.

Besides the ERT surveys, three different geophysical metal detection surveys were carried out in the cave. The cave floor was surveyed with 100% spatial coverage. Niches, walls, and low areas of the roof were also surveyed. No indications of even minor amounts of Ferro (iron and steel) or non-ferromagnetic (copper, bronze, silver, tin, etc.) metal were identified. This information is essentially categorical evidence that no large hoards of metal objects exist within the top three meters or so beneath the cave floor. Nevertheless, small objects such as coins may still be scattered at various depths in the roof collapse rubble. Very small objects in the very near surface (top 10 cm) such as coins were not identified by the geophysical instruments. Such small, near surface artifacts are better searched for with inexpensive, small coil, low powered metal detectors.

INTRODUCTION

The Cave of Letters was a refuge for Jewish rebels and their families toward the end of a failed rebellion against the Roman occupation of Palestine[1] which occurred from 132 to 135 of the Common Era (C.E.). Since the period of occupation of the cave was about 1873 years ago, it is believed that a significant thickness of roof collapse debris has accumulated.[2] The ability to map the interface between post-rebellion accumulated rubble and the underlying material may assist archaeologists in better focusing their excavation efforts. Additionally, any applications developed in the Cave of Letters may be useful in other similar environments.

In field programs in both 1999 and 2000, ground penetrating radar (GPR) was utilized as a reconnaissance tool in the concurrent intrusive archaeological investigations.[3] While GPR was successful in identifying various anomalous targets, and mapping various subsurface interfaces, it was felt that data from a second geophysical technique would be beneficial in supporting the GPR data. Additionally, while GPR provides reflection sections of unparalleled resolution, it is not always clear what causes a GPR reflection. GPR reflections are due to an assorted mixture of physical properties including electrical terrain conductivity, bulk dielectric constant, bulk porosity, bulk magnetic susceptibility, grain size, etc. Finally, as with any non-intrusive and indirect method of geophysical investigation, using multiple geophysical techniques and joint interpretations is often a benefit as it may greatly facilitate the final interpretation of what one is actually viewing under any particular site.

This paper describes the application of a relatively new near surface geophysical exploration technique, electrical resistivity tomography (ERT), as used in imaging the 'Bar Kokhba' floor at the Cave of Letters. The technique is described in detail below. Briefly, ERT is a technique for mapping the distribution of subsurface electrical resistivity in a cross-sectional format. As applied to the Cave of Letters, the goal of the ERT survey was to image the contact between roof collapse debris, which would have a large percentage of air-filled void space (and a correspondingly high electrical resistivity), and the underlying unweathered limestone bedrock which would have a different, and likely somewhat lower electrical resistivity. This predicted lower resistivity would likely be due to the much smaller amount of air-filled pore space, and the presence of clays and other minerals which may increase the rock matrix conductivity (i.e., decrease resistivity). Knowledge of the thickness of roof collapse debris would provide archaeologists with information on the maximum and minimum depths to which they could probe or dig.

Additionally, it was hoped that ERT would image an occupation level, loosely referred to in this article as the 'Bar Kokhba floor.' It is reasonable, and observed in other caves, that long term occupants of caves along the Dead Sea coast would rearrange large boulders and create packed earth floors so as to have more comfortable living quarters. Irrefutable evidence exists that the Cave of Letters was occupied during the Second Jewish Revolt.[4] Nevertheless, while geophysical techniques may be capable of identifying such an earthen floor buried beneath limestone rubble, only archaeological and archaeometric methods can assign uses and dates to such a feature.

The survey conditions for ERT are, quite clearly, particularly unfavorable in caves. Because ERT depends on the transmission of galvanic (DC) current through an electrode into the ground, it was initially thought that the likelihood of a successful ERT survey in the Cave of Letters was very small. The coarse surface material and lack of moisture in the very near surface created hostile conditions for putting current into the ground. Also, ERT survey lines must be linear, regardless of the cave geometry. Largely for these reasons, it is believed that ERT surveys have never before been conducted within a natural cave.

Besides the ERT survey, three types of geophysical metal detection surveys were carried out in the Cave of Letters. These included high resolution time domain metal detection (Geonics M61), terrain conductivity mapping (Geonics EM38), and magnetic gradiometer mapping (GEM Overhauser Walking Gradiometer).

ERT METHODOLOGY

ERT is a surface geoelectric technique for mapping the distribution of subsurface electrical resistivity in a cross-sectional format. Data are collected through a linear array of electrodes coupled to a DC resistivity transmitter/receiver and an electronic switching box. The collection process is driven by a computer that is also used for data recording. The entire data set is then inverted using a two-dimensional (2-D) finite difference, smooth inversion routine. The final product is a 2-D geoelectrical cross-section, plotting 'true' resistivity (in ohm-m) versus 'true' depth. The acquisition hardware used in this program was the ABEM Lund Imaging System. Below, the technique is described in greater detail.

BACKGROUND TO DC RESISTIVITY SURVEYS

All resistivity methods employ an artificial source of current which is introduced into the ground through point electrodes. The measurement of voltage potential at other electrodes in the vicinity of the current flow, combined with measurement of the input current, allows the determination of an effective or apparent resistivity of the subsurface.[5] Variations in resistance to the input current flow at depth cause variations in the potential difference measurements at the surface which, in turn, provide information on the subsurface electrical properties.[6] It is well known that the major problem with resistivity methods is their high sensitivity to minor variations in the electrical resistivity of the near surface layer, particularly in the vicinity of the current and voltage electrodes. With only a trace of soil moisture content, nowhere is this truer than in the Cave of Letters, a natural cave inset 100 m below the top of a 300 m overhanging cliff situated in the hyper-arid Judean Desert.[7]

The physics of 2-D resistivity (i.e., ERT) are no different than those of 1-D resistivity as initially investigated by Conrad Schlumberger in France and Frank Wenner[8] in the United States in 1917 and 1918. A known current moves in the subsurface from one electrode to an electrode return at some measured distance away (Figure 6.1). The potential drop that is, the force (i.e., voltage) needed for the current to overcome the resistance of the subsurface earth materials, is measured by two potential electrodes. If we measure the current we put into the subsurface, and we measure a voltage difference across two electrodes at surface, then with Ohm's Law we can calculate a resistance (voltage divided by current). Given the geometry of our four electrode array, we can then calculate an apparent resistivity (the resistance of a material across a standard length and cross-sectional area) of the subsurface. By changing the geometry of the electrodes, we can measure an apparent resistivity at a different depth in the earth. Until the 1970's, the interpretation of log-log sounding curves of 1-D resistivity surveys were done with overlays from books of type curves. By the 1980's, computer generated curves were used to more conveniently produce closer model fits to the data (Figure 6.2). The problems with 1-D resistivity, though, remained. The interpretation of any single 1-D resistivity sounding is inherently non-unique, with infinite models providing an identical solution to the curve fit. And of course, the solutions assume an often non-representative 1-D layered earth model.

Instead of four electrodes as with 1-D resistivity soundings, a 2-D resistivity array can be constructed of any number of electrodes, though in practice this typically ranges from 42 to 256. The electrodes must be collinear and evenly spaced, though 3-D arrays are also possible. While any one measurement still involves two current and two voltage measurements, all possible quartets of electrodes can be addressed by the transmitter/receiver via a patchbox filled with solid state relays, similar to those found in local neighborhood telephone routing centers (Figure 6.3). Which electrodes are addressed, and when, is determined by software commands laid out by the geophysicist as governed by the objectives of the exploration program (how deep, flat lying or vertical structures, etc.). All electrodes are connected to the electronics via multicore (multiple live conductors in the core) cable. The survey proceeds in an automated fashion. With present day transmitters and receivers, up to 10 voltage measurements can be made simultaneously, with a single measurement requiring as little as a fraction of a second. In practice, about 200 voltages (and apparent resistivities) are measured in an hour. Stacking, signal discrimination, and other quality control measures are carried out to ensure the recording of reliable data.[9] The product of a survey is hundreds or thousands of apparent resistivity measurements at various depths beneath the entire length of the electrode array, or what is known as a resistivity pseudosection (Figure 6.4).

Before the development of ERT, the 2-D survey described in the previous paragraph could have been acquired without solid state relays and with single wires dedicated to each electrode, so perhaps a better description of this modern approach is 'automated' 2-D resistivity surveying, for it is the automation of electrode selection that allows this type of survey to be efficient. Regardless of the number of data points and speed with which they are acquired, without numerical modeling, one is left with a pseudosection of apparent resistivities plotted at somewhat arbitrary depths. Numerical modeling techniques developed in the early 1990s by Loke and Barker[10] now allow one to invert large pseudosections consisting of several thousand field measured apparent resistivities into 'true' geoelectric sections (plotting 'true' resistivity versus 'true' depth) in a few hours. The size of the model blocks, and hence, the resolution of the technique is limited by current spreading and practical limitations on how much data can be acquired. While the size of the model blocks dictates that 2-D resistivity results may never approach the vertical resolution of a GPR survey, the ability to invert for true resistivities means that in certain geologic contexts, horizons described by different lithologies can be precisely and confidently identified.

The depth of investigation is dependent on the resistivity structure, the separation of the electrodes, and the geometry of the electrode array used.[11] Typical electrode arrays used in exploration geophysics are listed below:

- Wenner array (excellent vertical resolution, good horizontal resolution and deep penetration)
- Schlumberger array (excellent vertical and horizontal resolution, moderate penetration)
- Dipole-dipole array (poor vertical resolution, excellent horizontal resolution, poor penetration)

In the Cave of Letters survey, the Wenner array was used due to its excellent vertical resolution and depth of penetration. The geometry of the Wenner array is such that the two voltage electrodes are inside the two current electrodes, and all electrodes are identically spaced from the adjacent current or voltage electrodes.

ERT DATA PROCESSING

Resistivity Inversion Overview

Resistivity inversion is the process of converting measured apparent resistivity to true earth resistivities. For the Cave of Letters surveys, the 2-D inversion software RES2DINV was used to perform the two-dimensional inversion of the ERT data.[12] In the inversion process, initially, the apparent resistivity data are sorted and displayed in a pseudosection. The 2-D model used by the inversion program consists of rectangular blocks. The arrangement of the blocks is loosely tied to the distribution of the datum points in the pseudosection. The distribution and size of the blocks is automatically generated by the program with the constraint that the number of blocks does not exceed the number of datum points. The depth of the bottom row of blocks is set to be approximately equal to the equivalent depth of investigation[13] of the datum points with the largest electrode spacing.

A forward modeling subroutine is used to calculate the apparent resistivity values, and a non-linear least-squares optimization technique is used for the inversion routine.[14] The program supports both the finite-difference and finite-element forward modeling techniques.

INVERSION THEORY

As discussed above, the 2-D model used by this program divides the subsurface into a number of rectangular blocks. *The aim of 2-D inversion is to determine the resistivities of the rectangular blocks that will produce an apparent resistivity pseudosection that agrees with the actual measurements of the field survey.* The thickness of the first layer of blocks is set to some fraction (0.3–0.5) of the electrode spacing, and the thickness of each subsequently deeper layer is increased by a set amount (5% to 25%). The optimization method basically tries to reduce the difference between the calculated and measured apparent resistivity values by adjusting the resistivity of the model blocks. A measure of this difference is given by the root-mean-squared (RMS) error. In general, the most prudent approach is to choose the model at the iteration after which the RMS error does not change significantly.

The inversion routine used by the program is based on the smoothness-constrained least-squares method.[15,16,17] A new implementation of the least-squares method based on a quasi-Newton optimization technique[18] can also be used. This technique is more than ten times faster than the conventional least-squares method for large data sets, and requires less memory.

The least-squares optimization routine used in this software allows for the incorporation of a damping factor (smoothness control) and flatness filter (vertical/

horizontal bias). In the present survey, flatness filters were set to favor horizontal structures that would best represent a hypothetical floor of occupation.

ERT DATA COLLECTION

The locations of the three ERT sections surveyed are presented in Figure 4.7. ERT Line 1 was collected in July of 2000. ERT Line 1 was an 80 m line running from Entrance #2 across Hall 'A', through the AB Crawlway, and then along the west wall of Hall 'B'. A minimum electrode spacing of 1 m was used, providing a maximum depth of investigation of approximately 12 m below ground surface (mbgs).

ERT lines 2 and 3 were collected in July of 2001. ERT Line 2 was an 80 m line running from the Hall 'A' entrance to the AB Crawlway (overlapping Line 1), through the AB Crawlway, across the center of Hall 'B', and along the Passage 'BB'. A minimum electrode spacing of 1 m was used, providing a maximum depth of investigation of approximately 12 m (mbgs.).

ERT Line 3 was a 40 m line running from the northeast corner of Hall 'B', due north through the Hall 'C' Connection, and across the southwest corner of Hall 'C'. The length of Line 3 was constrained by the geometry of the cave. A minimum electrode spacing of 0.5 m was used, providing a maximum depth of investigation of approximately 6 m (mbgs.). While smaller minimum electrode spacing decreased the depth of investigation of Line 3, as compared to lines 1 and 2, it increased the spatial resolution.

In order to facilitate good electrode contacts (i.e., lower the contact resistance), approximately 200 ml of water was poured around each electrode placement. It is believed that the large amount of halite present on the surface,[19] when saturated with water, significantly assisted in decreasing the contact resistance making the ERT survey practicable in this extremely arid cave environment. Each of the electrodes used in each ERT spread were surveyed for relative relief along the line. The inverted (modeled) geoelectric sections account for any changes of relief along the lines.

METAL DETECTION SURVEYS

Three different metal detection surveys were carried out in the cave, using three different instruments. The Geonics EM38 terrain conductivity meter is a fixed frequency electromagnetic device that measures subsurface conductivity to a depth of approximately 1.5 m below ground surface (mbgs), with a peak response at a depth of 0.4 mbgs. Changes in terrain conductivity can be caused by both Ferro (iron and steel) and non-ferromagnetic (bronze, tin, silver, etc.) metal, as well as variations in clay content and salt concentration in pore water. The instrument has a resolution of 1 milliSiemen per meter (mS/m). Ohm-meters (ohm-m), as measured by the ERT system, are converted into units of mS/m by dividing into 1000. For instance, 200 ohm-m is equivalent to 5 mS/m, and 20 mS/m is equivalent to 50 ohm-m.

The GEM Instruments Overhauser magnetic gradiometer carries out high resolution measurements of vertical variations in the earth's magnetic field. While a gradiometer has no theoretical limit to its depth of investigation, practically speaking, it is a useful archaeometric exploration tool for the top few meters of the subsurface. The measured effect of a buried object drops off to the fourth power of the distance of the object to the sensors. As such, a small object in the very near surface may create a stronger anomaly than a larger object at significantly greater depth. A magnetic gradiometer is only sensitive to ferromagnetic (iron and steel) objects.

The Geonics EM61 time domain high resolution metal detector is the most sophisticated metal detector of its kind. The EM61 is only sensitive to metal, and is not affected by increased clay content, salt content, moisture content, or high magnetic susceptibility minerals. It was designed specifically for the detection of unexploded ordnance (UXO) and associated shrapnel. In order to work in the close confines of the cave, the EM61 handheld sensor was used. The EM61 can detect an object the size of a 55 gallon drum at a depth of 3.5 m. It can detect significantly smaller objects as the depths become shallower. Nevertheless, from experience in the Cave of Letters, it is clear that very small coil, inexpensive, low powered metal detectors are superior for locating small objects such as coins buried in the top few centimeters (cm). It is believed that there is no existing instrument that can detect an object the size of a coin at depths exceeding 20 cm.

For all the metal detection surveys, data were collected every meter on gridded lines spaced 1 m apart. As such, the cave floor was surveyed with essentially 100% coverage with all instruments. In addition, the EM61 sensors were carried into various niches and crawlways around the perimeter of the cave, as well as being dragged along the cave walls and lower portions of the ceiling.

RESULTS AND INTERPRETATION

ERT LINE 1

The location of ERT Line 1 is presented in Figure 4.9. The inverted (i.e., modeled) and interpreted section is presented in Figure 6.5. The geoelectric section is corrected for relief as surveyed in the cave.

Figure 6.5 clearly identifies two distinct layers. The upper layer (pink color), varying in thickness from about 3.5 m at Entrance #2, to less than one meter thickness along the AB Crawlway, displays extremely high resistivities generally exceeding 20,000 ohm-m. This layer is believed to be largely composed of coarse, gravel to boulder-sized roof collapse debris with a large percentage of air-filled void space. Given the lack of soil and moisture, the extremely high resistivities are reasonable. The deeper and thicker layer (blue) is believed to be unweathered limestone bedrock. The resistivities of this layer, while significantly lower than the upper layer, are still moderately high at 1,000 to 7,000 ohm-m. This layer continues to the bottom of the section.

Lying on the upper, highly resistive layer (i.e., at surface) are thin (<1 m thick), discontinuous layers of lower resistivity (green) material. These lenses are due to packed finer grained material at various locations on the surface along the line, including at Entrance #2, in Hall A, and in Hall B. The several meter wide, moderately resistive intervals (blue) bounding the 'AB Crawlway' are indicative of limestone bedrock subcropping a few centimeters below the cave floor. In other words, the cave floor found in the contemporary AB Crawlway is probably the same floor that existed 1873 years ago in the cave this assertion is supported by the fact that various artifacts were found at surface in the AB Crawlway, including a wooden comb.[20,21,22]

The upper, highly resistive layer (pink) varies in thickness along the section. In fact, and one would expect, this thinning and thickening approximately correlates with variations in the roof height of the cave, as this layer is largely the result of roof collapse rubble (see Figure 4.8). Hence, the rubble is relatively thick at the entrance (i.e., the cliff face) where the roof is high, thins along the passage into Hall A, thickens in Hall A, becomes very thin in the AB Crawlway, and thickens again in Hall B.

There is no well-defined internal layering through the roof collapse debris imaged in ERT Line 1. If existing as a packed earth floor, such layering would be expected to appear as a substantially less resistive layer slicing horizontally though the roof collapse debris. One explanation for the lack of such a horizon is that no such layering exists. A second explanation may be that if such an earth packed floor exists, it may be too thin and indistinct to be imaged by ERT. A third explanation, however, may be that such a layer did exist, but was destroyed during Yigal Yadin's excavation and trenching through Hall A.[23] While no such trenching was done in Hall B, ERT Line 1 ran along the western wall of Hall B, rather than across the center of Hall B, and therefore it may have bypassed any existing buried floor.

ERT LINE 2

The location of ERT Line 2 is presented in Figure 4.9. The inverted (i.e., modeled) and interpreted section is presented in Figure 6.6. The geoelectric section is corrected for relief as surveyed in the cave.

As with ERT Line 1, Line 2 images two major layers, including a highly resistive (>20,000 ohm-m) layer of roof collapse rubble (pink) overlying a moderately resistive (1,000 to 7,000 ohm-m) layer interpreted to be limestone bedrock (blue). The roof collapse rubble reaches a thickness of over 6 m in Hall B. As with Line 1, the thickness of the roof collapse layer approximately mimics the height of the roof, with the debris layer being thin to non-existent in the AB Crawlway, absent at the entrance and exit of the AB Crawlway, thickening in Hall B, and thinning in the BB Passage.

In striking contrast to Line 1, however, Line 2 appears to have imaged a fairly continuous, lower resistivity horizon, cutting almost completely across Hall B and partially through the debris in the BB Passage. The depth of this horizon varies from just below the surface to approximately 3 m below the cave floor. This horizon is hypothesized to be the anthropogenically packed 'Bar Kokhba Floor.' That this horizon is visible in Line 2, and not in Line 1 which also traversed Hall B, may be explained by the fact that Line 2 crossed the center of Hall B from the southwest corner to the northeast corner, while Line 1 traversed the western wall of Hall B where debris is considerably thinner, and a packed floor may not have been developed.

ERT LINE 3

The location of ERT Line 3 is presented in Figure 4.9. The inverted (i.e., modeled) and interpreted section is presented in Figure 6.7. Similar to the other two ERT lines, the geoelectric section is corrected for relief, based upon an elevation survey in the cave.

As with ERT Line 1 and Line 2, Line 3 images two layers, including a highly resistive (>20,000 ohm-m) layer of roof collapse rubble (pink) overlying a moderately resistive (1,000 to 9,000 ohm-m) layer, interpreted to be limestone bedrock (blue to green). Only a relatively small thickness of the limestone bedrock is imaged, because the depth of investigation for the 40 m long section is only 6 m below the cave floor. The roof collapse rubble reaches a thickness of about 6 m in Hall B and in the Hall C Connection. The rubble reaches at least 5 m below the cave floor in Hall C.

Similar to Line 2, Line 3 images a fairly continuous, lower resistivity (less than 7,000 ohm-m) horizon, cutting almost completely across the entire 40 m long section. The depth of this horizon varies from the surface to approximately

1.5 m below the cave floor. Again, it is hypothesized that this horizon is the 'Bar Kokhba Floor'. While the ERT section suggests that the floor is exposed at the surface for approximately 3.5 m, there was no immediate, visible evidence of this at the time of the survey in 2001. As with the remainder of the cave floor, the Hall C Connection was covered with boulders, smaller pieces of rubble, fine sediment, and bat guano. However, removal of a few boulders revealed, at the surface, a densely packed surface of fine grained material which required a trowel to excavate. Excavation of this layer in Locus 27 (Figure 4.9) revealed more than 1200 pieces of Roman period pottery from at least 14 different vessels, a Vespasian coin, and a wooden shaft that may have been part of a weapon.[24] It should be noted that the Vespasian coin, while clearly Roman, is dated at 70 to 71 C.E. Possible explanations for the presence of the coin include: (1) that it was left in the Cave during the First Jewish Revolt (66 to 73 C.E.), (2) the coin having been legal tender during the Second Revolt, or (3) the coin having been saved for re-striking during the Second Revolt.[25,26]

Metal Detection Surveys

All terrain conductivity measurements were less than 1 mS/m. As all ERT measured resistivities in the near surface exceeded several thousand ohm-m, such low conductivities are reasonable. As such, no indications of buried metal or other discrete anomalous features of any sort were identified. All EM61 measured responses approximated the background calibrated response of 0 mV. No indications of buried metal of any sort were identified, and the magnetic gradiometer did not identify any discrete anomalies, which indicates that no significant mass of buried Ferro-magnetic metal is present in the cave.

CONCLUSIONS

Three two dimensional electrical resistivity tomography (ERT) sections were collected in the Cave of Letters. The three sections, including 200 m of total line length, together traversed Entrance #2, Hall A, the AB Crawlway, Passage BB, the Hall C Connection, and the southwestern end of Hall C. Generally, a common interpretation follows each section with very highly resistive (greater than 20,000 ohm-m) roof collapse rubble overlying likely unweathered, moderately resistive (less than 7,000 ohm-m) limestone bedrock. The maximum thickness of roof collapse rubble appears to be about 6 m. A thin, moderately conductive horizon can be traced across most of the 120 m of ERT lines 1 and 2, traversing Hall B, the Hall C Connection, and 12.5 m of Hall C. It is hypothesized that this thin horizon, generally less than a meter in thickness, is a densely packed earthen floor which would have served as a relatively comfortable surface on which refugees in the cave would have dwelled. This thin horizon is found from the surface to a maximum depth of approximately 3 m below the cave floor. The likelihood that this horizon dates from the Roman period, and the possibility that this horizon is the 'Bar Kokhba Floor' (i.e., a surface utilized during the period of the Second Jewish Revolt), is given some support by the large number of Roman period finds from loci 63y and 27 in the Hall C Connection. Exact dating and a description of utilization of this surface can only be provided by archaeologic and archaeometric techniques, and not by any technique applied in the geophysical program. While this interpreted earthen floor does not appear in ERT Line 1, it is hypothesized that the layer may have been destroyed in Hall A by trenching carried out under the supervision of Yigael Yadin during his 1961 expedition. Trenching is known to have been carried out by Yadin as described and photographed in his book *Bar Kokhba*.[27]

Terrain conductivity, magnetic gradiometry, and time domain high resolution metal detection identified neither Ferro- nor non-ferromagnetic bodies. While these instruments are not sensitive enough to find coins in the near surface, or perhaps a few bronze keys at a depth of 1 m, they are certainly sensitive enough to identify and locate a substantially sized hoard of metal within the top 3.5 m. Even after multiple surveys of 100% coverage of the cave floor not even the slightest suggestion of buried metal was noted; this likely indicates that additional caches of metal objects, similar to those previously discovered by Yadin, will not be found. Nevertheless, scattered discrete small objects including coins, keys, weapons, etc. may still be present both in the very near surface and at depth.

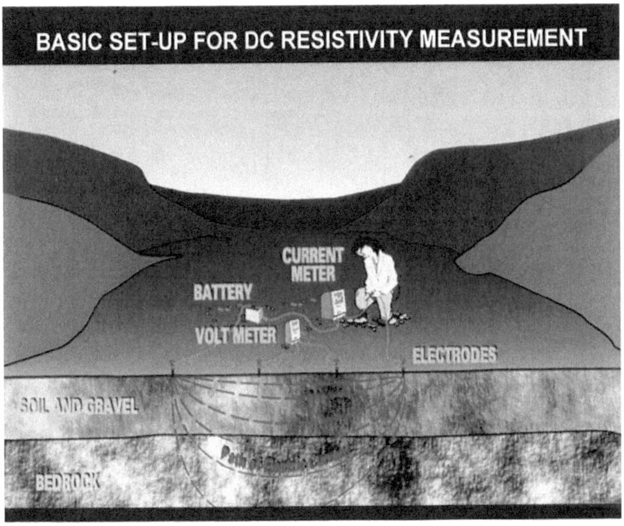

Figure 6.1: The basic set-up for 1-D resistivity soundings.

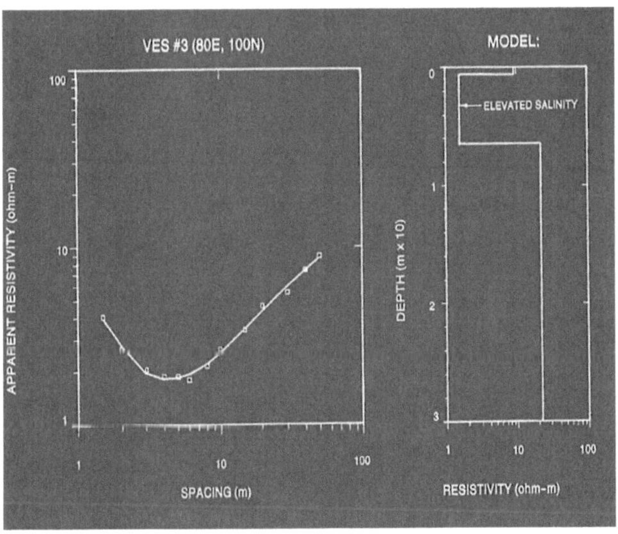

Figure 6.2: Computer curve matching and modeling of 1-D resistivity data.

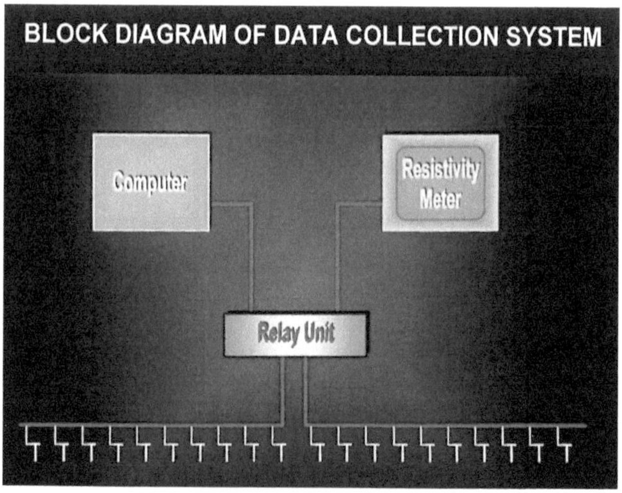

Figure 6.3: A generalized block diagram of an automated 2-D resistivity survey system.

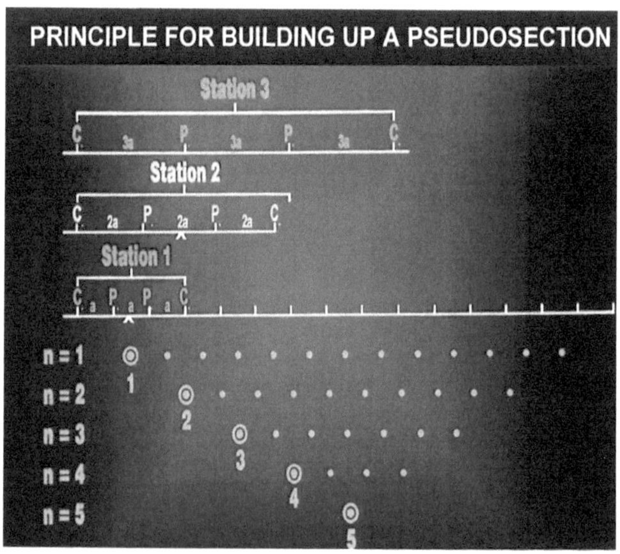

Figure 6.4: The construction of a pseudosection showing how the more broadly spaced quartets of currents and voltage electrodes define apparent resistivities at greater depths.

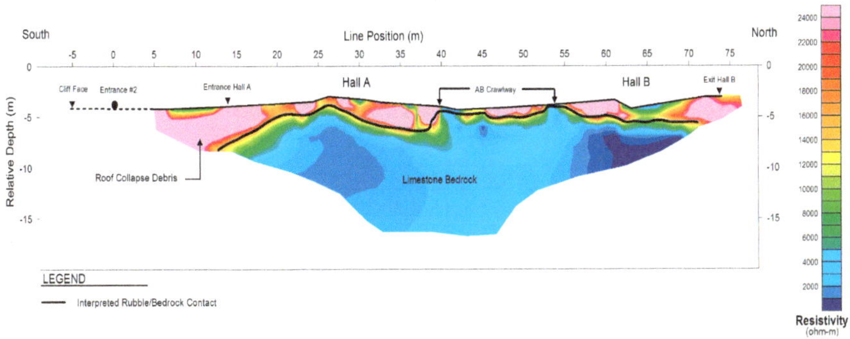

Figure 6.5. The geoelectric section of ERT 1 corrected for relief as surveyed in the cave.

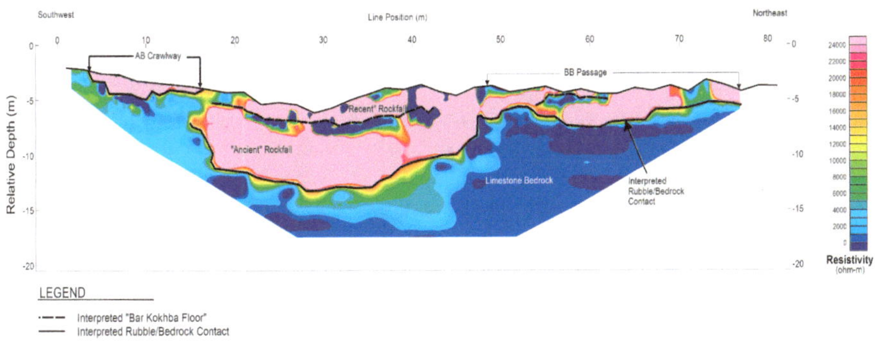

Figure 6.6. The geoelectric section of ERT 2 corrected for relief as surveyed in the cave.

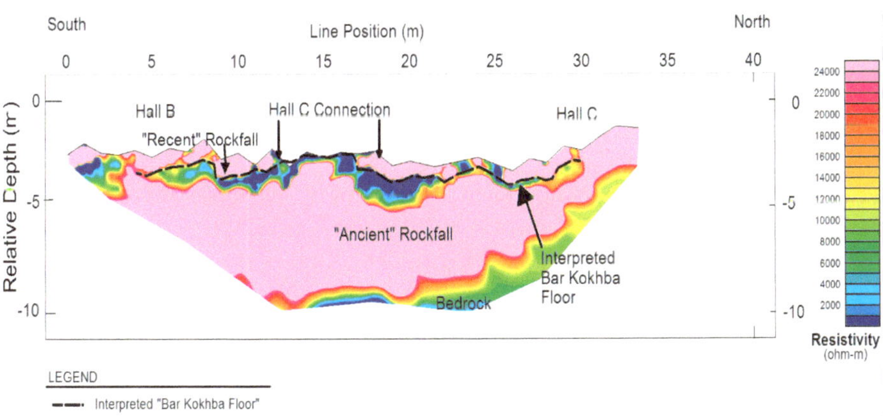

Figure 6.7. The geoelectric section of ERT 3 corrected for relief as surveyed in the cave.

NOTES

1. Yigael Yadin, *Bar-Kokhba; the Rediscovery of the Legendary Hero of the Last Jewish Revolt Against Imperial Rome* (London: Weidenfeld and Nicolson, 1971).
2. Richard A. Freund, *Secrets of the Cave of Letters: Rediscovering a Dead Sea Mystery* (Amherst, N.Y.: Humanity Books, 2004).
3. H.M. Jol, "Ground Penetrating Radar Subsurface Imaging Within the Cave of Letters" (Forthcoming).
4. Yadin, *Bar-Kokhba; the Rediscovery of the Legendary Hero of the Last Jewish Revolt Against Imperial Rome.*
5. W.M. Telford, L.P. Geldart, and R.E. Sheriff, *Applied Geophysics* (Cambridge, England: Cambridge University Press, 1990).
6. H. Robert Burger, *Exploration Geophysics of the Shallow Subsurface* (Englewood Cliffs, N.J.: Prentice Hall, 1992).
7. Philip Reeder, "Mapping of the Cave of Letters" (Forthcoming).
8. Burger, *Exploration Geophysics of the Shallow Subsurface.*
9. Torleif Dahlin, *On the Automation of 2D Resistivity Surveying for Engineering and Environmental Applications* (Department of Engineering Geology, Lund Institute of Technology, 1993).
10. M.H. Loke and R.D. Barker, "Rapid Least-Squares Inversion of Apparent Resistivity Pseudosections by a Quasi-Newton Method," *Geophysical Prospecting* 44, no. 1 (January 1996): 131–52.
11. Stanley H. Ward, "Resistivity and Induced Polarization Methods," in *Investigations in Geophysics* (Stanley H. Ward; Tulsa, Okla.: Society of Exploration Geophysicists, 1990), 147–90.
12. M.H. Loke and R.D. Barker, "Rapid Least-Squares Inversion of Apparent Resistivity Pseudosections by a Quasi-Newton Method."
13. L.S. Edwards, "A Modified Pseudosection for Resistivity and IP," *Geophysics* 42, no. 5 (August 1977): 1020–36.
14. C. deGroot-Hedlin and S. Constable, "Occam's Inversion to Generate Smooth, Two-Dimensional Models from Magnetotelluric Data," *Geophysics* 55, no. 12 (December 1990): 1613–24.
15. C. deGroot-Hedlin and S. Constable, "Occam's Inversion to Generate Smooth, Two-Dimensional Models from Magnetotelluric Data."
16. Yutaka Sasaki, "Resolution of Resistivity Tomography Inferred from Numerical Simulation," *Geophysical Prospecting* 40, no. 4 (May 1992): 453–63.
17. Yutaka Sasaki, "Resolution of Resistivity Tomography Inferred from Numerical Simulation."
18. M.H. Loke and R.D. Barker, "Rapid Least-Squares Inversion of Apparent Resistivity Pseudosections by a Quasi-Newton Method."
19. Jack Shroder, personal communication (2000).
20. Yadin, *Bar-Kokhba; the Rediscovery of the Legendary Hero of the Last Jewish Revolt Against Imperial Rome.*
21. Carl E. Savage, "First Century Occupation in the Cave of Letters? Results From" (Forthcoming).
22. Freund, *Secrets of the Cave of Letters: Rediscovering a Dead Sea Mystery.*
23. Yadin, *Bar-Kokhba; the Rediscovery of the Legendary Hero of the Last Jewish Revolt Against Imperial Rome.*
24. Carl E. Savage, "First Century Occupation in the Cave of Letters? Results From" (Forthcoming).
25. Freund, *Secrets of the Cave of Letters: Rediscovering a Dead Sea Mystery.*

26. Arieh Kindler, "Coins from the Cave of Letters of the Bar Kokhba," unpublished paper (2001).
27. Yadin, *Bar-Kokhba; the Rediscovery of the Legendary Hero of the Last Jewish Revolt Against Imperial Rome.*

BIBLIOGRAPHY

Burger, H. Robert. *Exploration Geophysics of the Shallow Subsurface.* Englewood Cliffs, N.J.: Prentice Hall, 1992.

Dahlin, Torleif. *On the Automation of 2D Resistivity Surveying for Engineering and Environmental Applications.* Department of Engineering Geology, Lund Institute of Technology, 1993.

DeGroot-Hedlin, C., and S. Constable. "Occam's Inversion to Generate Smooth, Two-Dimensional Models from Magnetotelluric Data." *Geophysics* 55, no. 12 (December 1990): 1613–24.

Edwards, L.S. "A Modified Pseudosection for Resistivity and IP." *Geophysics* 42, no. 5 (August 1977): 1020–36.

Freund, Richard A. *Secrets of the Cave of Letters: Rediscovering a Dead Sea Mystery.* Amherst, N.Y.: Humanity Books, 2004.

Jol, H.M. "Ground Penetrating Radar Subsurface Imaging Within the Cave of Letters," forthcoming.

Kindler, Arieh. "Coins from the Cave of Letters of the Bar Kokhba," 2001.

Loke, M.H., and R.D. Barker. "Rapid Least-Squares Inversion of Apparent Resistivity Pseudosections by a Quasi-Newton Method." *Geophysical Prospecting* 44, no. 1 (January 1996): 131–52.

Reeder, Philip. "Mapping of the Cave of Letters," forthcoming.

Sasaki, Yutaka. "Resolution of Resistivity Tomography Inferred from Numerical Simulation." *Geophysical Prospecting* 40, no. 4 (May 1992): 453–63.

Savage, Carl E. "First Century Occupation in the Cave of Letters? Results From," forthcoming.

Shroder, Jack. Personal communication, 2000.

———. "First Century Occupation in the Cave of Letters? Results From," forthcoming.

Telford, W.M., L.P. Geldart, and R.E. Sheriff. *Applied Geophysics.* Cambridge, England: Cambridge University Press, 1990.

Ward, Stanley H. "Resistivity and Induced Polarization Methods." In *Investigations in Geophysics,* Stanley H. Ward, 147–90. Tulsa, Okla.: Society of Exploration Geophysicists, 1990.

Yadin, Yigael. *Bar-Kokhba, the Rediscovery of the Legendary Hero of the Last Jewish Revolt Against Imperial Rome.* London: Weidenfeld and Nicolson, 1971.

CHAPTER SEVEN

Microclimatic Variations and Patterns of Cave Use at the Cave of Letters

PHILIP REEDER
Bayer School of Natural and Environmental Sciences,
Duquesne University
Pittsburgh, PA 15282, USA

INTRODUCTION

Research in 1960–61 by Yigael Yadin, in the Cave of Letters (COL), yielded a priceless collection of artifacts. Jewish rebel commanders, and their families, who sought refuge in the cave near the end of the Second Jewish Revolt against the Romans left these objects. Documents, mostly military correspondence, discovered in the cave bore the name of Simeon Ben Koseva, the historical figure known as Bar-Kokhba, leader of the Second Jewish revolt. Also discovered was the personal archive of Babatha, a young Jewish woman. These documents chronicle everyday life for a Jewish family at this important time in history.

After Yadin's expeditions to the Cave of Letters, little research was conducted at this site between 1961 and 1999. In July 1999 and July 2000 research permits were obtained from the Israel Antiquities Authority to conduct research in the cave that utilized state-of-the-art technology, as well as traditional geologic, geographic, archaeologic and surveying techniques, to add new information to the base of knowledge about the Second Jewish Revolt, and the Cave of Letters.

Part of the research design for the geographic component of the project involved the compilation of a micro-climatic database for the Cave of Letters. The creation of this data base is important because: (1) no micro-climatic data exists for the cave, (2) within the overall framework for the project, the geographic component seeks to establish relationships between human occupation of the cave and cave micro-climate, and (3) similarly, the geographic component seeks to establish spatial patterns of cave use that are dictated by cave's micro-climate.

Much of the literature associated with cave micro-climate research is focused on how human activities affect cave climate. Much of this research is framed within the context of developing caverns for tourism, and what effects these development projects will have on the cave micro-climate and related processes, and organisms. A micro-climate study was conducted in Kartchner Caverns in Arizona (USA) to provide the data necessary for determining the nature and the magnitude of micro-climatic changes that may result from the development of the cave[1]. Due to Arizona's arid climate, which is similar to the Dead Sea Region of Israel, the air outside the cave is usually less moist than the air in the cave. As a result, air exchange from outside the cave, with air in the cave, can have a drying effect on the cave's micro-climate, with this effect most pronounced near the cave's entrance(s). In single entrance caves, the effect of outside air on the cave's micro-climate is diminished as you move further from the entrance. In the case of the Cave of Letters, there are two entrances, but they are located in very close proximity to each other (about 8 meters apart). Other studies of note, which assess relationships between cave micro-climate and cave use include research conducted at Carlsbad Caverns[2,3] and Torgac Cave, both in New Mexico (USA).[4] The research in Torgac Cave determined that temperature and relative humidity curves generally track each other, indicating that as temperature falls, so too does relative humidity. This trend is the opposite of what one would expect, because normally temperature and relative humidity are negatively correlated, therefore, as one variable increases the other variable decreases or vice versa. The explanation for the positive correlation between temperature and relative humidity in Torgac Cave is that the air outside that cave are both cooler and drier then the cave air, and when pulses of this air enter the cave they cause both a cooling and drying of the cave atmosphere. This may also be the case for the Cave of Letters, which is located in an area with a similar type of regional climate.

Cave micro-climate studies have also been conducted in association with the conservation and protection of cave fauna, in many cases to assess micro-climate before and after cave gate construction to protect roosting colonies of bats.[5,6,7] The North American Bat Conservation Partnership (NABCP) has conducted research in various caves in the mid-western United States to assess relationships between cave micro-climate and bat habitat. More precisely, they have focused on quantifying the micro-climate of the caves prior to gate construction to protect the roosting bat populations, and post-construction, to quantify changes in cave micro-climate and the effect on the bat populations.[5]

Cave micro-climate has also been studied within the context of human impacts on artifacts in the cave, especially rock art in caves subjected to large scale tourism,[8] as well as the effects that natural fluctuations in cave micro-climate have on the preservation of rock art.[9] In the context of tourist caves, and preservation of archaeological remains (especially rock art), mass tourism has seriously deteriorated cave art to the point that limits have been put on the number of visitors and the time of the year that the caves can be visited.[10] In terms of the natural micro-climatic variations of the cave atmosphere, relationships between air flow, relative humidity and temperature fluctuations have been noted to adversely affect archeological remains, especially cave art. In terms of the micro-climatic research we conducted at the Cave of Letters, the literature is scant with respect to research that has quantified the temporal and spatial nature of cave micro-climate, and their influence on patterns of cave utilization. Hence, this phase of the research will add important new information, not only to the existing base of knowledge regarding the Second Jewish Revolt and the Cave of Letters, but to the base of knowledge regarding the general spatial and temporal aspects of cave micro-climate, and micro-climatic influences on human activities in caves.

RESEARCH DESIGN

Monitors were placed in the cave to record changes in percent relative humidity and temperature over short-term (2.5 days), medium-term (7.5 months) and longer-term (12 months) periods of record. The cave was monitored for temperature and relative humidity during the two-week expedition in the summer of 2000. This phase of the research began with a series of 24-hour studies that were intended to initially characterize the short-term variations in the cave's micro-climate(s). The monitors were programmed to collect data continuously for these 24-hour periods at six locations in the cave, as well as on the ledge outside the cave. The data was retrieved from the monitors and they were next programmed to take continuous readings for a 48-hour period. The 48-hour study from July 7th to 9th spanned a weekend which was a period when no research was occurring in the cave. The July 11th to 13th study spanned a period when research was occurring in the cave. Between 9 AM and 4 PM on these dates, 10–15 researchers were present in the cave at any given time. This pattern of collecting data for periods of 48 hours was established during the 2000 expedition to gain an understanding of (1) the short-term variation in microclimate in the cave, (2) how the presence of researchers working in the cave effect the micro climate, (3) how the micro-climate varied during periods when no one was in the cave (Friday afternoon to Sunday morning), (4) how the micro-climate of the cave was related to climate pattern outside the cave, and (5) if there was a spatial pattern in micro-climatic variations in the cave. The compilation of these data bases was also intended to determine if the climate monitors functioned adequately within the parameters of a cave-based study.

The July 2000 to March 2001 study was intended to assess longer-term variations in the cave atmosphere, as well as any seasonal patterns of variation, during a period when there was no human presence in the cave. The July 2001 to July 2002 study was intended to assess even longer term variations in the cave atmosphere, as well as the pattern of seasonal fluctuations over a full-year.

Understanding variations in the caves microclimate is important in quantifying the relationship between the cave and the physical environment that surrounds the cave. These temporal micro-climatic changes have implications in terms of both the modification of the cave passages by dissolution corrosion associated with fluctuations in cave relative humidity, and the human implications associated with the Bar Kokhba partisans, who were trapped in the cave, and their effect on the cave atmosphere, and thus the living conditions in the cave. Fluctuations in barometric pressure, and differences in barometric pressure between the cave environment and the outside environment, control whether the cave takes in or expels air. This has direct anthropologic relevance in that the partisans/rebels living in the cave in the second century built fires for cooking and perhaps firing pottery. This is evidenced by the hearth that was excavated in 1999 and 2000 in Hall "B" near the location where the AB passage enters this chamber (Figure 7.1). Hence, the patterns of airflow needed to be established so that a better understanding of the use of fire in the cave could be gained. By establishing a cave climate data base of various temporal dimensions, seasonal, annual, and longer term patterns in cave climate were established.

This research sought to answer the following questions:

(1) Are there short-term (daily) variations in the cave's micro-climate?
(2) Does the presence of the researchers in the cave affect the cave's micro-climate?
(3) Are the micro-climatic fluctuations in the cave related to climate variations outside the cave?
(4) Is there spatial variability within the temperature and relative humidity fluctuations within the cave?
(5) Is there a temporal variability within the temperature and relative humidity fluctuations within the cave for the short, medium and long term studies?
(6) Will airflow patterns in the cave indicate a pattern of human utilization?

PLACEMENT OF MONITORS IN THE CAVE

Six of the seven monitors were placed at various locations throughout the cave, and the seventh monitor was placed on a cliff ledge outside of the cave. In all there were four different periods of data collection between July 2000 and July 2002.

During the two-day monitoring periods (July 7–9 and July 11–13) the monitors were placed as depicted in Figure 7.1. Specifically, these locations were: (1) on a ledge approximately 10 meters below and 20 meters west of Entrance #2, (2) on a ledge in the room adjacent to Entrance #1, (3) in the southeast corner of Hall "A", (4) in the rubble on the eastern portion of the mid-section of the AB connection passage, (5) in Hall "B" on a ledge at the intersection of Hall "B", Passage BB, and the Hall "C" Connection passage, (6) at the entrance to the West Extension Passage, and (7) in the back portion of Hall "C". For the July 2000 to March 2001, and the July 2001 to July 2002 collection periods, monitor #7 was removed from the back of Hall "C" because the July 7–9/11–13 data interpretation indicated very little variation in this section of the cave. It was decided to place this monitor in the Entrance #2 passage for the July 2000 to March 2001 and July 2000 to July 2001 collection periods.

METHODOLOGY

HOBO Pro Series temperature and relative humidity monitors were used for this study.[11] The monitors were placed at various locations in the cave as indicated in the "Placement of Monitors in the Cave" section of this paper. The monitors were programmed and launched using a laptop computer and Onset Boxcar Pro 4 (version 4.3) software. In the July 7th-9th and July 11th-13th studies the monitors were programmed to collect temperature and percent relative humidity data every 30 seconds. In the July 2000 to March 2001 study the monitors collected data every eight minutes, and in the July 2001 to July 2002 study the monitors collected data every hour. For the July 7th-9th and 11th-13th studies, the monitors were launched and the data were downloaded in the cave. At the conclusion of the study that ended on July 13th the monitors were launched in the cave for the study that was to end in March 2001. The monitors were collected from the cave in March 2001 and were shipped to the United States and the data was downloaded on March 19, 2001. The monitors were launched again in the United States prior to departing for Israel in early July 2001. They were placed in the cave on July 15th and they were collected from the cave on July 13, 2002. They were shipped to the United States and the data were downloaded on July 22, 2002.

The data downloaded from the monitors was exported from the Boxcar Pro 4 spreadsheet program to Microsoft Excel 2003. Line graphs of the various data sets were then created using Excel. The line graphs were then compared visually to assess both temporal and spatial variations in the temperature and percent relative humidity data.

Airflow patterns were also determined by conducting a series of smoke tests within the cave. The hearth area in the southeast corner of Hall "B" was the focus of this research. A smoke producing device was ignited on two separate occasions in July 2000. The dispersion and movement of the smoke was visually monitored and sketched onto a map of the cave.

RESULTS AND DISCUSSION

48-Hour Micro-Climate Surveys

Two 48-hour studies were completed to assess the effect that human activities in the cave had on cave micro-climate (temperature and percent relative humidity). The study completed from July 7th to July 9th spanned a period when researchers were for the most part not working in the cave. The monitors were launched in the following order: Hall "C" (1:10 PM); West Extension (1:09 PM); Hall "B" (1:18 PM); AB Connection (1:26 PM); Hall "A" (1:36 PM), Entrance #1 (1:52 PM), and outside the cave (as the research team departed the cave at 4:12 PM). The data was downloaded on July 9th in the same order as they were launched and approximately at the same time. Hence, approximately the first three hours after launching the monitors and the last four hours before the monitors were downloaded reflect periods when researchers were present in the cave. The period from approximately 4 PM on July 7th to 9 AM on July 9th (41 hours) is the interval when researchers were not present in the cave.

The seven graphs for the July 7th to 9th collection period are presented in Figure 7.2. The temperature and percent relative humidity variations outside the cave have an inverse relationship, as expected, with temperature highs more or less corresponding with humidity lows, with the maximum temperature of 48 degree centigrade corresponding with a humidity low of about 8%, and the maximum relative humidity high of 43% corresponding with a temperature low of 27 degree centigrade. This inverse relationship occurs because cooler air has less of a capacity for holding water vapor, and it thus becomes saturated more easily, and visa versa with respect to warmer air and its ability to hold water vapor. This pattern also holds true for Entrance #1, although there appears to be an approximately four-hour lag between the temperature low and the humidity high. The humidity high of 47% occurred at about 10 AM on July 9th, which was just after the research team had re-entered the cave. Hence, this elevated reading reflects the human activities in the Entrance #1 area of the cave. In Hall "A" the temperature remained more or less constant at 24 degrees centigrade, even during the periods when researchers were in the cave. Elevated relative humidity readings also occurred while researchers were in the cave, but fluctuations also occurred when the cave was not occupied. Of note is a slight increase in relative humidity that occurred outside the cave at about 6 PM, and in Entrance #1 and Hall "A" at around 7 PM on July 7. This elevated reading was not detected deeper in the cave. Once the researchers departed the cave, and before our return, percent relative humidity and temperature were very steady deeper in the cave, beyond Hall "A".

The seven graphs for the July 11th to 13th collection period are presented in Figure 7.3. Researchers were present in the cave on these days between approximately 9 AM and 4 PM. The temperature and percent relative humidity variations outside the cave again have an inverse relationship, but the pattern is not as pronounced compared to the July 7th to 9th data set. The temperature high recorded by the data logger was approximately 38 degrees centigrade, which did not correspond with the percent relative humidity low of approximately 13%. There appears to be a lag of approximately three-hours between the temperature high and the corresponding percent relative humidity low. Percent relative humidity high values do correspond with the lowest temperature values, with little or no lag. This pattern also holds true for Entrance #1, and there again appears to be a lag, this time of approximately four-hours, between the temperature low and the humidity high. The humidity high of 57% occurred at about 6 AM on July 13th, which was prior to the research team re-entering the cave. The peaks of the percent relative humidity curves for Entrance #1 are wider when compared to the curve peaks when researchers were not in the cave, which probably reflects the activities of the researchers in the Entrance #1 area of the cave. In Hall "A" the temperature remained more or less constant at 25 degrees centigrade, even during the periods when researchers were in the cave. Elevated relative humidity readings also occurred while researchers were in the cave, with the average percent relative humidity increasing by approximately 15%. Increased percent relative humidity values were also noted in the AB Connection Passage, Hall "B" and the West Extension passage for the period while researchers were present in the cave. Hall "C", the passage most distant from the cave entrances, and not the focus of research activities, was very similar in pattern to the two-day data set when researchers were not in the cave.

Seven-Month Micro-Climate Survey

Six of the seven monitors utilized in the 48-hour surveys were placed in similar locations within the cave for a seven-month study. The monitor that was placed near the terminus of Hall "C" was placed in a new location, in Entrance #2 (see Figure 7.1), because the 48-hour data indicated no variations in Hall "C". The new placement in Entrance #2 was intended to better facilitate understanding the exchange of air between outside the cave, the two entrances, and Hall "A" in the cave.

Figure 7.4 depicts the variations in temperature and percent relative humidity at the seven locations where the monitors were placed (six in the cave and one location outside the cave). The monitors were launched in the cave on July 20, 2000 and were collected on March 16, 2001. The monitors were not downloaded in the cave, but were instead shipped to the United States and were downloaded on March 19, 2001. The substantial data variations late in the period of record are fluctuations that occurred during and after the collection and shipping process and are thus irrelevant to this study.

The period of record for this phase of the study started in mid-summer and extended through fall, to late winter. The range of the temperatures recorded outside of the cave was approximately 35 degrees centigrade (between 10 and 45 degrees), while the percent relative humidity range was 63% (between 37 and 100%). The temperature and percent relative humidity patterns for Entrances 1 and 2 were very similar for the period of record, as was the pattern for Hall "A". The amplitude of temperature fluctuation was less in Hall "A" compared to the two entrances, but a similar pattern was noted for all three locations. The temperature decreased slightly (about two-degrees centigrade) through the period of record in the AB Connection passage, and temperatures were generally consistent in Hall "B" and the West Extension passage. In terms of percent relative humidity, a consistent pattern can be detected at all seven locations, but the amplitude of the fluctuations decreases with distance into the cave. Of note is a steady ten-day decrease in percent relative humidity that begins in late October, and reaches a minimum value of approximately 40% in early November (range = approximately 45%). This pattern is clearly discernable, with no lag, when comparing the data from outside the cave, with the data from Entrances 1 and 2, and Hall "A". This similar pattern holds true for the rising limb of the relative humidity climograph, with the percent relative humidity increasing from the minimum value of approximately 40% around November 7th, to a maximum of approximately 77% in the cave at these three locations (range = 37%), and 100% outside the cave, around December 7th. The same pattern is detected in the AB Connection passage as well, with the temporal pattern being similar to the other three cave locations, but the range is 19%, compared to 37% for the other locations. This same pattern is also detected deeper in the cave in Hall "B" and the West Extension Passage. In the case of Hall "B" the temperature is more or less constant throughout the period of record, and the minimum percent relative humidity is similar in temporal pattern to the portions of the cave described above, that are closer to the entrances. The minimum percent relative humidity of approximately 52% occurred on November 17th, which is approximately ten days later than the areas of the cave closer to the two entrances. Hence, there appears to be a ten-day lag for the minimum relative humidity compared to the previously discussed areas of the cave. This same ten-day lag exists in the rising limb of the relative humidity climograph, with percent relative humidity increasing from the November 17th minimum, to the maximum of 60% on December 17th. The range and pattern are also similar for the West Extension Passage, but there is an additional ten-day lag compared to Hall "B", and a 20-day lag compared to the entrance areas of the cave and Hall "A".

A seasonal (temporal) pattern related to percent relative humidity also appears in this seven-month long data base, as does a spatial pattern with respect to the range of fluctuations in the cave as related to distance from the cave's entrances. The temperature followed a general pattern of continual decrease through the period of record, with the magnitude of the decrease becoming less and less, deeper in the cave, with the decrease being barely detectable at the monitor farthest from the entrances (West Extension Passage). As previously noted, a late fall decrease, followed by an increase to pre-decrease levels about 20-days later, was noted at all the data collection points. This decrease was most notable outside the cave, at the two entrances, and in Hall "A", and the trend is also evident at the three other data collection points (AB Connection, Hall "B" and West Extension). The range in the variation at these three locations decreased, as distance from the entrances increased. This temporal pattern, as previously mentioned, was lagged 10 or 20 days depending upon the location of the monitor with respect to distance to the entrances. The decrease in percent relative humidity during this period is probably attributable to the pattern of air exchange between the cave and the environment outside the cave. Caves breathe either drawing air in through the entrance or expelling air out the entrance, depending on the dynamic between barometric pressures in the cave versus the pressure outside the cave. If the barometric pressure outside a cave is higher than inside the cave, air moves into the cave. When a weather front moves over the area and the outside barometric pressure falls, the cave expels air (breathes out). Even slight air pressure changes due to diurnal air temperature fluctuations can be reflected in a cave's airflow. Consistently lower barometric pressure outside the cave caused the cave to expel air thus reducing the barometric pressure of the air in the cave. After about ten days the pattern shifted and the cave consistently drew in higher humidity air from outside the cave as the barometric pressure outside the cave surpassed the barometric pressure of the cave atmosphere, with the net result being an increase in relative humidity in the cave.

One Year Micro-Climatc Survey

The temperature and percent relative humidity data for a year-long period of record from mid-July 2001 to mid-July 2002 is depicted in Figure 7.5. The monitors were launched in the United States and then shipped to Israel, hence the first ten-days of data are conditions during shipment and prior to deployment in Israel, and this data should not be considered. The same is true for the last ten days of data in that it reflects a period after the monitors were collected from the cave, shipped to the United States, and then downloaded.

The monitors were placed in the same location as the monitors for the seven-month study (see Figure 7.1). The relative humidity sensor in the monitor that was placed outside the cave malfunctioned about three-months into the study in that it consistently recorded percent relative humidity values above 100% in mid-November 2001, and it eventually stopped recording relative humidity data in mid-February 2002, with the exception of one errant reading on March 23, 2002. The temperature sensor worked correctly for the entire period of record.

The 12-month temperature trends were similar to the seven-month data base, following the trend, as one would expect, of maximum temperatures in the summer months, and minimum temperatures in the winter months. In the seven-month data base the minimum temperature outside the cave and in the sections of the cave better connected with the outside (Entrances 1 and 2 and Hall "A") occurred in late February (see Figure 7.4). In the 12-month study, the minimum temperature outside the cave, and in the better connected areas of the cave occurred in early January, and the temperatures outside the cave and in the connected areas of the cave, increased from the minimum of 10 degrees centigrade outside the cave, to 16 degrees in Entrances 1 and 2 and Hall "A". The minimum temperature in the AB Connection Passage was 21 degrees centigrade, and in Hall "B" and the West Extension Passage the temperature remained more or less constant at 24 degrees centigrade. The maximum temperatures of 41 degrees centigrade occurred outside the cave in mid-July 2002. The temperature maximum of 25 degrees centigrade in Entrance #1, 27 degrees in Entrance #2, and 24 degrees in Hall "A" occurred in late July as well, as did the maximum of 23 degrees in the AB Connection Passage.

The trends in percent relative humidity were similar to the trends in the seven-month data set, and a similar pattern of decreased fluctuation in relative humidity with distance from the caves entrances were also found to exist. Maximum percent relative humidity values were all noted to occur in late August and early September in the better connected areas of the cave (Entrances 1 and 2, and Hall "A"). The same lag detected in the seven-month data set also occurred deeper in the cave in that the maximum relative humidity values occurred in the AB Connection passage Hall "B" in mid-September and in the West Extension Passage in late September. Also, the steady, lagged decrease in relative humidity from the front to the back areas of the cave, which was detected in the seven month data set, was not clearly discernable in the 12-month data set.

Air Flow Patterns in the Cave

During the limited excavation that took place in the cave during the 1999 expedition, an area in the southeast corner of Hall "B" was determined to be the location of a hearth. Beneath the 0.8 meters of rubble that covered this area, caches of firewood, charcoal, olive and date pits, and the remains of a clay oven were discovered. This area was designated Locus4/99 (Figure 7.6).

As part of the research related to the hearth and surrounding area, an experiment was conducted to test the air flow patterns in the cave, and their relationship to the placement of the hearth in the southeast corner of Hall "B". The hearth is situated at the point where the AB Passage, an on average one meter high crawlway, merges with Hall "B", which is on average has a width of 11 meters, and a ceiling height of six meters.

As previously discussed, airflow patterns in caves are often controlled by interactions between barometric pressure, the volume of air in the cave, and the air outside the cave. These interactions are what causes the cave to expel air though the entrance(s) (breathe out) or draw air into the cave via the entrance(s). In multiple entrance caves, like the Cave of Letters, air can also be exchanged between entrances in a set of processes that are separate from those controlled by changes in barometric pressure.

Research question #6, which was posed in the Research Design section, asked "will airflow patterns in the cave indicate a pattern of human utilization?" To answer this question, first the airflow patterns of the cave had to be quantified. This was accomplished by completing two separate smoke tests, wherein a "smoke candle" was ignited at the hearth in Hall "B" and the smoke was tracked throughout the cave. Figure 7.6 depicts the pattern of smoke movement associated with the smoke tests. The smoke from the device, almost in its entirety, moved from the hearth area, through the AB Connection Passage and into Hall "A". It is estimated that approximately 5% of the smoke released from the device drifted in Hall "B" away from the hearth, rather than flowing through the AB Passage. The small quantity of smoke that did drift in Hall "B" quickly dissipated, and after 3 or 4 minutes moved into and through the AB Connection. The entire volume of the smoke moved through the AB Connection Passage in approximately 5 minutes. The smoke moved out of the AB Connection at the single exit point where the passage meets the northwest corner of Hall "A". The smoke traveled along the north wall of Hall "A" and then out of the cave via Entrance #1. The smoke essentially hugged this wall rising no higher than 2 meters off the floor, and extending no further than three meters away from the cave wall. The remainder of Hall "A" was essentially smoke-free. Additionally, on both days that smoke tests were conducted, air flowing into the cave from Entrance #2 quickly exited the cave via Entrance #1, thus adding to the flow of air, and smoke, exiting the cave.

It appears that the physical makeup of the cave, and the physics involved with the movement of air through an open system, like a cave with two entrances, play a very important role in the removal of smoke from the ancient hearth discovered in the cave. As previously discussed, air will naturally move in and out of the cave based upon changes in barometric pressure. During intervals when air was moving out of the cave, the air/smoke from the hearth area would be drawn through the AB Connection. Due to the natural flow of air toward the entrances, and the decreased volume of the AB Connection, as compared to Hall "B", there was a natural "chimney effect" created that drew the air from the hearth, and through the

AB Connection. Once the air (smoke) was in Hall "A", the multiple influences of Entrances 1 and 2 drew the air along the north wall towards the entrances. Since Entrance #1 is larger, and thus better connected to the area outside the cave, it drew the smoke down into and through the passage that connected Hall "A" with Entrance #1. As previously noted, air was also being circulated within Hall "A", with air coming into the cave via Entrance #2, and exiting the cave, along with the smoke, via Entrance #1.

There is also cave use issues associated with the airflow pattern in the Cave of Letters. As already established, the natural airflow patterns in the cave influenced activities associated with the hearth, and how smoke moved through and out of the cave. Based upon archaeological investigations in the cave[12,13] it was determined that the airflow patterns also influenced how the Jews who sought refuge in the cave near the end of the Second Jewish Revolt against the Romans utilized the cave. Utilizing the natural airflow patterns in the cave, the hearth was sited at a location that would provide sufficient drainage of air to prevent the cave from filling with smoke. The hearth could have been sited in the southwest corner of Hall "A", where the maximum movement of air out of the cave was determined to exist, but this location was reserved for another purpose. During Yadin's research, his locus 57, yielded a trove of bronze artifacts, and it was also determined to be the site of the latrine for the inhabitants of the cave. I do not feel that this is happenstance, but rather a calculated effort to minimize the aesthetic and health concerns associated with the latrine. It was sited in the most ventilated location in the cave, where air is exchanged between the two entrances. It is also located in the path the smoke created by the hearth follows to exit the cave.

DISCUSSION AND CONCLUSIONS

Six research questions were presented in the Research Design. The conclusions related to each of the questions will be discussed separately. The first research question asks "are there short-term (daily) variations in the cave's micro-climate?" The answer to this question is most certainly yes. As depicted in Figure 7.2, which is data from when researchers were mostly not in the cave, a pattern of daily variation occurs not only outside the cave, as one would expect, but in Entrance # 1 and Hall "A", as well. The largest variation occurred in percent relative humidity, with the greatest variation (around 18%) occurring in Entrance #1. The elevated relative humidity values at the beginning and end of the graphs reflect the periods of time when researchers were actively working in the cave.

Figure 7.3 depicts data collected in the cave on a similar temporal scale as in Figure 7.2, but during this time period researchers were actively working in the cave. Similar to Figure 7.2, the beginning and end of each graph represent a period when researchers were working in the cave. Different from Figure 7.2, the center portion of each graph also represents an approximately eight hour period when researches were in the cave. The same daily variation noted in Figure 7.2 can be detected in Figure 7.3. Additionally, the researchers working in the cave noticeably affect cave relative humidity, beyond expected normal fluctuations. This is particularly evident in Hall "A", the A-B Connection passage and Hall "B". Slight variations are also detected in the West Extension Passage.

Research question #2 asks "Does the presence of the researchers in the cave affect the cave's micro-climate?" Based upon the short term cave micro-climate data that was collected, this question can be answered yes, without reservation.

Research question #3 asks "are the micro-climatic fluctuations in the cave related to climate variations outside the cave?" The answer to this question is yes as well. It has already been established that caves take in air from outside or expels air from the cave in an attempt to balance the dynamic equilibrium between barometric pressure in the cave and the atmosphere outside the cave. If the barometric pressure outside the cave is higher than inside the cave, air will move into the cave. When a weather front moves over the area and the outside barometric pressure falls, the cave expels air (breathes out). Even slight air pressure changes due to diurnal air temperature fluctuations can be reflected in a cave's airflow pattern. Interaction between the cave's atmosphere and the atmospheric conditions outside the cave were detected in Entrance #1 and Hall "A" in Figure 7.2, when researchers were not in the cave (note the center portion of the graphs). The variations in Entrance #1 and Hall "A" are related to exchanges of outside and cave-based air occurring in this portion of the cave. This pattern can also be detected in Figure 7.3, but as previously discussed, these variations are also affected by human-activities occurring in the cave. In the longer duration data base depicted in Figure 7.4, similar trends exist when comparing the graph from outside the cave with the graphs from Entrances 1 and 2, Hall "A" and the AB Connection Passage, thus indicating longer temporal relationships between air in the cave and the atmosphere outside the cave. It is difficult to make this same assertion for the year-long data base (Figure 7.5) because the relative humidity sensor in the data logger on the ledge outside the cave ceased to function after about six months. The temperature sensor did continue to collect data, and a similar pattern can be noted between the atmosphere outside the cave, and Entrances 1 and 2, and Hall "A".

Research question #4 asks "is there spatial variability within the temperature and relative humidity fluctuations within the cave?" Based upon the conclusions and discussion already presented for research question 1, 2 and 3, the answer is certainly yes. Entrances 1 and 2 are clearly associated with the atmosphere outside the cave, as is Hall "A". Moving deeper into the cave, further from the entrances, these relationships are diminished. In the short term data depicted in Figure 7.2 (no researchers in the cave), no clear relationship is visible in the A-B Connection

Passage, Hall "B", the West Extension Passage and Hall "C". If the temporal scale is expanded as in Figure 7.4 (seven months), and Figure 7.5 (one-year), similar longer-term variations are detected in the variables from the in-cave sampling stations, as compared with outside the cave.

Research question #5 asks, "Is there a temporal variability within the temperature and relative humidity fluctuations within the cave for the short, medium and long-term studies?" The answer to this question is yes. As previously discussed as part of the conclusions for research questions 1–4, temporal variations exist in the data collected both outside the cave, and in the cave, for all the temporal scales (two day, seven months and one year). The patterns of temporal variation were most evident outside the cave, and within the cave in the areas with the most direct linkages to outside the cave (Entrances #1, Entrance #2, and Hall "A"). Temperature did not vary beyond the A-B Connection Passage in all of the studies, and percent relative humidity varied to a much more limited extent, as compared to the front portions of the cave.

Lastly, research question #6 asks "will airflow patterns in the cave indicate a pattern of human utilization?" Based upon the smoke-tests conducted in the cave, it was determined that there was a pattern of airflow associated with the "hearth area" in the southeast corner of Hall "B", Hall "A" and Entrances 1 and 2. The hearth was placed in this location because of the natural chimney effect the A-B Connection Passage has on air moving from Hall "B" toward the caves entrances. The smaller size (volume) of the A-B Passage causes the air (smoke) moving toward Hall "A" to accelerate as it moves through the passage and when the air/smoke enters Hall "A" its is then effected by air patterns in Hall "A" which are being influenced by the air being exchanged between the two entrances. The air/smoke is drawn through Hall "A" along the north wall, and it then exits Hall "A" through the passage that connects Hall "A" to Entrance #1. As reported by Yadin,[14] the latrine for the inhabitants of the cave was located near where the Entrance #1 passage enters Hall "A". The air being exchanged between the two entrances would pass by the site of the latrine, thus keeping this area better ventilated than other parts of the cave. Based upon the location of the hearth and the latrine, the cave's inhabitants did take into account the air flow patterns in the cave in terms of locating these important features of their day-to-day existence in the cave.

With regard to the micro-climate of the Cave of Letters, based upon the research design developed for this part of the project, it was possible to successfully answer the six research questions put forth. Therefore, additional information has been added to the existing base of knowledge regarding the physical and cultural aspects of the micro-climate of the Cave of Letters. Furthermore, the knowledge gained from this segment of the project can be utilized by other areas of the project, such as the archaeological component. The results can assist with interpreting the physical constraints and controlling factors associated with archaeological discoveries, as well as relationships that exist between the cave's physical environment and cave utilization.

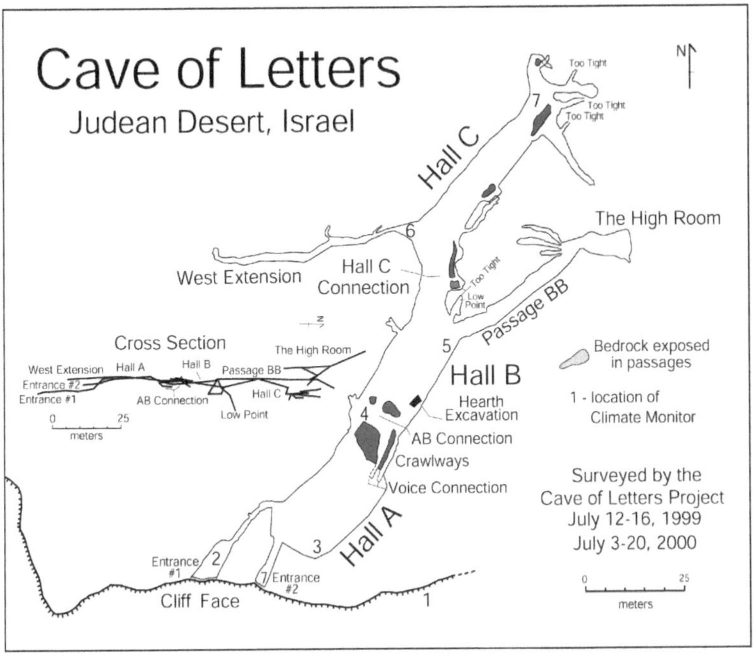

Figure 7.1: Location of the temperature and relative humidity data loggers placed in the Cave of Letters.

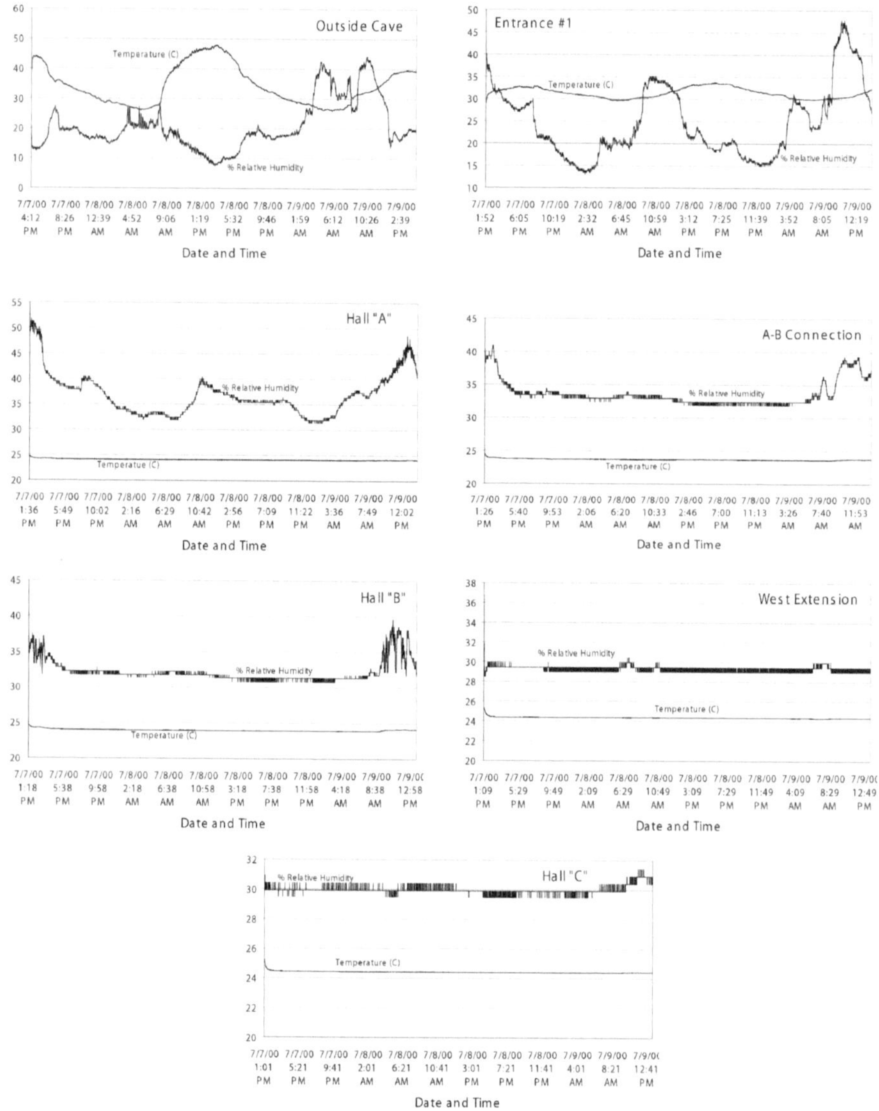

Figure 7.2: Temperature and relative humidity data collected in the Cave of Letters from July 7th to July 9th when researchers were generally not in the cave.

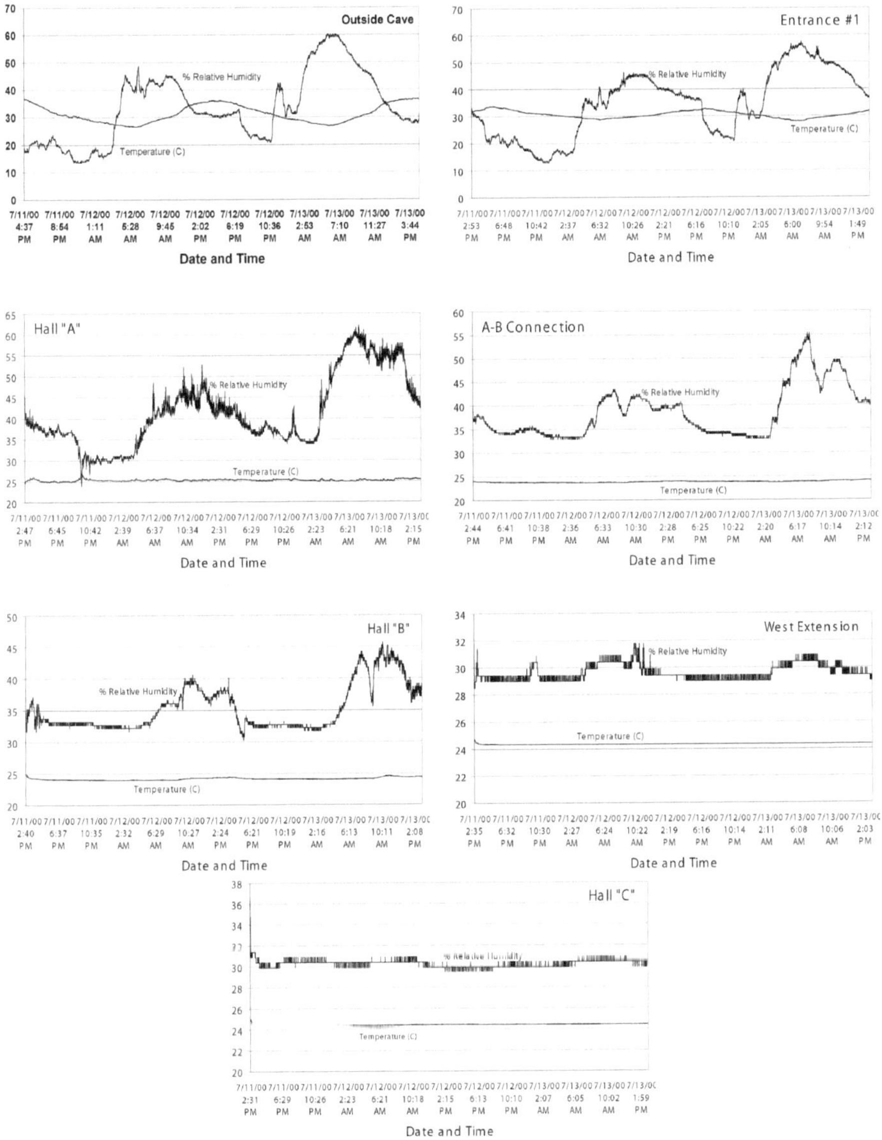

Figure 7.3: Temperature and relative humidity data collected in the cave from July 11th to July 13th when researchers were working in the cave.

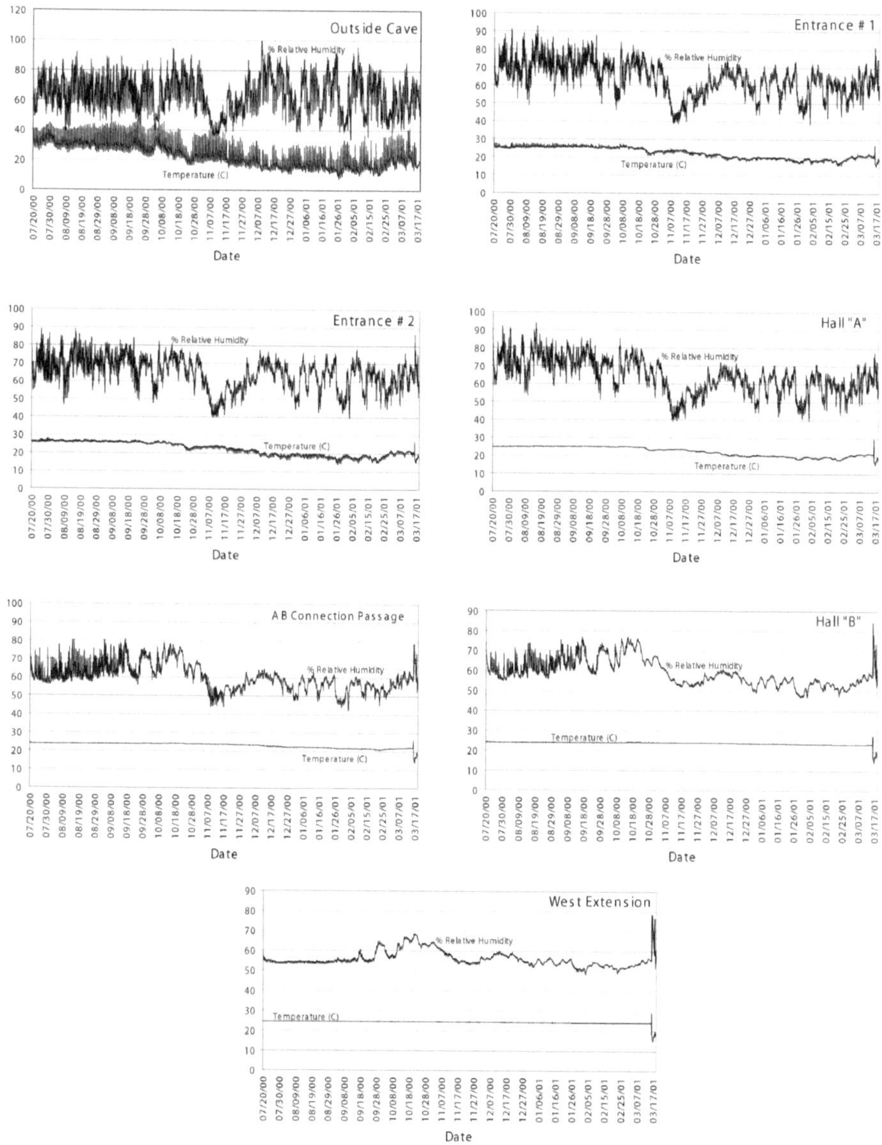

Figure 7.4: Temperature and relative humidity data collected in the Cave of Letters between July 2000 and March 2001.

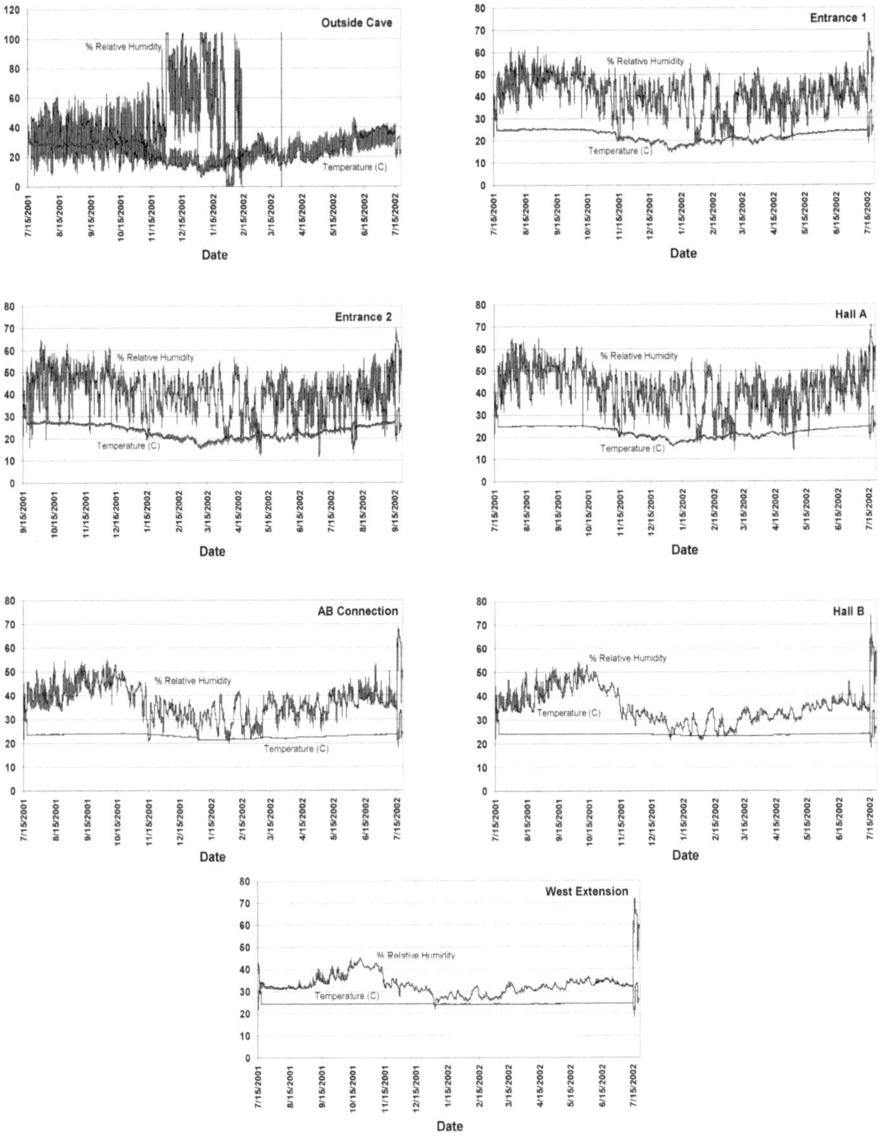

Figure 7.5: Temperature and relative humidity data collected in the Cave of Letters between July 2001 and July 2002.

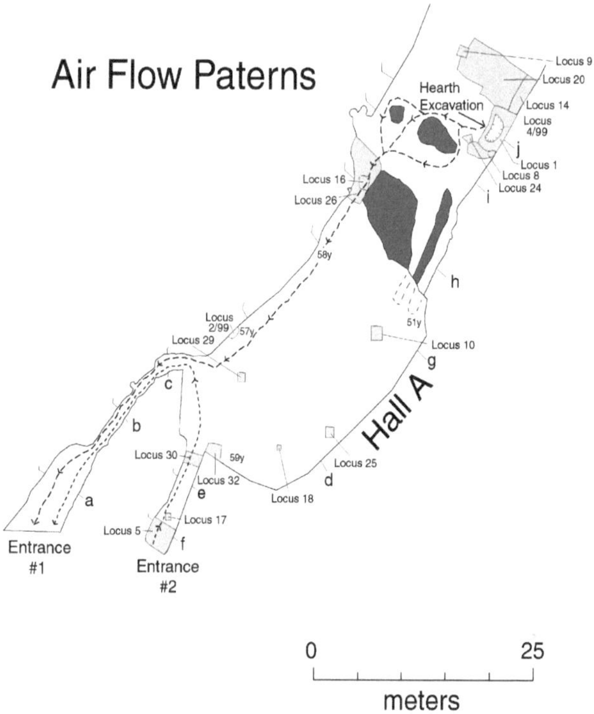

Figure 7.6: Air flow patterns in the Cave of Letters as determined by a series of two smoke tests conducted in southeast corner of Chamber "B" during July 2000.

NOTES

1. Buecher, R.H., "Microclimate Study of Kartchner Caverns Arizona." *Journal of Cave and Karst Studies* 61(2) (1999): 108–120.
2. McLean, J.S., "Factors Altering the Microclimate in Carlsbad Cavernes: New Mexico." *U.S. Geological Survey, Open File Report National Park Service Research Project CACA-N-1a, Albuquerque, NM* (1976).
3. Mclean, J.S., "The Microclimate in Carlsbad Caverns, New Mexico", *U.S. Geological Survey, Open File Report, 76–171, Albuquerque, NM* (1971).
4. Forbes, J., "Air Temperature and Relative Humidity Study: Torgac Cave, New Mexico." *Journal of Cave and Karst Studies* 60(1) (1998): 27–32.
5. Kennedy, J., "Pre- and Post-Gate Microclimate Monitoring. Proceedings of the Bat Gate Design: Technical Interactive Forum", edited by Kimery Vories and Dianne Throgmorton, U.S. Department of Interior Office of Surface Mining, and Coal Research Center, Austin, TX, March 2002.
6. Hall, L.S., "The Effect of Cave Microclimate on Winter Roosting Behavior in the Bat, Miniopterus schreibersii blepotis." *Australian Journal of Ecology*, 7 (1982): 129–136.

7. Postawa, T., "A Cave Microclimate as Modeled by External Climatic Conditions and its Effect on a Hibernating Bat Assemblage B: A case of the Jaskinia Pod Sokila Gora Cave." in *Proceedings of the VIIIth EBRS, Approaches to Biogeography and Ecology of Bats*, Edited by B. W. Woloszyn, 199–218.
8. Hoyos, M., Soler, V., Canaveras, J.C., Sanchez-Moral, S., and Sanz-Rubio, E., "Microclimate Characterization of a Karstic Cave: Human Impact on Micro-Environmental Parameters of a Pre-Historic Rock Art Cave (Candamo Cave, Northern Spain)." *Environmental Geology* 33(4) (1998): 231–242.
9. Dragovich, D., "Cavern Microclimates in Relation to Preservation of Rock Art." *Studies in Conservation*, 26(4) (2008): 143–149.
10. Gerjevic, V.D., "Monitoring of Caves Microclimate in Skocjan Caves, Slovenia." *Geophysical Research Abstracts*, 8-08161 (2006).
11. Onset Computer Corporation. Bourne, MA, http://www.onsetcomp.com/data-logger.
12. Yigael Yadin, *The Finds from the Bar Kokhba Period in the Cave of Letters.* (Judean Desert Studies; Jerusalem: Israel Exploration Society, 1963).
13. Freund, R.A., *Secrets of the Cave of Letters.* (Humanity Books, New York, 2004).
14. Yadin, *The Finds from the Bar-Kokhba Period in the Cave of Letters.*

CHAPTER EIGHT

Return to Locus 57y

A Re-Examination of Some of the Bronze Artifacts from Hall A of the Cave of Letters

RICHARD A. FREUND
The Maurice Greenberg Center for Judaic Studies,
University of Hartford, West Hartford, CT 06117, USA

Archaeology is based upon comparisons of materials from other sites for dating and authenticating. We date architecture, pottery and other elements by comparison with other dated sites and a relatively clear time line has been developed that allows us to assess these elements from site to site because we have established comparative models. On May 7, 1996, a bronze, short-handled incense shovel was found in an extremely disturbed Hellenistic-early Roman layer of occupation in Area A square G54, locus 152 at Bethsaida (Et-Tell).[1] The shovel was found under debris in close proximity to a large Iron Age Bit Hilani style palace structure, close to the city gate complex, and 9 meters away from of the southwest corner of a structure which measures 20 meters by 6 meters and apparently was a Roman style temple at Bethsaida (Figure 8.1). The incense shovel was immediately recognized as bearing a startling resemblance to an incense shovel found in the Cave of Letters by Yigael Yadin's excavations in 1960. The Bethsaida shovel is especially similar to the one designated by Yadin as incense shovel no. 5 (and slightly similar to no. 6-see Figure 8.2). The main reasons are: size, shape, design, decoration, and functionality suggested by the other elements. The shovels are both made out of bronze and have five small concentric circular details in the four corners of the pan and in the center.[2] My research concluded that the incense shovels were not for burning incense but rather for carrying incense to an incense burner or altar for burning. The short handled incense shovels[3] especially made it difficult to think that they had been used to transport hot coals on them and there is no sign of burning present on the shovels. So startling in fact was the resemblance between the Cave of Letters shovel and the Bethsaida shovel as to raise the question of whether they were both cast from similar molds. In the end, it became clear that they were both the product of a similar process but that they were not from the same casting. This discovery raised questions about the religious meaning(s) and use(s) of the incense shovel in general and at Bethsaida in particular. Bronze incense shovels of this type and quality are exceedingly rare in Israel, and the discovery and examination of the entire archaeological context of these incense shovels in Israel is rarer still. In the Bar Kokhba "Cave of Letters", Yigael Yadin found three incense shovels together in a basket in locus 57y (see Map 1 in Chapter 1).[4] The incense shovels were found together with other bronze objects and while Yadin agreed that the manufacture of the shovels could be assigned to the first century CE, he insisted their original use was from a pagan, second century context.[5] According to Yadin, this cache of bronze artifacts was "the largest collection of Roman metal vessels found to date in Palestine and the neighboring region."[6] It still is. Together, the incense shovels found in the caves were of different sizes which may suggest similar but distinct purposes (Figure 8.2). Little more than 135 shovels have been identified and classified worldwide and most lack an exacting archaeological context.[7] The documented shovels appear in and around what might be called Hellenistic and Roman ritual sites and in burial caves. Most of the shovels come from the eastern parts of the Roman Empire and the Bethsaida shovel is possibly the only one in Israel that comes from a documented archaeological city context and not from a hoard, a random discovery, or a private collection. The meaning and purpose of these shovels has been debated because the incense shovel is prominently found in synagogue iconography of the third through sixth centuries. Yadin's impression, for example, was that the incense shovels he found in the Cave of Letters were part of the booty taken by Bar Kokhba's troops. He states rather ambiguously: "…of the units of the Roman Legions or the Auxilia, which carried them about for ritual purposes."[8] He was not sure what the ritual purposes were, but he, similar to other investigators who had written on these bronze shovels, assigned them to some ambiguous pagan ritual. In addition, although Yadin himself pointed out that the incense shovels figure with a number of other well-known Jewish symbols such as the lulav (palm-branch), ethrog (citron), and Menorah (candelabrum) in synagogue iconography of the third through sixth centuries, he does not address the point of why synagogue iconography would adopt a pagan ritual object for presentation in a Jewish building. But incense services (and the requisite artifacts to transport and burn it) necessitated a number of artifacts, some of which are mentioned in the Bible while others are mentioned only in later Rabbinic texts. My research[9] has concluded that there was some type of incense service that may have continued from the Temple Priestly service to the Rabbinic synagogue ritual and continued through the 8th century CE.

THE INCENSE SHOVEL: FROM THE TEMPLE TO THE SYNAGOGUE

The incense shovel is found depicted in the third century synagogues of Kohav HaYarden and Peqi'in. Later synagogues at Hammat Tiberias, Bet Shean, Bet Alpha, Capernaum, Huldah, and Husifah and Sepphoris, all throughout the Jordan Valley, Galilee, and Golan have the symbol. A total of 16 different sites display the incense shovel either in mosaic or incised at a synagogue from the third through the sixth centuries CE. It was a very prominent symbol that was generally combined with the other prominent symbols of the period, the Menorah, palm branches, citrons, ram's horns, and the Torah Shrine. It is, of course, possible that some or *all* of the symbols—Menorah, palm branches, citrons, ram's horn, and incense shovels—were originally pagan symbols popular in Jewish and non-Jewish circles[10] and that Jewish artisans knowingly adapted their use in synagogues. The use of the seven branched Menorah described in the Bible in synagogue art, was deemed problematic in some Rabbinic literature and its appearance in synagogue iconography may be indicative of the debate since the symbols do not appear on all synagogues in the Galilee and Golan.[11] Often, small changes were made to the Menorah so that it was similar to but not a copy of the one in the Temple. The synagogue was not intended to be a replacement for the Temple. The symbols of the synagogue were only intended *to remind* worshippers of the Temple not duplicate the items in the Temple. This provides a clue as to why the incense shovel, a major symbol of the Imperial Cult, may have become a representative symbol of the incense shovel used in the Temple.

Part of the problem of identifying the incense shovels and their use in the synagogues of late antiquity is the fact that Jewish institutions were using an artifact that had been used in the Temple of Jerusalem and therefore did have some Jewish lineage, but it was also used in pagan institutions of the same period and therefore could be assigned to "pagan" lineage. The incense shovel found at Dura Europos[12] in the excavations from the Hellenistic and Roman period, which was similar to the ones found at Bethsaida and the Bar Kokhba Cave of Letters, was assigned as a pagan incense shovel and assumed to be a "brazier." In the same excavations from the Hellenistic and Roman periods At Dura Europos, two glazed clay incense burners (one conical) in the Temple of Atargatis adorned with the sacred animal, the deer, were discovered as well as a bronze "brazier."[13] The Dura-Europos incense shovel is a free-standing artifact with decorations in an ornate fashion. The presence of the two clay incense burners suggests that the "brazier" was only a shovel for transporting incense and not a receptacle for the burning of incense. When one compares the Dura Europos incense artifacts to the ones found by Yadin in the Cave of Letters the issue of usage is clearly the line of demarcation. Despite Yigael Yadin's perception of slight traces of burning on the shovels from the Cave of Letters (which cannot be confirmed-even microscopic analysis does not reveal any burning on the metal), incense burning does not appear to be the primary purpose of the incense shovel from the Cave of Letters or from Bethsaida. These shovels were used for transporting incense to another location where the incense was burned. In the time that there was a Jewish Temple standing in Jerusalem this incense shovel was used to deliver the incense to an incense altar. But what happened after the incense altar was no longer in existence? If we assume that the incense shovel continued to have a life in the synagogue ritual, how was it used? I think that the incense shovel still was used to deliver incense but this time to a smaller, incense burner in the synagogue. In the synagogue, it appears to have been some type of bronze container that would allow the burning to take place and the incense shovel was still just a transporter of incense from one location to another. This is virtually a synagogue re-creation of the incense service in the Temple. Archaeological evidence exists for a free standing bronze incense burner in the early Hellenistic period in locations all around the Mediterranean. As synagogues began to proliferate throughout the Mediterranean, it would have been natural that this type of artifact would be found there. One of these Hellenistic incense burners, or censers, is described as having been made "…in the form of a cup with a splaying foot, one side of the cup being raised up and surmounted with knobs. It seems to act as a sort of shield from heat, while the handle is in the form of an ibex standing facing outwards on a plinth."[14] Although the examples are not normally associated with a synagogue ritual in the earliest synagogues, it is clear that by the fourth century synagogues, the incense burner was in use.

Examples of bronze incense burners are found in synagogue excavations and in iconography in Israel. In the fourth century CE synagogue of Beth Shean, which features a mosaic of an incense shovel, archaeologist Nehemiah Zori found a two part conical bronze incense burner[15] that had four feet and could be free standing (although it had a bronze chain connected to the upper part for hanging). The decorated sides of the "pan" or receptacle were ornate (bronze lattice work) and resemble the side panels of the Dura-Europos Hellenistic pan. Zori, who excavated the synagogue, says that the side panels of the burner are designed to look like animals. Zori also states that this incense burner is similar to those some found in Syria[16] and dated the incense burner to the sixth or perhaps seventh century. This is the same synagogue where a mosaic with an incense shovel was found in what is known as Beth Shean A.[17] It is interesting that two incense services related elements were found in this one location. An incense shovel is featured in the synagogue mosaic while an entirely different type of incense burner was found in another part of the excavations. The incense burner and the mosaic are clearly from different periods. The earliest stratum of the Beth Shean mosaic is assigned to the fourth century, while the incense burner found in the sixth/early seventh century remains of the Beth Shean synagogue resembles those which were a part of the Byzantine church service. One explanation of the discrepancy is that the two may represent different developments of the same contemporary incense rituals. Another explanation is that the two elements were two parts of the on-going incense ritual in the synagogue. The shovel may have continued to be used only for transporting dry incense to a conical incense burner. While this is not exactly the same way the incense shovel had been used in the Temple of Jerusalem (transporting incense to an incense altar), it still allowed the artifact to play a part in a synagogue ritual after the destruction of the Temple in much the same way that the shofar, the palm branch/citron and Torah Shrine and Menorah all had a part in synagogue rituals.

All of this suggests that incense burning was not in fact an original "Christian" invention but rather a re-invention of a common Jewish synagogue practice which continued through the fourth century CE. The Byzantine Church did not at first accept the use of incense in services, apparently because of the Imperial Cult use of incense in the pre-Christian period (but also perhaps because it was so connected with the Temple and synagogue service). The Church Fathers, Trypho and Irenaeus in the second century CE interpreted the biblical concept of incense as a totally spiritual metaphor. Irenaeus writes that "The incense is the prayers of the saints."[18] This trend continued in the Church Fathers into the third century, as witnessed in the writings of Tertullian, who, perhaps under the influence of the Marcion controversy, attacked many Hebrew Bible practices including the use of incense.[19] In addition, during this period, trade in incense by Christians was forbidden.[20] Other Church figures, including Arnobius in his assaults against pagan practices in the Church denigrates the use of incense in the late third and early fourth century.[21] Although it is known that some early Christians did use incense at graves, the use of incense was not a sanctioned Church ritual until the fifth and early sixth centuries.[22] A Byzantine incense shovel from Sardis, decorated with a large cross[23] very similar to the ones described above demonstrates that the shovel was used by the fourth century CE in Christian contexts. The fourth century shovel shows the development of a two artifact incense service (shovel and censer-paralleling what was done in the Temple and synagogue) in Christian ritual but by the sixth century only a bronze conical censer is found in eastern and western churches-the shovel has basically disappeared. By the sixth century, incense is an authorized part of Church services with liturgical blessings taking place. By the ninth century the use of incense is a near canonized part of eastern and western Church rituals.[24]

The late Roman and Byzantine period synagogue iconography, therefore, implies that an incense burning ritual survived the end of the Temple service into the synagogue service of the rabbis. Because the Menorah, Torah shrine, lulav, ethrog, and the shofar did survive as important active ritualized parts of the synagogue service (and not just passive symbols that people remembered from the days of the Temple of Jerusalem) an incense burning ritual of some kind in the synagogue in late antiquity is implied. Although it is not clear how they used the bronze incense shovel in the ritual. Whether or not it was performed with an actual burning of incense in a conical burner (found in synagogues from late antiquity) or was displayed, shaken, and/or used for dry incense scenting in synagogues on a bronze incense shovel is not known. The symbol on the mosaics seems to imply that there was some type of active incense ritual symbolized by the incense shovel on the mosaic. The lulav and ethrog were actually used in synagogues on Sukkot, a shofar was used in synagogues on Rosh HaShanah and Yom Kippur and most synagogues had a Torah shrine and Menorah displayed in the synagogue. It would stand to reason that the incense shovel iconography suggests that the shovel either symbolized or was a part of an active synagogue incense ritual.

There is literary evidence of an incense burning ritual in the literature of the rabbis. Very little is known about rituals of the synagogue in the Amoraic-Geonic (late Roman and Byzantine) period for a number of reasons. One reason is that the standardization of rabbinic practice took place at the end of this period (10th to 11th century) and any practices discontinued in the medieval synagogue were not written about after this period. Remnants of variant practices can be found in Karaite polemics against rabbinic practices embedded in Geonic response literature from the 9th and 10th century. One of the main targets for this polemic rabbinic responsa literature is directed against the Karaites, who criticized the Rabbis for introducing practices into the synagogues that were not included in the Bible. Some of these practices continued into medieval rabbinic Judaism, while others did not. For example, the introduction of the practice of the ritual lighting (and blessing) of Sabbath candles in the period of the Geonim survives; Rabbi Saadiah Gaon took the final action of obligating Jews to kindle the Shabbat light and say a blessing over its light apparently in reaction to the Karaite Sabbath practice of not lighting candles or any other fire during the Sabbath.[25] While the Rabbis and the Karaites were at odds with one another over theological and religious authority issues, they were most prominently separated by the rabbinic development of ritual not mentioned specifically in the text of the Bible and Temple rituals directly transferred to the synagogue—clearly enhancing the position of the synagogue and the Rabbis. Both seemed improper to the Karaites. In the rabbinic prayerbook developed by the Rabbis in the period of the Tannaim and Amoraim, the incense service of the Temple figures prominently both in the opening and closing of the morning services. The ritual of incense is clearly suggested again in a polemic against the Karaite Daniel al-Qumisi by Rabbi Saadiah Gaon. Al-Qumisi's writings (as many of the controversies and the rabbinic responses) are preserved in manuscript form and express the major differences between the Rabbis and the Karaites. The Gaon Natronai ben Nahshon, for example, composed a treatise refuting the Karaite views on the Jewish calendar, and was discovered only in manuscript form.[26] Once the controversy was finished and the Karaites were no longer a force to be reckoned with, the rabbis apparently felt no need to preserve the original content of the arguments. This is the case of the incense ritual. In a Cambridge Geniza manuscript, Al-Qumisi writes about his dispute with Saadiah Gaon:

> "…If they say that we should burn incense and light candles because they (synagogues) are a 'Temple in miniature', we will say to them: it is not as you argue, even more so, God forbade that the incense and the lighting of candles in synagogues because there is in this pagan worship, because they are placing the synagogues in the same level as the Temple of God (in Jerusalem) in the holiness and the doing of these things. …this is forbidden to put the synagogues in the same level with the Holy Temple in regards to the bringing in of impurities, because synagogues are not holy. And every one who lights candles in them or burns incense in them because they are holy is transgressing the commandments of God and is making holy that which is not holy and is not preparing his heart for the God of his fathers…but those who obligate the lighting of candles and the burning of incense in synagogues are the Rabbis, and they bring evidence (to support this view) of the holiness of the synagogue from the writing that it shall be a 'Temple in miniature'"…[27]

My assessment that the bronze incense shovel continued to be used beyond the destruction of the Temple in Jerusalem is found in earlier research. Professor Goodenough in his assessment of the incense shovel[28] presents a number of sources that also seem to imply that the burning of incense continued in the period following the destruction of the Temple. Evidence of open-top square box incense burners associated with Jewish burials may reflect refinements of the incense shovel burner. Goodenough's anthropological approach to the use of the incense burner among Jews found that it continued in the East but not in the West. Evidence for incense burning in synagogues and at home can be found among Ethiopian Jews, Syrian Jews, and especially among Yemenites up until the modern period.[29] They are mostly incense burners similar to the Eastern Orthodox tradition of Christianity. Although the incense burning may have survived in eastern/oriental Jewish rites, the Western and Eastern European Jewish rites removed it entirely perhaps because it was too similar to medieval Church ritual. Jewish liturgy continued to have a disproportionately large number of readings on incense in the daily, Sabbath, and holiday services. The mystical medieval Jewish tradition also incorporated elements of the incense service (and perhaps even the conical incense burner) into a new/old ritual called *Havdalah* performed with an ornate "spice box" (*Hadas*) at the end of the Sabbath.[30] Similar to other parts of the synagogue service, the rabbis in Eastern and Western Europe seemed to move to a new level of symbolic understanding of the mysteries of the Temple—from one physical symbol of the Temple service to another physical symbol of the remembrance of the Temple in liturgy and ritual (Figure 8.3).

The Bronze Patera and Thetis in the Cave of Letters

Because the rest of the artifacts Yadin discovered in the Cave of Letters were seen by him to be either Jewish or used by Jews, it is very strange that the bronze artifacts were declared pagan from the outset. One of the reasons for this designation was the pagan symbols on the cup handles and the image of Thetis on the bronze patera. The Thetis and the Sea Creature decorations were found on a patera that can be dated to by the prominent use of such imagery in the first century CE.[31] I have written elsewhere that the Cave of Letters bronze horde, which Yadin argued was from second century CE "... units of the Roman Legions or the Auxilia, which carried them about for ritual purposes," is more likely artifacts from the Temple in Jerusalem that were brought to the Cave and buried in the first century CE.[32] Yadin remarked that, based upon the wear and tear, the utensils in Cave of Letters were at least a century old[33] when they were hidden, placing them squarely in the period of the Second Temple. This would make these artifacts either relics being carried around by this Auxilia in Ein Gedi or they were from another period and buried and have nothing to do with the story of the Second Revolt. Although Yadin held that the utensils were originally pagan and not for Jewish use, he also left open the possibility that they may have served Bar Kokhba because of the slight defacement of the images and that was the reason why they buried.[34] It is interesting to note that the famous Thetis-Sea Creature scene found on the patera from the Cave of Letters was a prominent scene on many different artifacts from the first century. It may also explain the defacement or wear and tear on the patera and on some of the bronzes. They were in use for such a long time that they had natural wear and tear which included the defacement through use. In fact not all of the bronzes are defaced. Yadin does not fully address the second half of his theory, which is that if the followers of Bar Kokhba had defaced these mythological images on the bronze objects, they could use them again, but for what purpose? If they were objectionable enough to be defaced, Bar Kokhba and his followers could just have easily melted them down for the metal and not buried them. Yadin merely states that "...according to the *halakhah*, this rubbing was sufficient to render the vessel usable, without having to deface the other parts of the relief."[35] But what possible purpose might these vessels be used for (presumably later when they were dug up) if they were ritual objects? Whether or not Yadin was correct concerning the purposeful defacement (or whether the apparent defacement is the result of regular wear and tear) does not matter. If they were purposely defaced (and not melted down), it was an attempt to render these vessels fit for use and their placement in the caves was still for their protection so they could be used in the future (presumably ritual purpose).

In Yadin's time, the controversy of Jewish art was just getting underway. He would not have seen these items as potentially Jewish given the state of the question in the 1950s. The main claim against the Arch of Titus and the Jewish use of the Cave of Letters patera are that the mythological images somehow disqualified these objects as Jewish ritual objects. This claim is undermined by the mythological iconography of many synagogues in the Roman and later Byzantine period. Yadin was an advocate of the "sudden" change theory that took place in Rabbinic Judaism with regards to the Roman iconography. As Yadin knew, many synagogues beginning in the third and fourth centuries were decorated with Greek symbols such as Greek idols naked (Hammat Tiberias) or clothed (Beth Alpha). Also, Hercules (Chorazin) and carved images of Zeus are found on Jewish graves in Beth Shearim. So the argument that Jews did not use Greco-Roman mythological images is difficult even though all of these synagogue mosaics are from the late Roman and Byzantine period[36] and not the first century CE. Even the doors of the Temple, as they are rendered on the third century CE Dura Europos Synagogue, have mythological figures. The assumption is that this is an anachronism.[37] We can assume that they are depicting some type of historical information, that some historical tradition is being represented, and that this is not a total invention of the artist. In the Dura Europos synagogue, in particular, we find both a dolphin and a Capricorn/Thetis-like character on a number of ceiling tiles as well.[38] In addition, amongst the second century CE "Letters", in the Cave of Letters, two small seal impressions with mythological images were found sealing the letters. One seal has a motif of Hercules killing the Nemean lion, while the other appears to be another mythological figure (perhaps a mermaid) holding a torch.[39] Yadin himself could not understand how Bar Kokhba, a symbol of the fight against Rome, would choose such a symbol for his letters. It is similar to other Roman (non-Jewish) seals of the same period. It is likely that the seals' symbols in their absolute form symbolized the very fight of Bar Kokhba against the formidable Roman army. Even if it looks like Hercules, it may have been understood by the Jews as just a symbol of battle or Judaized in the mind of the Jews as a biblical hero such as Samson who battled a lion or Daniel. In this way, we approach the problem of Thetis (Figure 8.4).

THETIS ON BASE OF THE JERUSALEM TEMPLE MENORAH

The original Arch of Titus, built at the curved end of the Circus Maximus in Rome after the victories of Vespasian and Titus, no longer exists. The current "Arch of Titus," found at the end of the Via Sacra between the Roman Forum and the Colosseum, was completed in the period of Domitian, Vespasian's second son. It was rendered in the mid-80s CE either as a tribute to Domitian's father's victory in Jerusalem or in memory of his brother (perhaps to deflect criticism circulating in Rome that he had poisoned his brother). This so-called "Arch of Titus" is built of imported white marble around a concrete core and has panels showing Titus' carrying away the treasures of the Temple of Jerusalem.[40] The artifacts shown on the Arch were probably kept in the Forum of Vespasian, built to identify the new dynasty with the blessings of peace after the period of strife both within the western and eastern empires. This is confirmed by the statements of Josephus in the Jewish Wars, 7.5.5–7. In June 455, King Gaiseric, king of the Vandals, sacked Rome, took the Temple treasures (as well as other treasures), presumably including the Menorah and this is the last reliable literary reference to the artifacts.[41]

The Menorah is depicted on the Arch of Titus. One of the main questions about the Menorah is its base, upon which is found the mythological figures (including Thetis). Unlike other scholarly understandings of the base as a foreign ornamentation or a Roman stabilizing device, I maintain that the base is original and known to the Rabbis. The base of the Menorah in the Arch of Titus is one smaller octagonal block stacked on a larger octagonal block, which resemble steps that are approximately nine inches thick when the height of the Menorah is compared with the height of the individuals carrying it on the Arch of Titus. Although many point to the rabbinic period (mosaic) tradition of the base of the Menorah having three legs, only Rashi specifically says that the Exodus 25:31 passage about the *yerekh* of the Menorah, "it is the bottom foot which is made in the form of a box, with three feet coming out from the bottom." It is widely recognized that the rabbinic era mosaics are from a period much after the destruction of the Temple and unreliable as witnesses to the original Menorah.

Other rabbinic interpreters, including the Rashbam (Rabbi Shemuel Ben Meir, the grandson of Rashi), for example, simply states that the *yerech* is the wide footing at the bottom of the Menorah (without any reference to the tripod of Rashi).[42] Rashi's interpretation may be based upon the versions of the Menorah he had seen in his own times (or mosaics from earlier Byzantine representations) but is not necessarily the earliest version of the base. *Yerech* generally means "thigh" in the Bible, not base, and therefore (in the Bible) would be referring only to the stem or shaft of the Menorah and not the base.[43] The post-Temple/rabbinic versions of the Menorah with the tripod foot-like base alluded to by Rashi might be the earliest attempts *not* to present the Menorah as it existed in the Temple. This ruling is in accord with rabbinic law which forbade the creation of any exact representation of Temple vessels in the synagogue after the destruction of the Temple.[44]

The stepped base of the Menorah depicted in the Arch of Titus may actually have been known to the Rabbis. According to the Mishnah Tamid, 3.9, "There was a stone before the Menorah on which there were three steps; on this the priest stood to trim the lamps. He left the oil jar on the second step and came away." This curious arrangement of a carved stone step ladder with three steps is not found again in Tannaitic literature. According to Rabbinic traditions in BT Menahot 28b and elsewhere, the height of the Menorah is supposed to be 18 *tephahim* (handbreadths)[45]; 4 1/2 feet high, which is similar to the Menorah's sizing on the Arch of Titus. BT Menahot 29a further clarifies this tradition with an early third century insight of Rab:

> Rab said, The height of the candlestick was nine handbreadths. Thereupon R. Shimi b. Hiyya raised the following objection to Rab. We have learnt: There was a stone before the Menorah in which were three steps; on this the priest stood to trim the lamps. He answered, You, Shimi! I meant only from the point where the branches begin [to rise] and upwards.

If the Arch of Titus Menorah is to scale, the Menorah would appear to be 8 handbreadths (2 feet) for the base and 10 handbreadths (2 1/2 feet) for the top. If this third century CE tradition is taken seriously, the middle portion of the shaft is approximately one handbreadth and the branches are 9 handbreadths. If we take the ornamented bottom attached to the octagonal base as 1/2 foot (6 inches), each of the two parts/steps of the octagonal base are nine inches each or 1/2 cubit in height. Each one of the steps of the step ladder would need to be 1/2 cubit high with the third just above the pedestal attached to the octagonal base.[46] 1/2 cubit is in fact the height of all of the steps in the Temple Mount and would, therefore, match the conformity of the design for other steps. The question remains, why did the Priests need a three stepped stone step ladder at all for a Menorah which was only 4 1/2 feet high? It might be that some help would be necessary to insure that a 5 1/2 foot (average height) tall Priest could trim the wicks and clean the Menorah with dignity, but such a stone step ladder itself cumbersome and extremely unwieldy for moving around the octagonal base. The purpose is not stated other than a further clarification that the holy oil cruse was left on the second step after the trimming and filling of the oil dishes (Mishnah Tamid 3.6).

One possible solution to why the Menorah has this three stepped stone ladder for trimming may relate to the base itself and the decoration on it. Many scholars note that the base is decorated by Greco-Roman mythological characters. The various explanations of why many of the rabbinic versions of menorah have a three footed bottom is related to explaining the decorative plates on the base as well. One of the earliest publications on the subject argued that the entire base of the Menorah was created by the Romans to give it stability for the move to Rome. Another view that many relied upon was that the Menorah on the Arch of Titus was not the real Menorah from the Temple at all; the real Menorah had been hidden away and when the Romans arrived at the Temple in Jerusalem in 70 CE either they took what was an available copy (BT Hagigah 26b-27a) or another Menorah similar to the one Herod gave to Rome (Wars 6:388) but not "the" Menorah from the Temple. All of this because of the mythological designs decorated the base.

But the idea that the Menorah base on the Arch of Titus depiction is different from all of the other presentations of the Menorah is not altogether correct. Some of the Rabbinic period representations have the three footed model; some of the earliest depictions of the Menorah have what looks like a block base with steps, similar to the Titus Menorah-but without the decorations. On a ring from 3–4th century CE[47] and many of the "Rabbinic" period menorahs on synagogues show the base as a combination flat block base without feet.[48] The three earliest representations of the Menorah also show the base as blocks: The first century CE Arch of Titus appears to be two octagonal stacked blocks. On the Mattathias Antigonus coin of 37 BCE (the only time that the Menorah appears on a Jewish coin at all in antiquity), the Menorah base seems to be a single block (perhaps slightly tiered).[49] Finally, on the earliest known representation (found in the Jerusalem excavations) on plaster, the Menorah base appears to be triangular but with no feet.[50] One of the most famous and complete mosaic representations found at the synagogue of Naaran has the stepped base of the Arch of Titus. It quantitatively appears that while many late Roman and Byzantine Jewish depictions of the Menorah have a tripod bottom, many others have a block or slightly tiered footing as well.

The upper part of the Menorah on the Arch of Titus is assumed to be the Menorah of the Hasmoneans by most scholars. It fits artistic renderings found in both post and pre-destruction contexts, but the mythological figures are problematic. The Menorah was, according to the Talmud, remade in Hasmonean times of iron overlaid with tin (or wood overlaid with iron). "Then they grew richer they made it out of silver, and when they grew richer, they made it out of gold" (Rosh Hashanah 24b; Avodah Zarah 43b). Although the books of Maccabees do not clearly state that the Hasmoneans remade all the Holy Vessels, including the Menorah, I Maccabees 4.49 states:

> "They also made new sacred vessels, and they brought the Menorah and the altar of incense and the Table into the nave."

The Maccabees tradition may allude to only some new vessels while the Menorah, Incense Altar, and Table were brought back or recaptured from Antiochus. Talmudic tradition assumes that only in Herodian times were the Temple vessels restored to their original glory. Josephus, however, states that when the Menorah as remade by the Hasmoneans, it was made of gold from the beginning (*Antiquities* 12.5.4). He says that Antiochus carried off "…the vessels of God, the golden lampstands and the golden altar and table…" and that it was necessary to remake the Holy Vessels. When Pompey and his men saw the gold menorah when they visited in the mid-first century BCE it was clearly gold and not in the intermediate stage of development. (Antiquities 14:7).

But the question is whether the images on the base of the Menorah were acceptable "Jewish" symbols or just Herodian apologetic attempts at impressing his benefactors. Although eagles, for example, are not generally considered Jewish images (but were clearly a symbol of Rome), eagles are found throughout the Herodian-Roman periods in clearly Jewish/Rabbinically sanctioned contexts.[51] Rabbinic burials in Beth Shearim as well as Hellenized Jews throughout Israel used the eagle despite the obvious connection to astral and immortality symbolic connotations in the Near East and despite its direct connect with Greek myths.[52] Dolphins[53] and sea-creatures[54] also appear in synagogue and other Jewish ornamentation. Their appearance on the Arch of Titus is therefore not as unusual as was once suspected. Perhaps even from the early Hellenistic period these were acceptable mythological figures, even before Herod. In addition, the source of the Menorah and its base may actually be clarified from our understanding of how some of the artifacts in the Temple came to be there. Although most of the vessels were crafted by Jewish artisans, some of the artifacts apparently in use in the Temple were donations from non-Jews. In one sequence in *Wars*, Josephus recounts that some of the Temple vessels were once seized by the anti-Roman Zealots. In Josephus' The *Wars*, Book V.562–564 and in Philo's *Legatio ad Gaium* 157,319, we learn that there were many treasures in the Temple that were destroyed by Jewish Zealots. Although the Zealots are not necessarily in accord with rabbinic standards, they sometimes exceeded rabbinic standards in their antipathy toward Roman life and culture. Philo notes that while no statues or other outstanding objects of the Imperial Cult had been placed in the Temple in Jerusalem, statues of Caligula as well as other dedicated items, including golden shields, crowns, pillars and inscriptions, had been placed in Diaspora synagogues (*Legatio*, 20–23, 140–158). Certain artifacts were donated by foreign leaders in the Greco-Roman period for the Temple in Jerusalem by the Ptolemies, Augustus, and Tiberius. Emperor Augustus and Livia in the early first century CE, in particular, had donated such items that would still have been in use during the rebellion and would have been the subject of controversy given the views of the Zealots. Some of these items presumably were ultimately melted down by John of Gischala, the Jewish rebel leader who continued to fight in Jerusalem up until the destruction of the Temple in July, 70 CE.

> "When there was nothing left that John could extort from the people, he turned to sacrilege and melted down many of the offerings in the Sanctuary and many of the vessels required for services, basins, dishes, and tables, not even keeping his hands off the flagons presented by Augustus and his consort. For the Roman emperors honored and adorned the Temple at all times."

According to Josephus, the smaller vessels seem to have been the first vessels to be melted down by the Zealots. These included presumably the holy vessels constructed by Jewish artisans and the smaller vessels donated by the Emperor Augustus. They were both considered "holy" and honored in this period. One of these donated items may have been the base for the Menorah. In addition, the PT Megillah 24a reports that even later, Antoninus (Caracalla) apparently donated a Menorah to a synagogue in Sepphoris during the period of his great period of friendship with Rabbi Yehudah HaNasi in the third century CE, apparently indicating that the tradition of accepting gifts for ritual Jewish use continued beyond the period of the Temple.

The base of the Menorah includes an artistic rendering of one of the enduring Greek myths which was used by a variety of peoples throughout the Greco-Roman world: the story of Thetis, Achilles' mother, riding on a sea creature to bring weapons to her son Achilles.[55] The sea creature usually has a human upper body torso with the body of a fish extending from the waist and ending in a tail which curls upward. The sea creature can be depicted facing either direction. Usually Thetis rides on the end of the creature's tail and large fish or dolphins are usually nearby. Often the sea creature turns backward to look at Thetis but sometimes he gazes in the opposite direction. This type of scene is found in pottery, reliefs, paintings, seals, mosaics, coins, and even on sarcophagi from the first centuries CE.[56] A similar sea creature is seen on seventeen ceiling tiles at the Dura Europos synagogue, a frieze in the synagogue at Capernaum, a second century column base from Beit Shean, on the ceiling of the Jewish Catacomb of Vigna Randanini and possibly on the Hammam Lif synagogue in Tunisia.[57]

CONCLUSIONS: WHY THE BRONZE ARTIFACTS WERE BURIED IN 57Y

The bronze artifacts of the Cave of Letters remain the finest examples of Roman period bronze artifacts ever found in the Land of Israel. They were found by Yigael Yadin using a new technology, the metal detector, in 1960 during his excavations. Yadin theorized that they were buried after they were taken from a pagan Auxilia in the second century CE. This assignment has always been problematic because the Second Revolt refugees would have needed the bronze for many different pressing uses. The bronze artifacts were apparently brought to the cave from a location where bronze artifacts of this quality would be used (not an Auxilia unit in the desert). It is my conclusion that the bronzes were buried at a period earlier than the second century CE. The burial of the artifacts near entrance #1 of the cave, in what clearly became the latrines of the refugees, may indicate that the second century CE refugees had no idea that the bronzes were buried in this area. There was one more test which we performed on the bronzes that seem to suggest that they were buried in the first century rather than the second century. One of the two large bronze bowls that was discovered in the palm branch basket in 57y had a rope tied to the handle and the rope was then fastened to the basket to close the basket (Figure 8.5). Yadin's bowl #2[58] is identified also at the first century sites of Pompeii and Herculaneum and it has birds' heads on the one *omega* handle. The rope around the handle of the bronze bowl (RTT 3860-IAA 286) was C14 tested and can be dated to the older extreme of the dates (see the introduction chapter for a full description), 70 CE, based on an evaluation of the C14 data.

Figure 8.1: The incense shovel from Bethsaida.

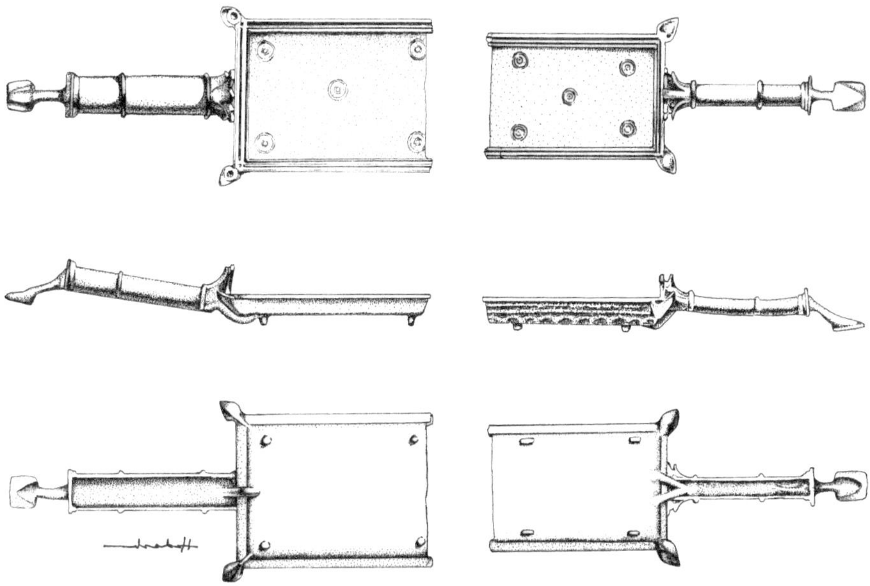

Figure 8.2: The Incense Shovel of Bethsaida (right) and the Cave of Letters (left).

Figure 8.3: The mosaics of the synagogues of Israel.

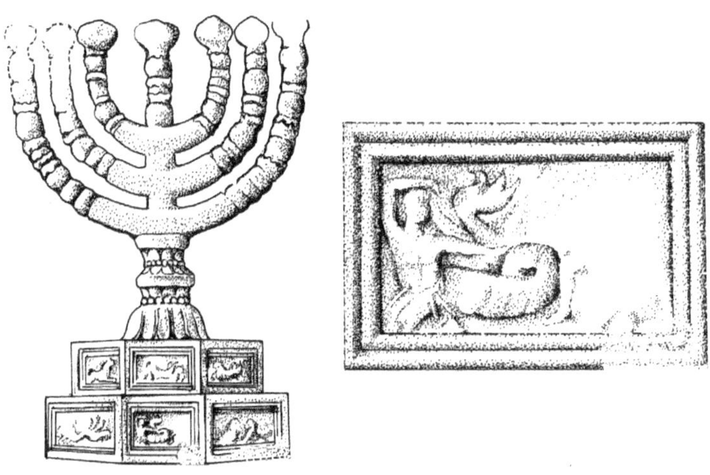

Figure 8.4: The Thetis Image from the Patera of the Cave of Letters and on Menorah of the Temple in Jerusalem.

Figure 8.5: The Bronze Bowl and the Rope.

NOTES

1. This is all referenced in the following articles: *The Dead Sea Scrolls: Fifty Years After Their Discovery*, (Jerusalem: Israel Exploration Society, 2000), edited by Lawrence H. Schiffman, Emanuel Tov, and James C. VanderKam, in my article entitled: "A New Interpretation of the Incense Shovels of the Cave of the Letters," 644–660 and in Rami Arav and Richard Freund, *Bethsaida: A City by the North Shore of the Sea of Galilee,* Volume II, (Kirksville, MO: Truman State University Press, 1999), articles by Rami Arav, "Bethsaida Excavations, Preliminary Report, 1994–1996," 31–44 and in Richard Freund, "The Incense Shovel of Bethsaida and Synagogue Iconography in Late Antiquity," 413–459. In the same locus above, below and around the find was found in the preceding days—May 2–6, 1996: Hellenistic cooking pots, casserole bowls, juglets, jars, decorated ware, so-called Galilean bowls, Roman glass, a bow spouted Herodian lamp, and Eastern Terra Sigillata pottery shards among hundreds of other small finds. This discovery was recorded in the pages of *Biblical Archaeology Review*, January/February 1997, 23.1, 32 as a *Prize Find.*
2. Yigael Yadin, *The Finds from the Bar Kokhba Period in the Caves of Letters,* (Jerusalem: Israel Exploration Society, 1975) 53, No. 5—57.23 (P1. 16; Fig. 14). [Yadin, *Finds* hereafter].
3. Yadin, *Finds*, 53. Yadin's incense shovel no. 5's handle is 12.3 centimeters long; the overall length of this shovel is thus 23 centimeters. The incense shovel of Bethsaida is 20.5 centimeters in length and 6.7 centimeters in width and it has a handle which is 11.1 centimeters long.
4. My book, *Secrets of the Cave of Letters* (Prometheus, 2004) provides a much wider insight into the question. Yadin, *Finds,* 55–57 lists 9 incense shovels in Palestine, one in Gaul, 3 from Pompeii and Herculaneum, 4 from Lebanon and Syria (indicating that others are to be found in in the National Museum in Damascus); but this is not a complete list. Most of the shovels which have been discovered come from the eastern parts of the Roman empire and the Bethsaida shovel is one of the few that comes from a documented archaeological context and not from a hoard, a random discovery or from a private collection. The incense shovel does not seem to have received as much attention as it should in archaeological artifact lists. Most recently in Manns, Frederic

and Eugenio Alliata, eds., *Early Christianity in Context: Monuments and Documents*, (Jerusalem: Franciscan Printing Press, 1993). Plate VIII. (found between pages 312 and 313) an incense shovel is listed among the finds in excavations from Capernaum but is not found in other publications on Capernaum. In Yadin's section marked "Varia," for example, he lists only one shovel in the Rockefeller Museum and a significant number from other collections. The one from the Rockefeller that Yadin lists bears the inscription: "Good luck to the purchaser", which may indicate private use. Two others are presently found in the collection of the Rockefeller Museum. One apparently comes from the vicinity of Hebron, and is said to have been found together with a coin struck by a procurator in the first half of the first century CE. Other shovels listed in Yadin's work are no longer in their original locations. This list of shovels found in Yadin's work is, therefore, far from exhaustive.

5. Yadin, *Finds*, 18–22. On Friday, April 1, 1960, while working in Niche 4, Hall A, locus 57 (57y on Map 1 in Chapter 1), after removing a medium-sized stone and an additional half-meter of dirt, a basket containing a cache of metal utensils was discovered. The basket contained 19 metal objects, including bowls, patera, three incense shovels and jugs. On Sunday, April 3, 1960, at a distance of several meters north of his locus 57 and near the metal utensils, a scroll fragment of Psalms was found. Yadin's assessment of the manufacture and locale is in Finds, 45.
6. Yadin, *Finds*, 44.
7. "An Herodian Tomb near Tell Abu Shusha", A. Siegelman, in B. Mazar, Ed. *Geva*, (The Israel Exploration Society, 1988), 36. The shovel from the Herodian Tomb is only 90 mm by 130 mm. He suggests that the shovel was used for "pagan" incense burning. A. Mazar, who found another shovel in a tomb at French Hill in Jerusalem, suggested it was used for moving the bones. Siegelman also cites the shovel found near the Hellenistic Temple from the Beer Sheva excavations, note number 84. Other authors have variously assigned different meanings to this type of shovel. M. Narkiss, M. Avi-Yonah and J. Braslavi concluded that the incense shovel was a Jewish symbol. M. Narkiss, "The Snuff Shovel as a Jewish Symbol," *Journal of the Palestine Oriental Society*, (1935), 14–28, M. Avi-Yonah, "The Meaning and Use of Incense Shovels", *Journal of the Palestine Oriental Society*, (1940), 87–97, J. Braslavi, *Symbols and Mythological Figures in the Early Synagogues in Galilee in All the Land of Naphtali*, (Jerusalem: Hebrew University, 1967), 115–117.
8. Yadin, *Finds*, 45.
9. Most recently in my book, *Digging Through the Bible* (Lanham: Rowman and Littlefield, 2008), 225–246.
10. See Jeffrey Rubenstein, *The History of Sukkot in the Second Temple and Rabbinic Periods*, Brown Judaic Studies 32, (Atlanta: Scholars Press, 1995) 94 ff. demonstrates that the symbol of palm-branches and citrons were non-Jewish as well as Jewish symbols.
11. BT Avodah Zarah 43a. "A person may not make…a menorah after the design of the [Temple's] menorah. One may, however, make one with five, six or eight branches, but with seven he may not make it even though it be made of other metals [than those used in the Temple's menorah]".
12. Yadin, *Finds*, 57. Goodenough, Symbols, O Cit., I, 174, III, no. 439.
13. M. Rostovtzeff, *Dura-Europos and its Art*, (Oxford: University Press, 1939) 46 and plate VIII.
14. See R. D. Barnett's article, "A South Arabian Ivory Vessel" in *Eretz-Israel*, Vol. 7 (Jerusalem: Israel Exploration Society, 1964), 5.
15. N. Zori, "The Snuff Shovel as a Jewish Symbol," 163. The bronze incense burner was found in room 10 with pieces from the 6th to early seventh century.

16. N. Zori, "The Snuff Shovel as a Jewish Symbol," 163. M.C. Ross, *Catalogue of the Byzantine and Medieval Antiquities in the Dumbarton Oaks Collection,* Washington, 1962, 42–45, Plates XXXII: 47, XXXIII: 48–49.
17. R. Hachlili, *Ancient Art and Archaeology in the Land of Israel,* (Leiden: Brill, 1988), Figure 103.
18. Ante-Nicean Fathers, Vol. I, 1976 reprint, (Grand Rapids: Eerdmans, 1976), 574.
19. Ante-Nicean Fathers, Vol. III, 346.
20. Ante-Nicean Fathers, Vol. III, 67.
21. Arnobius, "Against the Heathen", *Ante-Nicean Fathers,* Vol. VI, 1976 reprint, 428–429.
22. Cited in Goodenough, *Symbols*, Vol. 4, 200.
23. Cited by Yadin, *Finds*, 57. note #36 as: G.M.A. Hanfmann, *Excavations at Sardis*, 1958, *BASOR*, 154 (1959), 5 ff., fig. 11.
24. Ante-Nicean Fathers, Vol. VII, 537, 552–556, 563–564.
25. L. Hoffmann, *The Canonization of the Synagogue Service,* (Notre Dame: University of Notre Dame Press, 1979), 86–89.
26. *Rabbi Saadya Gaon's Translation of the Torah*, Ed. M. Zucker, (New York: Feldheim, 1959), 5.
27. *Ibid.*, Zucker, 170–171 note 666. There is other evidence suggested there of this incense ritual as well.
28. Goodenough, *Jewish Symbols in the Greco-Roman Period,* Vol. 4, (New York: Columbia University Press, 1954), 199 ff.
29. Goodenough, *Symbols*, Vol. 4, 204–205.
30. B. Narkiss, "The Hadas in Jewish Art," in *Eretz Israel* 7, (1961), 55.
31. See R.A. Freund, *Secrets of the Cave of Letters.* (New York: Prometheus, 2004) for fuller explanations of this thesis.
32. Yadin, *Finds*, 45. I have argued in *Secrets of the Cave of Letters* that the claim that Bar Kokhba even started to reinitiate Temple worship can be found in a number of sources. See, for example, M. Avi-Yonah in his book, *The Jews Under Roman and Byzantine Rule*, (Jerusalem: Magness Press, 1984), 13, states: "Bar Kokhba established his government in Jerusalem and probably resumed the Temple ritual as far as possible."
33. Yadin, *Finds*, 45.
34. Yadin, *Finds*, 44.
35. Yadin, *Finds*, 60.
36. Even the fourth-fifth century rabbinic tradition of Rabbi Abun recorded in the Jerusalem Talmud (PT) Avodah Zarah 3, 42d (which is usually cited to show the "new" policy which made this normative) which made the practice of pagan mythological figures on synagogue mosaics permitted cannot be seen as a "sudden" development without precedent.
37. See Zev Vilnay, "Pictures of Jerusalem and its Holy Places", *Eretz Israel,* 6 (Jerusalem: Israel Exploration Society, 1960), 154. Although Dura Europos is from the third century CE, the rendering clearly has mythological figures on the main door and while this may represent a later standard, it is possible that it represents the earlier standard as well.
38. Goodenough, *Symbols*, Vol. 5, 11 ff., 4, Figure 1.
39. Yadin, *Finds*, 118–121 and footnote 24 on page 121.
40. J.B. Ward-Perkins, *Roman Imperial Architecture,* (New Haven: Yale University Press, 1994), 73.
41. For more on the use of the Menorah in Judaism, see my article: "The Mystery of the Menorah and the Star," in *Nationalism, Zionism and Ethnic Mobilization of the Jews in 1900 and Beyond,* Ed. M. Berkowitz, (Leiden: Brill, 2004), 125–147.

42. Others found a compromise. The Medieval commentary known as "*Daat Zeqenim MiBaalei HaTosefot*" stated that it was a wide, thick block at the bottom with three feet. Much of these materials are not observations but rather are derived hermeneutically from words in the text. The Rabbis, for example, derive the tradition that the Menorah was 18 handbreadths tall from the Gematria of the Hebrew word "This [is the work of the Menorah]": *hazeh*, which is 18. The fact that the Tabernacle had only one Menorah and the Temple had 10 (perhaps others for a total of 11) is to be found in the extra letter in Exodus 25.31 ("Make [a Menorah]") "Va'Asita". The extra Yud's value is 10.
43. The Pesiqta Rabbati 6.3 seems to think it is the mid-section (stem) of the Menorah and not the base.
44. BT Menahot 98a-b.
45. The view that they were much larger is based upon a Rabbinic tradition preserved in BT Sukkot 52b that the Menorahs were 50 cubits or 75 feet high (in need of four ladders!).
46. The steps listed in the Temple Court in Middot 2.3 are half a cubit (nine inches). Corresponding close to the size of the paneled stepped base of the Arch of Titus Menorah.
47. *Encyclopedia Judaica*, Vol. 11, (Jerusalem: Keter, 1972), 1363, Figure 9.#12.
48. Samaritan synagogue, El-Hirbeh, 4th century, see the listing of the Bet Shearim menorot.
49. *Encyclopedia Judaica*, (Vol. 11), 1357, Figure 9.#2. The bottom seems to be a single (perhaps tiered) base connected directly to the central shaft.
50. *Encyclopedia Judaica*, 1358, Figure 9.#4. also has a single base which connects directly the central shaft and flares out.
51. Rachel HaChlili, *Ancient Jewish Art and Archaeology in the Land of Israel* (Leiden: Brill, 1988): 81, 187, 189, 206–208, 329–330, 332–335, 365–366, 376–377, 393–394.
52. HaChlili, *Ancient Jewish Art and Archaeology in the Land of Israel* 334, [HaChlili, *Ancient Jewish Art* hereafter].
53. HaChlili, *Ancient Jewish Art*, 318, 330–332, 377, 387.
54. HaChlili, *Ancient Jewish Art*, 344.
55. Homer, *The Iliad*, beginning of Book 19.
56. Yadin, *Finds*, 60.
57. Goodenough, *Symbols*, Vol. 5, 4–12.
58. Yadin, *Finds*, 46.

CHAPTER NINE

THE Niche OF Skulls IN THE Cave OF Letters[1]

NICOLAE RODDY
Creighton University, Omaha, NE 68178, USA

INTRODUCTION

On the second day of Yigael Yadin's first season in the Cave of Letters—March 25, 1960, to be exact—a young volunteer excavator by the name of Pinchas Prutzky (later Porat) squeezed his way into the entrance of a narrow niche located near the center of the eastern wall of Hall C in the innermost recesses of the cave.[2] Described by Yadin as "a real maverick,"[3] the novice explorer had stolen away from the rest of the team in Hall C, eager to catch first sight of whatever treasures the cavern might be concealing.

His candle flickering in the blackness of those forbiddingly narrow confines, Prutzky inched his way along the niche,[4] later designated Locus 2 by the recent excavators (see Plate 3), only to stumble upon a horrifying and grisly sight. There among the silent, trembling shadows the young explorer saw an array of human skeletal remains, including a number of skulls piled together in woven baskets stacked one atop the other; the rest of the bones had been laid out lengthwise, wrapped in plaited mats. The macabre spectacle so startled Prutzky that he recoiled from the scene and inadvertently disturbed one of the stacks of baskets.

The following day, Yadin's team recovered additional bones and textile fragments from another niche in Hall C, designated Locus 3. A few days later, just two days before the end of the eleven-day season, additional basket and mat burials were found in Locus 8, a much smaller niche located in Passage BB.[5]

Who were these early inhabitants and what is one to make of their curious burials? The present study revisits the subject of the cave burials, offering fresh contextual analysis of Yadin's findings supplemented by new information gathered from the recent John and Carol Merrill Cave of Letters Expedition (1999–2001), including C-14 analysis of finds recovered from both expeditions. Viewed against the backdrop of current anthropological and archaeological research on burial customs and practices in Palestine during the Roman and early Byzantine periods, the evidence leads one to conclude that 1) the majority of the skulls and skeletons are indeed the remains of Judean rebels of the Bar Kokhba Revolt (133–135 CE) who had sought refuge in the cave; 2) their bodies were laid to rest in secondary burial alongside other skeletons, presumably also Judean, who had died in the cave roughly a half-century earlier; and 3) the burial assemblage, while curious, is neither as unique nor unorthodox as one might suppose. A related question, whether or not the people who survived long enough to arrange the bones in such a curious manner had intended ever to recover them, will also be addressed.

SETTING THE SCENE

The Burials

According to Yadin's report, excavators discovered colorful textiles covering a leather-lined basket that contained the skeletal remains of a child wrapped in a nearly complete tunic at the center of the burial niche at Locus 2. Four baskets were stacked on a large rock at the right-hand side of the niche. Each basket contained four to six skulls, many of which lacked jawbones. At the far right-hand corner of the niche lay a pile of bones, the missing jawbones among them. These and other smaller bones were contained in baskets, but the longer ones had been carefully enveloped in plaited mats and covered in various textiles. Atop the rocks lining the left-hand side of the niche, the excavators found fragments of textiles and tufts of hair. According to the report, upon moving a number of smaller stones near the entrance of the niche excavators found several additional textile fragments resting upon "a fairly thick layer of worm remains,"[6] indicating that at least some primary burials took place in Locus 2. In addition to the bones, textiles, and woven baskets and mats already mentioned, excavators found a few pieces of wood and shards from two cooking pots.

Yadin's team recovered the skulls and bones from the niche and turned them over to Hebrew University where they were analyzed by Dr. H. Nathan. According to Dr. Nathan, the skeletal remains comprised seventeen individuals: three males, 14–40 years in age; eight females, 15–30 years old; and six children, ranging in age from two to ten years of age.[7]

This burial assemblage was not the only one found in the Cave of Letters. Locus 8, a narrow cavity about three meters deep located in passageway BB, yielded a basket containing two skulls, as well as two, presumably corresponding skeletons wrapped in textiles. As in Locus 2, the lower jaws were missing from the skulls, but were found together with the rest of the bones. These remains also were analyzed by Dr. Nathan, who determined them to be those of a male, 40–45 years in age, and a female, roughly 20 years in age.[8] As in the case of Locus 2, "it was possible to discern much worm remains." Besides the bones and textiles, materials retrieved from Locus 8 include a piece of wood, likely a table leg, and the neck of a juglet.

Finally, yet another burial niche was discovered by Yadin at the far end of Hall C. Designated Locus 3, this niche was much smaller in size and showed obvious signs of having already been disturbed, perhaps by human scavengers.[9] In addition to human remains, the niche yielded textile fragments of wool and linen and a single oil lamp. Because the contents of the niche were so disturbed, making the finds more difficult to assess, apart from brief mention here, Locus 3 will be ignored for purposes of the present study.

Other Finds

According to Yadin's report, several of the crevices and cavities of Hall C, the innermost and largest of the chambers, indicated that deliberate attempts had been made to conceal their openings with rocks. Dismantling the rocks at the opening of Locus 7, located at the northern end of Hall C, Yadin's explorers discovered the waterskin bag that contained the famous Bar Kokhba correspondence, along with various other items. Also found in Hall C were the papers and possessions of Babatha bat Shimon; the so-called "Cache of the Net," which included two cooking pots and a Bar Kokhba coin; the "Cache of the Six Keys"; and very near the burial niche (Locus 2), the "Cache of Glass Vessels."

The burial niche itself yielded two fragments of a wooden bowl that were similar, if not identical, to two wooden bowls retrieved from Babatha's cache. The bowls were made of lathe-turned acacia wood and described as "shallow, with flattened, outsplaying rim."[10] In addition to this, woven baskets and mats recovered from both Babatha's possessions and the famous "Horde of Bronze Vessels," were of the same type found concealed in the burial niches. Finally, the textiles found wrapped in waterskins and containing women's articles (Loci 7 and 61) are "from the same period" and "identical in types" as those found in all three burial niches.[11] Finally, in Locus 2, some of the same textiles were found enmeshed with a thick layer of worm remains and lying under a rock, attesting to at least some primary burials among the Second Revolt inhabitants of the cave. The intact skeletal remains of a child wrapped in a tunic, laid in a leather basket, and placed in the center of the niche almost certainly belonged to this community as well.[12]

In sum, the material record indicates that the types of artifacts found intermingled with the human remains in the primary and secondary burials of Loci 2 and 3 are identical to the types found in association with the datable epigraphical materials found elsewhere in the cave, thus directly linking at least some of the humans laid to rest in the niches to the rebels and refugees of the Bar Kokhba revolt. On the basis of Yadin's report alone, there is insufficient evidence to identify all of the bodies as belonging to this group. The report indicates that the skulls and bones in baskets and mats appear to have been deposited at the same time, however Yadin was not certain about the burials on the niche floor.[13] Since that time, Carbon-14 analysis of bones and artifacts carried out in connection with the John and Carol Merrill Cave of Letters Expedition (1999–2001) yielded some intriguing and unexpected results. Before turning to these findings and drawing final conclusions about the curious burials, it will be helpful to delve briefly into the growing body of knowledge about Jewish burial customs and attitudes towards death in Roman-period Palestine to establish a social context for assessing the data.

JEWISH BURIAL PRACTICES AND ATTITUDES TOWARD DEATH IN PALESTINE DURING THE EARLY ROMAN PERIOD (100 BCE–100 CE)

R. Hachlili observes that research into burial practices and the material remains associated with them "is effective in reconstructing the history of a society, its religious beliefs and its social outlook."[14] Moreover, ethnographic research demonstrates that social attitudes toward death, co-arising with the rituals and that reflect and reinforce them, are among the most stable pillars of society; thus it will be worthwhile to explore rabbinic literature as a source of insight into Jewish burial practices and attitudes toward death in Roman Palestine. While the pitfalls associated with historical analysis of rabbinical literature are well known, the stability of social attitudes and practices regarding death permit the assumption that rabbinic literature's insights into pre-Mishnaic attitudes and practices concerning death can largely be trusted.[15]

Although burial practices shift only gradually over time, innovation and change may occur as a result of social disruptions caused by war resulting in occupation or exile, widespread natural disaster, or the lure of pervasive foreign culture. Despite these outward changes, a singular attitude toward death remains; namely, that even under normal circumstances, contact with the dead raises great societal anxiety in light of the dangers associated with contracting social or ritual impurity. This is especially true for Judaism traditionally, where under Jewish law contact with a corpse is affirmed to be the supreme cause of impurity (avi avot ha-tuma); thus it should come as no surprise that rabbinic literature has a great deal to say about Jewish burial practices and the social and religious considerations that govern them.[16]

Death and Burial under Ordinary Circumstances

Supplementing, and often challenging, information gleaned from rabbinic sources, a growing body of archaeological data and secondary studies dealing with death and burial in Roman Palestine has emerged over the past four decades.[17] As a result, one finds that rabbinic precepts are not always supported by facts on (or under) the ground. For example, rabbinic dictum that a grave (kokh) be four cubits long, seven spans high, and six spans wide seems hardly more than an instructive ideal seldom played out in the material record, as Judean burial arrangements appear randomly appointed.[18] A proper understanding of actual Jewish burial practices depends upon archaeology, for as R. Simeon ben Gamliel is said to have observed, everything depends on the nature of the rock (M. BB 6.8).

The material record shows that Jewish tombs in Roman Palestine are generally small, family-oriented, often multi-chambered caves cut into natural rock and stone, a practice in evidence in Jerusalem since the IA II period.[19] To date, nearly 1,000 such tombs have been discovered in and around Jerusalem. Rock-cut tombs newly hewn or renovated during the latter half of the Second Temple Period show an array of Hellenistic design and motifs. Carved niches (kokhim) in the tombs serve as final resting places for the deceased. These are of two basic types: the loculus, which is a shaft hewn perpendicular to the walls of the tomb chamber, and the arcosolium, an arch-shaped cavity carved along the walls of the chamber.[20] One may easily imagine that loculi would be more difficult to cut than arcosolia-type niches in requiring greater labor confinements for the removal of stone. However, loculus design offers greater spatial economy, for anywhere from six to nine loculi could be cut into the face of the rock in about the same space required to accommodate just three arcosolia.[21]

On the basis of the archaeological record, it would appear that the use of the rock-cut tomb was the most common Jewish burial practice in Roman Palestine; but one must keep in mind that this form of burial simply happens to be the most durable and not necessarily the only means. The simple trench grave, once thought to be uniquely Essene, or at least Qumranic, was employed widely throughout Judea, with simple demographic analysis filling the gap left by the lack of surviving hard evidence.[22] Jodi Magness, along with others, asserts that "only the wealthier members of Jewish society could afford rock-cut tombs," and that, "the poorer classes were buried in simple individual trench graves dug into the ground."[23] While this statement holds true as a general rule, the only real requirement is access to rock and stone in a family place designated for burial.

Another widespread funerary phenomenon of first-century Judea/Palestine, found in association with the rock-cut tombs, is the practice of ossilegium, or the gathering of human bones (Heb., liqqut asamot) for secondary burial in a stone container, or ossuary (Heb., aron; Gr., soros).[24] These receptacles are usually made of chalk, limestone, or wood, and most, if not all, supported lids.[25] Ossuaries may be decorated or plain and sometimes bear the name of the deceased, often crudely inscribed, suggesting a purpose more functional than decorative. Perhaps the best example of this is the tomb of the priestly family Caiaphas, discovered by Zvi Greenhut in December, 1990, which yielded a number of relatively modest ossuaries—not at all what one might expect from such an important and wealthy priestly family.[26]

The use of ossuaries in Judea appears to have begun during the late first century, no earlier than 20 BCE.[27] The practice appears to have died out by the beginning of the Byzantine period, with a slight revival in the third century. Rahmani attributed ossilegium to the rise of Pharisaic belief in individual bodily resurrection of the dead, for in addition to providing a repository for decomposition of the body and the expiation of sins, the bone box would ostensibly keep one's bones together and prevent them from mingling with the bones of others, thus preserving them neatly until the time the deceased's body is resurrected.[28] S. Fine and others have challenged this assertion on a number of points, which Magness concisely summarizes.[29] First, ossuaries often contain the bones of more than one individual; for example, the famous Caiaphas ossuary contained the remains of six individuals. Second, ossuaries sometimes contain incomplete skeletons. Third, rock-cut tombs containing ossuaries also yield skeletons deposited in loculi, pits, or charnel rooms. Fourth, the use of ossuaries all but disappears by the end of the first century C.E., only temporarily reviving itself in the third century. Finally, Magness observes, the greatest difficulty lies in the fact that sources for the period

> …associate the belief in individual, physical resurrection of the dead with the Pharisees (see, e.g., Josephus, *Ant.* 18.1.3). These same sources tell us that the Sadducees rejected this concept (Josephus, *Ant.* 18.1.4; Matt. 22:23). But there is no doubt that the monumental rock-cut tombs with ossuaries belonged to Jerusalem's elite, many of whom were Sadducees. In fact, some of these tombs and ossuaries belonged to high priestly families, such as the tomb of Bene Hezir and the tomb and ossuaries of the Caiaphas family. Ossuaries were used by the same members of Jerusalem society who rejected the concept of individual, physical resurrection of the dead.[30]

So to what is the appearance of ossuaries to be attributed? Magness and others have suggested that the widespread use among Romans of cineraria, or lidded boxes that preserved the ashes of cremated family members influenced the use of ossuaries. Others, Hachlili among them, downplay the notion of foreign influence in the use of ossuaries, citing significant differences in size and function, chiefly asserting that the cineraria hold ashes as opposed to bones. Magness disputes this assertion on basis of the fact that cineraria identical in shape and size have been found throughout the Roman orbit, including Rhodes and Asia Minor. Magness also witnesses to having viewed unpublished cineraria in the Archaeological Museum in Afyon, in west central Anatolia, of which she writes:

> Aside from the fact that they contain cremations, the Afyon urns are virtually identical to the plain Jerusalem ossuaries: the same size, shape, and with the same kind of lids.[31]

Similarities in decoration and inscription, along with parallels in later developments in Roman burial practices as reflected at Beth She'arim and elsewhere, argue strongly for accepting that the appearance of ossuaries in first-century BCE Palestine is not tied to the rise in the belief in resurrection, but likely the result of Roman cultural influences adopted by the Judean elite and filtering down to lower social levels.[32] The fact that bones instead of ashes are contained within them reflects cultural differences between Judeans and Romans concerning proper disposition of the dead; simply stated, Romans cremated, while Jews traditionally did not. In any event, the use of ossuaries all but disappears after the destruction of the Second Temple and dies out altogether by the end of the third century.[33] As Magness observes:

> If the use of ossuaries was connected with the concept of the individual, physical resurrection of the dead, they should have become even more popular after 70 C.E., when this belief became normative in Judaism.[34]

In practice, the use of ossuaries usually involved a family ceremony that marked the end of a one-year period of mourning.[35] According to Mishnaic sources, the process of decomposition over a twelve-month period was believed to purify the deceased from his or her sins and prepare the departed for resurrection.[36] Although the actual time it takes for a body to decompose depends upon such factors as temperature, moisture in the air or ground, and exposure to certain bacteria, insects, or larger predators, in most cases a year would allow sufficient time for decomposition to have occurred.

The rabbinic tractate Ebel rabbati (Major Mournings), which likely dates to the third century CE[37] and is customarily referred to by its euphemistic title Semahot (Rejoicings), reflects Jewish burial practice and related purity concerns in recounting the passing of Rabbi Eleazar bar Zadok's father:

> Rabbi Eleazar bar Zadok said: "Thus spoke father at the time of his death. 'My son, bury me at first in a fosse [i.e., a burial niche]. In the course of time, collect my bones and put them in an ossuary, but do not gather them with your own hands.' And thus did I attend him: Johanan entered, collected the bones, and spread a sheet over them. I then came in, rent my clothes for them, and sprinkled dry herbs over them. Just as he attended his father, so I attended him."[38]

In sum, the use of caves for burial has an extraordinarily long history in this region, but the use of ossuaries in Roman-period Judea appears to have been hardly more than a passing but widely popular cultural trend based on an imported technology. The evidence is sufficient to conclude that the use of ossuaries occurred irrespective of the diversity of religious and social concerns of the late Second Temple period. Except for a brief resurgence in the third century, the practice of ossilegium practically disappears sometime around the time of the Temple's destruction in 70 CE.

Death and Burial under Extraordinary Circumstances

The materials cited above deal with death and burial under stable social circumstances, but how might these norms have played out in situations like insurrection and war? Certainly these disruptive events created great anxiety for those directly affected; but because the ancestral tomb or burial ground is the proper place for family members to be laid to rest, the loss of sons and brothers to slavery or death on the battlefield would no doubt have generated additional anxiety. Survivors of rebels killed in combat, or refugees dying under long-term siege, would necessarily be forced to martial whatever resources they had at their disposal to deal with the crisis, at the expense of a minimum of compromise to the demands of custom or conviction. The rebels and refugees hiding in the Cave of Letters were no exception, forced as they were to strike a balance between culturally-reinforced attitudes towards death and the forbidding reality of a desperate situation.[39]

Obviously, the first concern in times of insurrection or war would be to find an acceptable burial place, one that would not imperil the lives or ritual well-being of the survivors. According to rabbinic sources, burial on a battlefield was not at all inappropriate, for transporting the bodies of the fallen would expose survivors to increased risk that might lead to the shedding of one's own blood. The dilemma arising from the conflict of crisis and custom is reflected in an insightful anecdote attributed to Rabbi Akiba, an early second-century contemporary of Bar Kokhba:

> This is how I began attending upon the Sages: Early one morning I arose and found a slain man [*met mitzvah,* or unattended corpse]. I carried him a distance of three Sabbath limits, until I brought him to a burial place and interred him. When I came and excitedly declared the matter to the Sages, they said to me: "Every single step that you took is accounted to you as if you had shed blood." I then applied to myself the principle of *a minori ad majus*: Having thought to do good, I sinned; had I thought to do less—how much more so![40]

Rabbinic instruction not only approves of burial on a distant battlefield, but also establishes a hierarchy of appropriate places for interring an unattended corpse (met mitzvah). For example, an uncultivated field is thought to be more suitable than a plowed field; a plowed field is deemed better than a sown field; a sown field is preferred over an orchard; an orchard is better than a vineyard; and so on.[41] Although there are no explicit rabbinic references concerning burials in a (non-family) cave per se, it is clear that under Roman siege, with military encampments on both sides of Naḥal Ḥever overlooking the Cave of Letters, the burial niches would not only have served their obvious practical function, but would have been considered rabbinically appropriate final resting places for Jewish corpses, even satisfying the purity law that requires that burials be removed at least fifty cubits (roughly 75 feet) from a habitation.[42] The situation in the Cave of Letters is complicated by the fact that the skulls were found separated from the bodies and gathered together in baskets, their jawbones missing and long bones wrapped up in mats.

The way the bones were gathered in the Cave of Letters may lead to speculation that survivors did not intend this to be a permanent inhumation. Under normal circumstances, once a body finally has been buried it is not appropriate to disturb it; but is it possible that the cave's inhabitants entertained a secondary-level concern for transporting the skeletal remains at some future time? According to rabbinic sources survivors who are able to recover Jewish corpses and transport them to their respective family tombs honor the deceased and his or her family by doing so:

> Neither a corpse nor the bones of a corpse may be transported from a wretched place to an honored place, nor, needless to say, from an honored place to a wretched place; but if to the family tomb, even from an honored place to a wretched place, it is permitted, for by this he is honored (Semahot 13.7)

Although relocation appears to have been legally permitted under certain circumstances, apart from third-century Beth She'arim and a few singular examples there is no real evidence that transportation of remains was practiced. The legal considerations that governed the transportation of human remains are as follows. First, according to rabbinic sources it is not permissible to relocate a corpse unless the mover is certain that the grave is a temporary one (Semahot 13.5). One cannot know for certain whether or not survivors in the Cave of Letters ever considered the niche to be a final resting place. Norms of both simple practicality and social sanctioning allowed this to be the case, but why the curious arrangement of skulls and bones?

Second, it is not permitted for the bones of a corpse to be forcibly taken apart, however the bones may be allowed to separate through the natural process of decomposition (12.6), a process generally thought to take a full year.[43] Given the environmental variables that govern the process of natural decomposition, it is not possible to tell from the reports whether any of the jawbones were "assisted" in the dislocation process.

Third, skeletal remains, like sacred scrolls, may be transported from place to place in a container, but they must never be allowed to rattle about loosely (Semahot 13.2). Under normal circumstances, intact jawbones simply could have been tied to prevent the mouth from falling open; thus one is still at a loss to determine the thoughts and intents of the last surviving inhabitants of the cave.[44]

Finally, although one imagines that anyone planning to remove the skeletal remains from the cave would want to recover all of the bones, it may have been apparent to the survivors that conditions would permit only certain parts of the skeletons to be removed at a time. Ideally, an entire body would be offered up for *ossilegium*, however, in cases where parts of the corpse are missing, it is essential that at least the spinal column or, better, the skull be counted among the remains. The period of mourning for survivors of a victim who had drowned or was devoured by a wild beast, for example, would not begin until the "greater part of the corpse"—

that is, until the spinal column or skull—had been recovered (12.10). Neither was it permitted that a body be carried on a bier unless the spinal column or skull was intact (12.13). Thus, practically speaking, any future removal of the bones from the cave would likely entail the lowering of the baskets of skulls first, perhaps by ropes tied to the basket handles, then followed, if possible, by removal of the long bones wrapped and in mats and secured with ropes.

On the basis of both evidence and reason, one may conclude that the rules governing burials in extraordinary circumstances were influenced by the same convictions that inform ordinary situations, namely the preservation of life and its sanctity, concerns for the maintenance of ritual purity, and the ideal importance of the family burial ground. In extraordinarily dire and urgent situations such as we find in the Cave of Letters, practical considerations govern these convictions in descending priority order. Burial in the family plot would be given up in favor of maintaining ritual purity; purity concerns, although closely related, would be secondary to the preservation of life and respect for the dead.

Carbon 14 Results

Radiometric analysis of organic materials such as wood or bone seldom produces the precise results investigators would like. As an indicator of age, the decay of the Carbon 14 isotope is both constant and measurable; however, any number of environmental factors can affect the process of analysis, resulting in a wide range of possible dates. Imprecision in C-14 dating led Magen Broshi, former curator of the Dead Sea Scrolls, to measure radiocarbon analysis against some of the dated epigraphical materials retrieved from Bar Kokhba caves in order to find out which aspect of the range of dates came closest to the actual date.[45] While the range of dates can be quite wide—as much as two centuries—Broshi was able to demonstrate on the basis of dated organic materials that the actual age is not found near the mean of the range, but almost always very near the older extreme. Needless to say, this discovery proved extremely valuable for dating materials recovered from Qumran, but when Broshi's research is applied to finds recovered from the Cave of Letters, an interesting pattern emerges. All dates are according to 2sigma from Elizabeth Boaretto, Weizman Institute, 2002 (see Chapter 7 in this volume). Also please note that the calibrated dates in Figure 9.1 in this chapter are slightly different than some of those presented in Chapter 7. The dates used in this chapter we based upon a slightly earlier interpretation.

The earliest samples from the Cave of Letters were taken from pieces of firewood found in an oven pit under a layer of collapsed roof. A sample taken from a piece of finished wood, possibly a table leg, was calibrated at 20 BCE–130 CE (RT 3916). Two other samples (RTT 3917 and RTT 3918) date even earlier, to

100 BCE–90 CE and 60 BCE–80 CE respectively.[46] The data strongly suggests that the oven was in use before, and perhaps even during the First Jewish Revolt.

In addition to these materials, the earliest bone fragments recovered from the burial niche at Locus 2 (RTT 3726–1.1 and 3726–1.2) were calibrated to 20–250 CE and 50–340 CE respectively, roughly contemporary with the use of the oven. Two other bone samples recovered from the burial niche (RTT 3800 and 3726) were calibrated to 120–390 CE and 130–350 CE, witnessing to a later habitation. Finally, a piece of textile from the niche was also tested and calibrated at 80–330 CE (RTT 3862).

In sum, radiocarbon analysis of specimens taken from bones and related materials recovered from the burial niche at Locus 2, represent a wide range of dates that appear to polarize into earlier and later habitations. At least two occupations lasted long enough to witness the arrival of death among their respective communities, necessitating proper disposal at least on the part of the latter group. The fact that material remains associated with the dated epigraphical materials are found intermingled with the human remains from the latter period habitation not only adds support to the conclusions of Broshi's research, but suggests that the later inhabitants of the cave buried their own dead in the same places as the earlier inhabitants Since the skulls and long bones found by Yadin are not available for testing (they were reburied in the 1980s in another location) it is difficult to say whether or not the baskets containing these bones were from the earlier period. Certainly nearby residents, especially those from En Gedi, were already well aware of the protection the caves offered in times of distress. In any event, whether the wide spectrum of occupation was continuous or intermittent, its continuity as a place of refuge extends to witnesses of the First Jewish War (66–70 CE), as well as the rebels and refugees of the Bar Kokhba rebellion (133–135 CE).

CONCLUSION

The material record of textiles and utensils found in the burial niches—in many cases intermingled with the human remains—indicates a primary relationship with similar items associated with the epigraphical evidence in the main hall of the cave. However, C-14 dating of organic materials provides a wide range of dates, which Broshi's Law narrows to at least two occupations within a hundred-year period, overlaying both the first and second Judean uprisings. It is reasonable to conclude that skeletal remains found in Locus 2 of the Cave of Letters were gathered by second-century Jewish rebels and refugees of the Bar Kokhba revolt, and that most of the bodies were members of their own community.

But what of the curious burial arrangement? Other unusual death and burial assemblages have been found at nearby sites, the nearest located just across Naḥal Ḥever , in the so-called Cave of Horrors excavated by Yohanan Aharoni in the early

1960s.[47] Named for its own grisly yield of human remains, the macabre assemblage of human remains led its discoverers to conclude that the cave's inhabitants, believed also to be refugees from the Second Jewish Revolt, had died painfully of dehydration and starvation with no one to bury them. Cave 32, designated by Aharoni as the "Cave of Skulls," yielded a secondary burial assemblage containing seven skulls piled together, with other bones gathered alongside.

At nearby Nahal David, six small caves were excavated by N. Avigad, also in the early 1960s.[48] The caves yielded both primary and secondary burials dated to the first century BCE. Among the burials several wooden coffins were found, one of which was found to contain seven skulls (Cave 4). Last but not least, nine tombs uncovered at 'En Gedi, all badly eroded, indicated a variety of burial practices dating around the cusp of the Common Era.[49]

The curious arrangement of burials in the Cave of Letters invites speculation that survivors of the siege might later return to recover the remains, along with the documents and household vessels hidden elsewhere throughout the cave. Woven mats and baskets would certainly offer the most practical means for temporary burial and later removal, especially given the treacherous adit and exit of the cave. The jawbones, which would normally be tied to prevent the mouth from falling open, could have been forcefully removed and placed with the rest of the bones despite rabbinic protest against it. However, quite to the contrary, there is simply no reason to maintain that a future retrieval of bodies had been planned.

The rebels and refugees of the Bar Kokhba revolt were forced to strike a balance between well-established, second-century Jewish beliefs and customs and the crisis of a horrific new situation; but practical considerations aside, even from a rabbinic legal perspective, the burials were appropriate exactly as they had been arranged, with baskets and textiles serving as receptacles for efficient secondary burial. How much more would the burials have been appropriate from a practical perspective in light of the inhabitants' constraints of space and the fear and stress of their unfortunate and apparently lengthy siege?

It may be concluded that the burial arrangement in the Cave of Letters was a practical one based on the principle of spatial economy and respect for the deceased, with no identifiable theological or legal concern in evidence. One must carefully avoid retrojecting imagined cultural or social scenarios based on idealized sources and personal expectations on Judean beliefs and practices during the period of Roman occupation. Although the burials in the Cave of Letters may strike most people as strange, perhaps even shocking, one finds that such assemblages are not unique to the Judean desert region of the early centuries CE—or other periods, for that matter! The archaeological record shows that skulls have been gathered together for various reasons in the Judean desert region at every age: Neolithic and Chalcolithic, Bronze and Iron, Hellenistic and Roman; and even in Byzantine and Orthodox monasteries into modern times.[50]

In the end, the fate of those who arranged the Cave of Letters burials cannot be known, let alone being able to discover their real intentions in doing so. Perhaps the last refugees starved to death, piteously and painfully huddled in some dark corner of the cave, their remains dragged away by predators. Perhaps they surrendered or were captured by watchful Roman soldiers encamped some three hundred feet overhead, atop both side of the wadi. The only thing that is truly certain is that none among them ever returned to retrieve the items of great value they left behind.

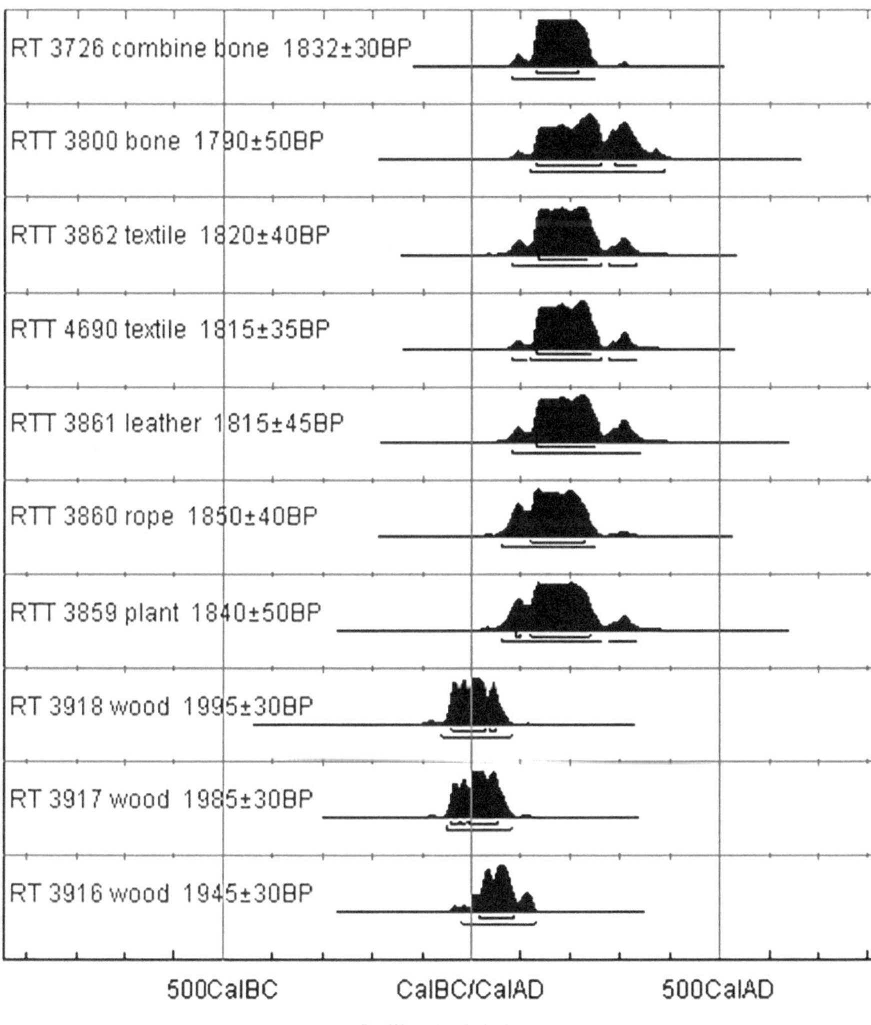

Figure 9.1: C-14 Results, John and Carol Merrill Cave of Letters 2000 Expedition.

NOTES

1. The preliminary research for the present article was presented at the International Meeting of the Society of Biblical Literature, hosted by the Pontifical Biblical Institute and Gregorian Institute in Rome, Italy, July 8–12, 2001. In addition to the incorporation of additional data, I wish to acknowledge insights gleaned through personal conversations and/or e-mail correspondence with Gideon Avni, Richard Freund, Rachel Hachlili, Jodi Magness, Lawrence Schiffman, and Joseph Zias, none of whom shall share blame for any faults contained herein.
2. Yigael Yadin, *The Finds from the Bar Kokhba Period in the Cave of Letters* (Jerusalem: Israel Exploration Society, 1963), 18. A recent interview with Prutzky, who long ago had changed his name to Porat, was conducted by the PBS Nova series just before his death in 2002, and is available at http://www.pbs.org/wgbh/nova/scrolls/porat.html.
3. Yadin, Bar Kokhba, *The Rediscovery of the Legendary Hero of the Last Jewish Revolt against Imperial Rome* (London: Weidenfeld and Nicolson, 1971), 60.
4. The entrance to the niche is extremely narrow, roughly half a meter wide. The crevice itself slopes downward about twelve meters in length and at some points is very constrictive vertically and horizontally.
5. These latter finds were left in situ until the following season. The remains eventually would be buried, along with skeletal remains recovered from the Cave of Horrors, in an Israeli state-sponsored ceremony held near Naḥal Ḥever on May 11, 1982, on Lag B'Omer, a Jewish holiday with roots in rabbinic traditions surrounding R. Akiba and Simon bar Kokhba.
6. Yadin, *Finds from the Bar Kokhba Period*, 31–33.
7. Yadin, *Finds from the Bar Kokhba Period*, 33–34.
8. Yadin, *Finds from the Bar Kokhba Period*, 35–36.
9. Some of the bones from Locus 3 found scattered throughout Hall C had been picked up by Aharoni's previous expedition in 1953.
10. Yadin, *Finds from the Bar Kokhba Period*, 125–29.
11. Yadin, *Finds from the Bar Kokhba Period*, 171.
12. The placement of the child in the center of the burial space is curious. Primary burials took place on rock ledges along the sides of the niche, while secondary burials appear also to have been arranged with an eye toward spatial economy, placed as they were along the sides and in corners. In the absence of additional clues, it is reasonable to imagine that the child was the last of the refugees to be placed in the niche and, on the basis of the child's adornment and placement, that this was a primary burial occurring towards the end of the group's habitation.
13. Yadin, *Finds from the Bar Kokhba Period*, 32.
14. R. Hachlili, *Jewish Funerary Customs, Practices and Rites in the Second Temple Period* (Leiden: E. J. Brill, 2005), xxxv.
15. For fuller treatment of attitudes towards death in rabbinic literature, see D. Kraemer, *The Meanings of Death in Rabbinic Judaism* (New York: Routledge, 2000). Unfortunately, Kraemer fails to treat the rabbinic tractate *Ebel rabbati*, euphemistically referred to as *Semahot*, in his otherwise thorough study. The tractate is dated by most to the third century, but Kraemer regards it as post-Talmudic, outside the scope of his study.
16. See A. Steinsaltz, *The Essential Talmud* (New York, 1976), esp. ch. 23, "Ritual Purity and Impurity," 193–98.
17. Earlier studies include Saul Lieberman, "Some Aspects of Afterlife in Early Rabbinic Literature," in *Harry A. Wolfson Jubilee Volume* (Jerusalem: *JAAR*, 1965): 495–532, and Eric M. Meyers, *Jewish*

Ossuaries: Reburial and Rebirth (Rome: Pontifical Institute, 1971). For more recent studies, see L. Y. Rahmani, "Ancient Jerusalem's Funerary Customs and Tombs," in *Biblical Archaeologist* 44–45 (1981–82); P. Figueras, *Decorated Jewish Ossuaries* (Leiden: E. J. Brill, 1988); R. Hachlili, *Ancient Jewish Art and Architecture in the Land of Israel* (Leiden: E. J. Brill, 1988).

18. M. BB 6.8, Tos. BB 6.22; BT BB 100b-101a, cited in Hachlili, *Jewish Funerary Customs*, 56–57.
19. See Elizabeth Bloch-Smith, *Judahite Burial Practices and Beliefs about the Dead* (Sheffield: Sheffield Academic Press, 1992); A. Kloner, "Burial Caves from the First Temple Period in Jerusalem." *Avraham Biran Volume* (Jerusalem: Israel Exploration Society, 1992) 241–46. Most of the data for burial practices comes from the two main cemeteries at Jerusalem and Jericho, excavated most extensively by R. Hachlili and A. E. Killebrew (see Bibliography), as well as from smaller cemeteries around Jerusalem, near 'En Gedi (including Qumran), and in the Galilee.
20. For additional description and terminology associated with Roman-period Jewish burial assemblages, see Hachlili, *Jewish Funerary Customs*, xxxvii–xl.
21. Hachlili, *Jewish Funerary Customs*, 69.
22. See Jodi Magness, "Ossuaries and the Burials of Jesus and James," *Journal of Biblical Literature* 124.1 (2005): 121–54.
23. Magness, "Ossuaries and the Burials of Jesus and James," 121–54.
24. See L. Y. Rahmani, "Ancient Jerusalem's Funerary Customs and Tombs: Part Four," *Biblical Archaeologist* 45.2 (1982), 109–119; *A Catalogue of Ossuaries in the Collection of the State of Israel*, Jerusalem 1994. See also R. Hachlili and A. Killebrew. Jericho: *The Jewish Cemetery of the Second Temple Period*, IAA Reports 2, Jerusalem, 1997.
25. Stone sarcophagi are also included in this category, although it is not known whether these were intended for primary or secondary burials. Hachlili suggests that sarcophagi probably functioned as large ossuaries for secondary burial, but that they could also have served as primary burial for a significant person, as some sarcophagi have headrests. See Hachlili, *Jewish Funerary Customs*, 115.
26. Magness, "Burials," 130.
27. Rahmani, *A Catalogue of Jewish Ossuaries*, 21.
28. Rahmani, "Ancient Jerusalem's Funerary Customs and Tombs, Part One," *Biblical Archaeologist* 44 (1981): 175–76.
29. S. Fine, "A Note on Ossuary Burial and Resurrection of the Dead in First-Century Jerusalem," in *Journal of Jewish Studies* 51.1 (2002) 69–76. Magness, "Burials," 131–32.
30. Magness, "Burials," 132–33, 136.
31. Magness, "Burials," 134
32. G. Foerster and Lee Levine, seconded by J. Magness, "Burials," 133–34.
33. Magness, "Burials," 136.
34. Magness, "Burials," 136.
35. Thus the modern Jewish observance of marking a one-year anniversary of one's death, known since the western medieval period as Yahrzeit, has its roots in this Mishnaic practice.
36. M. Sanhedrin 6: 5–6. McCane, *Jews, Christians, and Burials*, 73–75.
37. Zlotnick, *The Tractate "Mourning,"* 9.
38. *Semahot* XII. 9, D. Zlotnick, trans. *The Tractate "Mourning"* (New Haven: Yale University, 1966) 82. In a statement attributed to R. Akiba, "The bones may not be gathered until the flesh has wasted away" (*Semahot* 12.7).

39. While it is common among scholars to refer to the participants in the First and Second Jewish Revolts as rebels and revolutionaries, the nature of the skeletal and remains and related artifacts compels us to view the at least some of the inhabitants of the cave as refugees.
40. *Semahot* 3.19; Zlotnick, 21–22.
41. *Semahot* 4.18. According to *Semahot* 8.9, when Rabbi Judah ben Betera and Rabbi Hanina ben Tardion learned that Rabbi Akiba had been executed in Caesarea, they took this as a sign that very soon "no place in the Land of Israel will be found where bodies of the slain have not been cast" (8.9).
42. M BB 2.9, see Hachili, *Jewish Funerary Customs*, 15–16.
43. In a statement attributed to Rabbi Akiba, "The bones may not be gathered until the flesh has wasted away" (*Semahot* 12.7).
44. Securing the jawbone in this manner is prescribed by rabbinic law (b. Shabbat 23.5) and is still practiced in some areas of East Europe, although sometimes for varying reasons; for example, I know from personal experience that rural Christians in the Transylvanian plateau tie the chin with a white cloth in order to prevent the deceased from talking. I suspect that entrenched popular custom necessitated rabbinic sanctioning of this ancient practice.
45. M. Broshi, "The New Radiocarbon Dates of the Dead Sea Scrolls and Their Significance," in *The Practical Impact of Science on Near Eastern and Aegean Archaeology* (London: Archetype Press, 1999) 108. See also Freund, *Secrets of the Cave of Letters*, 226–30.
46. C-14 tests were conducted under the supervision of Dr. Elisabetta Boaretto, at the Weizmann Institute of Science. I received preliminary results from Richard Freund, University of Hartford, through e-mail correspondence dated June 6, 2001, with a current update received on Nov. 16, 2006.
47. Y. Aharoni, "Expedition B: The Cave of Horrors," *Is rael Exploration Journal* 12.4 (1962) 186–99.
48. N. Avigad, "Expedition A," *Israel Exploration Journal* 12.3–4 (1962) 169–83.
49. Hachlili, *Jewish Funerary Customs*, 11.
50. See J. Zias, "Three Trephinated Skulls from Jericho," *Bulletin of the American Schools of Oriental Research* 246 (1982) 55–58; Baggati, Bellarmino. "Scavo di Monastero al Dominus Flevit," *Liber Annuus* 4 (1955–56): 240–70.

CHAPTER TEN

A First Century Occupation in the Cave of Letters?

Results from the B/C Passage

CARL E. SAVAGE
Drew University, Madison, NJ 07940, USA

INTRODUCTION[1]

The archaeological and geological survey and exploration of the Cave of Letters (reference numbers: 18270–09320, 18300–09405) upon which this study is based took place during July 3–20, 2000, nearly 50 years after the initial survey of Aharoni and Yadin's larger exploratory effort.[2] It was the second expedition carried out in the cave under the auspices of the Consortium for the Bethsaida Excavations Project which was led by the University of Nebraska at Omaha and the University of Hartford. The project itself was funded by a generous grant from John and Carol Merrill. This second expedition was devoted to several tasks that were tested preliminarily during the 1999 survey of the cave by the consortium, and this report focuses on one of the primary research goals. Because of the initial success of the ground penetrating radar [GPR] survey that was carried out in Hall B during the 1999 expedition to the cave, it was determined that GPR and electro-resistivity tomography [ERT] could be used to discover in the cave substructure where habitation layers might reside under the considerable debris fall that covers the floor throughout the cave.[3]

Figure 10.1 shows a printout of one of the GPR transect lines run in the B/C passageway. The dark black lines in the figure at about the 1 to 1.5-meter level correspond to a possible habitation layer. One can see on the right side of the figure a significantly disturbed area that may indicate possible artifactual remains. Figure 6.7 displays the ERT data from the Hall B to Hall C data line. The B/C passage (Hall C Connection) is marked just left of center in the illustration. The dotted line in the figure marks the boundary of a possible habitation layer. These two non-invasive survey techniques agree that there could be an occupation layer in this region of the cave. This determination would aid in identifying areas where "traditional" archaeology might be of service to investigate one of the research questions at the heart of this re-investigation of the cave: Was there a first-century CE habitation in the cave?

Yadin had already observed that the corridor between Halls B and C, the B/C passageway in my terminology,[4] was "covered by a thick occupational layer."[5] He records that he found many household utensils in this area that he designated as Locus 63 (see Figure 4.9 for the locus locations). His list of finds included a metal tool, some basket fragments, and some wooden fragments, but no pottery.[6] However, he states in his narrative diary that there was found by the team working in the B/C passageway "a large quantity of sherds."[7] Earlier Aharoni had documented that the same area appeared to have been partly dug by the Bedouin prior to his own survey. Aharoni indicates that he found "only sherds, cloth, wood, etc., and nothing else."[8] Of course, these were now the very items we were interested in recovering!

Yadin designated two other loci in the B/C passageway. Locus 4 was "an extremely long and narrow crevice between Halls B and C." It was filled with earth with some fragmentary finds that included two rim fragments of storage jars, listed in his finds catalog as 4.5 and 4.6.[9] The last of Yadin's B/C passageway loci was Locus 67 found at the mouth of the corridor in Hall C. He states that this locus "comprises one unit together with Locus 63," which I take to mean that he did not think there was an architectural or stratigraphical reason to distinguish between the two, but he simply differentiated between the two on the basis of the work in each area having been conducted on different days. He records that the finds from the two loci were packed together.[10]

The GPR data and ERT data collected by our investigations has confirmed what both Yadin and Ahorni had determined, the B/C passageway is an area of the cave where an occupation level was likely.[11] The "thickness" of it was also confirmed. The passage is an approximately 7.5 × 17.5-meter area that is relatively debris free that became an area of prime interest for our attempt to recover dateable artifacts that could be used to test the first century CE occupation hypothesis. The key then was to ensure that we were examining a portion of the passageway that had not previously been disturbed by the activities of prior investigators or looters.

Evidence was found of some "digging" in the B/C passageway situated nearly into the opening into Hall C on its western side. We determined that we would locate our sifting at this location since it had already been disturbed. This location was on the opposite side of the opening from where Yadin indicated his Locus 67 was. Presumably the west side disturbance may have been one such area that Aharoni had alluded to when he stated that he had found evidence of Bedouin activity or that he had dug in the corridor between Halls B and C, but since he only refers to that activity in a generic fashion—i.e., he states that they had dug everywhere possible—the origin of this disturbance could not be fully ascertained. Likewise, we were able to pinpoint Yadin's Locus 4, the crevice, on the eastern wall of the B/C passageway very near the opening into Hall B. With these two disturbed areas located; we looked for an area more toward the center of the passageway as a place that would allow us to systematically explore an occupational level or levels below the present surface that was not covered by rock fall.

It was determined by visual examination of the B/C passageway that proceeding from the east side of the corridor at coordinates BB-CC 33,[12] which location was determined to be just west of Yadin's Locus 4, the surface appeared somewhat uniform, clear of obvious signs of disturbance and without much roof fall debris. Locus 6 was begun there as a 2.5 × 2.5-meter square that was later extended across the passageway to the west. From Locus 6 were recovered over 400 pottery fragments, 20 pieces of rope, 17 wooden fragments, a tool, 3 fragments of reed basket, an animal horn, numerous textile fragments, 2 leather pouch fragments, a small piece of papyrus, and 2 beads. The diagnostic pieces of the pottery will be discussed in greater detail below.

Later Locus 23 was started at a location approximately 2.5 meters farther north than Locus 6 in the B/C passageway (Y 32), on the opposite side (west) from Locus 6. This locus provided much of the same type of material as Locus 6: 875 pottery fragments, 2 pieces of leather, assorted textile fragments (including one piece knotted to a stick), seeds, a few basket fragments including some with rope, and one coin located by means of a metal detector. Again, the diagnostic pieces will be discussed below.

One final locus was created in the B/C passageway when a second coin was found through use of a metal detector, Locus 27 at Y29 near the opening into Hall B. This was the only find recovered from that location.

Excavations were conducted to a depth of approximately 1 to 1.5 m and the soil was sifted to recover smaller artifacts that may have been overlooked in the restricted lighting that was available in the passageway. The beads, seeds and many of the textile and basket fragments were retrieved in this manner.

DIAGNOSTIC POTTERY AND COINS

Of the over 1200 pottery fragments that were collected from the B/C passageway during the 2000 expedition, only a small handful are what could be termed diagnostic pieces: rims, handles, bases, oil lamp fragments, etc. The majority of the rest were body shards of a wide range of household ceramic types, mainly cooking ware and storage vessels. Except for the juglet rim, and the possible juglet base, no clearly identifiable tableware pottery was recognized in this context. However, one of the wooden fragments appears to have been a small bowl, so it is perhaps possible that either meals were not eaten in this area or that the tableware may have been mainly wooden and did not survive.[13]

Figures 10.2 and 10.3 and their accompanying data in Figure 10.4 show the profiles of the diagnostic shards and give a description and detail of their characteristics.[14] Obtaining parallels from these rather poor specimens has not proved an easy task. In general, the pottery found in the B/C passageway is similar to that found in the other Judean caves.[15] For example, Figure 10.2:4 is similar to a jar found by Aharoni in the Cave of Horror.[16] Likewise, the jar in Figure 10.2:8 is similar to a form found in the Cave of the Pool by Avigad.[17] The oil lamp fragment, Figure 10.2:11, a discus type common in the 1st and 2nd century CE, are similar to ones found in the Cave of Horror[18] and in the Cave of Treasure.[19] The juglet rim, Figure 10.2:12, is of a type that parallels a form found in other Judean Caves and possibly at Heshbon in the Transjordan.[20] It could be dated in either the late first century CE or first half of the second century CE.

Two body shards, Figures 10.3:14 and 10.3:15 were included in the diagnostic pieces because of the irregular black markings found on them. These markings appear to be ink. There is some speculation that these could have been fragments of ostraca because of their striking similarity to pieces found in the Cave of Horror by Aharoni, although there is no decipherable writing.[21]

The remainder of the pottery illustrated in Figure 10.2 seems to consist of forms that are at home in a second century CE setting. It looks as if there are no late Hellenistic forms or clear first century CE forms present that one might expect to find if the assemblage were to be dated to a first century CE context. The assemblage seems to contain mostly unremarkable common ware typical to a household with no fine ware in attendance.

The coins recovered in the B/C passageway present another source for dating the habitation of the passageway, but again these cannot be straightforwardly used as data without some discussion. The coins, shown in Figures 10.3:16 and 10.3:17, were both of sufficient preservation that they were easily identified and thus are dateable. Figure 10.3:16 is a Vespasian silver tetradrachma possibly from Antioch. It is approximately 25 mm in diameter and weighs nearly 14 gm. The graphical axis is oriented the same on both obverse and reverse. On the obverse better preserved coins of this type read: ΑΥΤΟΚ [ΡΑΤΩΡ ΚΑΙΣΑΡ ΣΕΒΑΣΤΟΣ ΟΥΣ] ΠΑΣ [ΙΑΝΟΣ]. (The unbracketed letters indicate what can be read on this specimen.) This coin shows the laureated head of Vespasian facing right. On its reverse one can more clearly see an eagle facing left with outspread wings standing on a club. It has a wreath in its beak, and there is a branch in the field to the left of the eagle. The inscription [ΕΤΟVΣ VΕΟΥ ΙΕ] ΡΟΥ Α (= yr. 1 = 68/69 CE?) appears around the circumference of the coin. Again, only a few of the letters can be clearly seen on this specimen, shown as the unbracketed letters. Figure 10.3:17 is a small bronze Nabatean coin of Rabbel II and Gamaliah minted during his reign of 70–106 AD. The obverse and reverse are again oriented approximately along the same axis. The coin, however, is poorly struck leaving it unclear whether the obverse contained both the faces of Rabbel II and his wife and sister Gamaliah as is customary. The reverse indicates the crossed double cornucopia and the Aramaic inscription containing the names Rabbel and Gamaliah.

The two coins are thus clearly dated to a pre-Bar Kokhba minting. The silver Vespasian coin apparently dates to the period of the Great Revolt. If one adds to that the fact that it was found in a locus apparently devoid of pottery and distinctly south from a locus that can most plausibly dated to the second century CE, the Bar Kokhba habitation period, it may create a possibility that the coin could be an indicator of a prior habitation. However, there is a very plausible argument for placing this coin within the Bar Kokhba context. These Vespasian silver coins were over struck by the later rebels to create their own silver coinage.[22] The coin could therefore be explained as a coin collected for that purpose but never re-struck.

The Rabbel II Nabatean coin likewise needs some explanation since it too may span the relevant time frames between the Great Revolt and the Bar Kokhba revolt. This coin, however, unlike the silver tetradrachma, was found within the context of the pottery from locus 23 which adds some further perspective for deciding upon a proper temporal setting for the coin. In addition, it is well known that coins often circulate far beyond their minting, and one is left to more plausibly locate the coin in the second century CE context of the Bar Kokhba revolt. The fact that a Nabatean coin minted in the early second century would be found among the household items in the cave is made even more plausible when one considers that it could be connected to a group of rebels fleeing first from Nabatea, as did the famous Babahta whose cache of documents was one of the most prized finds of Yadin's campaigns in the Cave of Letters, and then from their first refuge in En Gedi to the cave. This becomes an easier explanation than one which would need to explain why a Nabatean coin might be among the items carried by persons fleeing the Romans from Jerusalem during the Great Revolt.

If the coins and pottery had been found together outside of the Cave of Letters context, one might be able to speculate that a late first century date for the collection of finds would be plausible given the indistinct temporal markers for most of the pottery finds. One might in that case give a greater weight to the absolute date found on the coins when deciding a likely time span for the collection as a whole and thus favor an earlier rather than later date. Nonetheless, when seeking to construct the most plausible explanation for this assemblage, given the whole of the discovery context, one is drawn to one conclusion: given the existing parallels for the pottery that can be relatively dated and the interpretive possibilities for explaining the presence of these particular late first century coins into the 130's CE, one must favor a second century date for the assemblage of the B/C passageway.

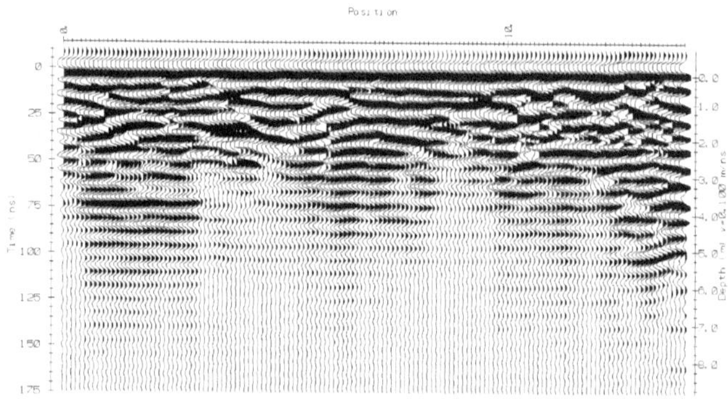

Figure 10.1: A GPR Transect Printout of the BC Passageway.

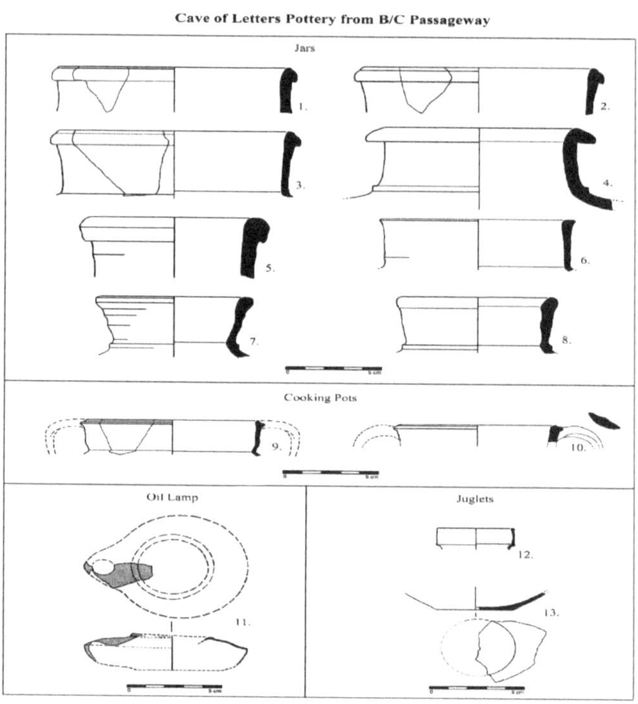

Figure 10.2: Cave of Letters pottery from the BC Passage.

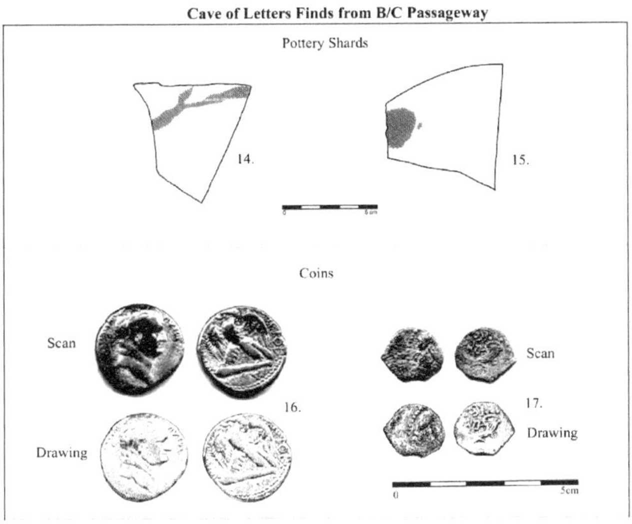

Figure 10.3: Cave of Letters finds from the BC Passage.

	Description	Date of excavation	Locus: Grid: Basket:	Special Features and Munsell clay colors
1.	Jar rim shard	13-07-2000	L6 AA32 160	12 cm mouth exterior whitish coating
2.	Jar rim shard	14-07-2000	L6 AA32 165	10 cm mouth exterior color: 5YR 6/6 reddish yellow clay core color: 5YR 6/6 reddish yellow very fine white grits medium high firing
3.	Jar rim shard	18-07-2000	L23 Y32 209	12 cm mouth exterior color: 7.5 YR 5/6 strong brown clay core color: 7.5 YR 5/6 strong brown
4.	Jar rim shard	17-07-2000	L23 Y32 205	10 cm mouth exterior color: 10YR 5/3 brown clay core color: 10YR 5/2 grayish brown few small white grits medium firing
5.	Jar rim shard	17-07-2000	BC=E L23 Y32 201	8 cm mouth exterior color: 7.5 YR 6/4 light brown clay core color: outer 5YR 5/6 inner 2.5Y 5/3

	Description	Date of excavation	Locus: Grid: Basket:	Special Features and Munsell clay colors
6.	Jar rim shard	16-07-2000	BC=E L23 Y32 182	10 cm mouth exterior color: 10YR 4/3 brown clay core color: 7.5YR 4/6 strong brown white grits medium high firing
7.	Jar rim shard	16-07-2000	L23 Y32 182	8 cm mouth exterior color: 10YR 5/2 grayish brown clay core color: outer-5YR 4/6 inner-5YR 5/1
8.	Jar rim shard	12-07-2000	L6 BB32 152	8cm mouth exterior color: 10YR 6/4 light yellowish brown clay core color: 10YR 6/4 light yellowish brown very fine white grits and voids
9.	Cooking pot Rim shard	11-07-2000	L6 BB32 146	9 cm mouth exterior color: 5YR 4/4 reddish brown clay core color: 5YR 2.5/1 black medium firing
10.	Cooking pot Rim shard	20-07-2000	L23 Y32 227	8 cm mouth exterior color: 10YR 3/2 very dark grayish brown clay core color: 5YR reddish brown
11.	Roman Disk Oil lamp (2) nose shards	13-07-2000	L6 AA32 163 & 165	red skip exterior exterior clay color: 2.5YR 4/4 reddish brown clay core color: 2.5YR 5/6 red evidence of burning at nose hole
12.	Juglet rim shard	18-07-2000	BC=E L23 Y32 209	exterior clay color: 2.5YR 4/6 red clay core color: 2.5YR red very fine clay body
13.	Juglet base shard	19-07-2000	L23 Y32 217	4 cm base exterior core color: 5YR 4/4 reddish brown clay core color: 5YR 4/4 reddish brown

	Description	Date of excavation	Locus: Grid: Basket:	Special Features and Munsell clay colors
14.	Pottery shard	19-07-2000	L23 AA32 217	exterior clay color: 5YR 6/6 reddish yellow clay core color: 2.5YR 6/6 reddish yellow few white grits, medium high firing; black ink on exterior surface
15.	Pottery shard	17-07-2000	L23 Y32 186	exterior clay color: 5YR 6/6 reddish yellow clay core color: 5YR 6/4 light reddish brown interior whitish coating; black ink on exterior surface
16.	Coin	17-07-2000	L27 Y29 193	Silver Vespasian Tetradrachma (See text for more details)
17.	Coin	17-07-2000	L23 Y32 207	Bronze Rabbel II (See text for more details)

Figure 10.4: Description of Cave of Letters Pottery from B/C Passageway.

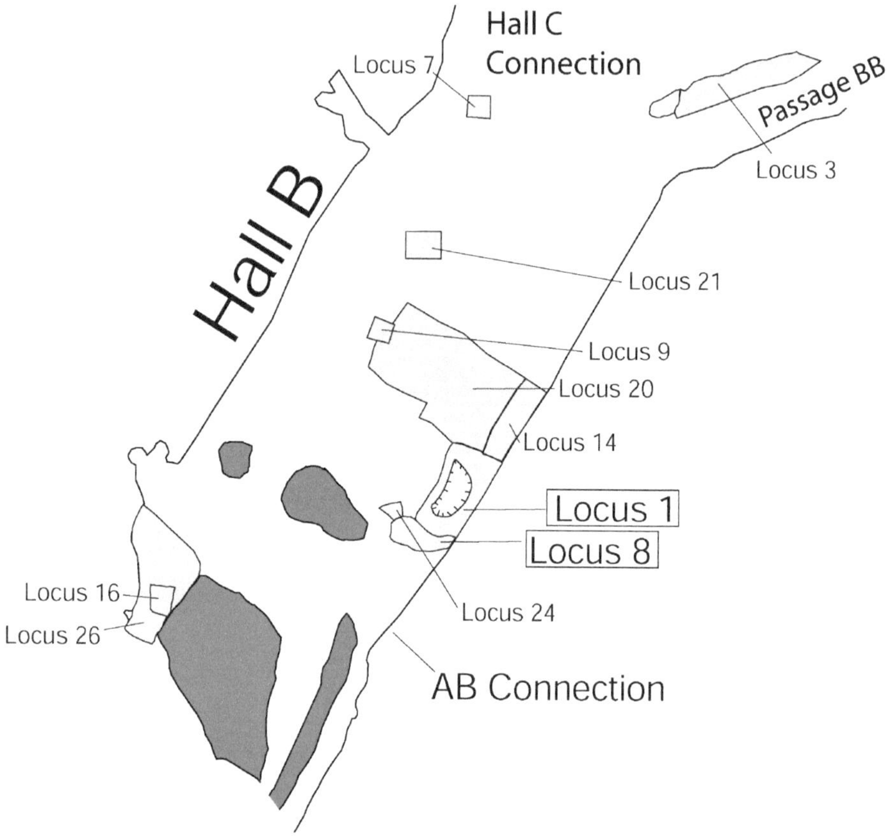

Figure 10.5: Printout of the Pen Top Field Map (Chris Morton).

NOTES

1. Some of this material was previously presented by the author at the 2000 Meeting of the American School of Oriental Research in Denver. However, the material has been much revised in the present form.
2. See Yohanan Aharoni, "The Caves of Nahal Hever," *Atiqot* 3 (1961): 148–62; Yohanan Aharoni and Beno Rothenberg, *In the Footsteps of Kings and Rebels* (Rarnat-Gan, Israel: Massada, 1960); Yigal Yadin, *The Finds from the Bar Kokhba Period in the Cave of Letters, Judean Desert Studies.* (Jerusalem: Israel Exploration Society, 1963); Yigal Yadin, "Expedition D—The Cave of Letters," *Israel Exploration Journal* 12, no. 3–4 (1962): 227–57; Yigal Yadin, "Expedition D," in *The Expedition to the Judean Desert.* (Jerusalem: Israel Exploration Society, 1960), 36–52. Later expeditions visited the cave between 1981 and 1996, but these were mostly to study the geology of the cave and were not focused on archaeological research. See Amos Frumkin, "The Cave of Letters Sediments—Indication of an Early Phase of the Dead Sea Depression?" *The Journal of Geology* 109 (2001): 79–90.

3. See Figures 5.3 and 6.7.
4. See Reeder's Figure 4.9 in "Mapping of the Cave of Letters" in this volume for his map of the passageway that includes the location of the loci from both Yadin's and the Cave of Letters Project.
5. Yadin, *The Finds from the Bar Kokhba Period in the Cave of Letters*, 40.
6. Yadin, *The Finds from the Bar Kokhba Period in the Cave of Letters*, 40
7. Yadin, *The Finds from the Bar Kokhba Period in the Cave of Letters*, 22.
8. Aharoni and Beno Rothenberg, *In the Footsteps of Kings and Rebels*, 131–32.
9. See Plate 34 (4.6) Yadin, *The Finds from the Bar Kokhba Period in the Cave of Letters* and Figure 43 (4.5) Yadin, *The Finds from the Bar Kokhba Period in the Cave of Letters*, 117.
10. Yadin, *The Finds from the Bar Kokhba Period in the Cave of Letters*, 22.
11. See again Figures 5.3 and 6.7.
12. The grid coordinates are from the rough pen top map that was used in the cave. This map was later incorporated by Reeder into his more sophisticated maps. See Figure 10.6 for an example printout of the pen top field map employed during the 2000 expedition.
13. Yadin notes quite a few wooden examples of tableware that were recovered in his Locus 61. See Yadin, *The Finds from the Bar Kokhba Period in the Cave of Letters*, 39, plates 39–40.
14. The drawings in Figures 10.2 and 10.3 are by DreAnna Hadash staff artist for the project.
15. Avigad notes in his report on the Cave of the Pool that he was able to use the finds from the Cave of Letters to make possible his reconstruction of vessels from his site. Nahman Avigad, "Expedition A—Nahal David," *Israel Exploration Journal* 12, no. 3–4 (1962): 175.
16. See Figure 3:3 in Yohanan Aharoni, "Expedition B—The Cave of Horror," *Israel Exploration Journal* 12, no. 3–4 (1962): 193.
17. See Figure 5:3 in Avigad, "Expedition A—Nahal David," 176.
18. See Renate. Rosenthal, Renee Sivan, and Miriam Schloessinger, *Ancient Lamps in the Schloessinger Collection, Qedem.* (Jerusalem: The Institute of Archeology, The Hebrew University, 1978), 85–89; Figure 32.G in Edna J. *Stern, Excavations and Surveys in Israel.* (Jerusalem: Isarel Department of Antiquites and Museums, 1998); also see Figures 4:17.15, 4:17.16 and Table 4:8 in Carl E. Savage, *Biblical Bethsaida: An Archaeological Study of the First Century* (Lanham, MD: Lexington Books, 2011), 104, 107–9.
19. See Figure 3:14 in Aharoni, "Expedition B—The Cave of Horror," 193; Figure 1:8 in Pessah Bar-Adon, "Expedition C," in *The Expedition to the Judean Desert.* (Jerusalem: Israel Exploration Society, 1960), 28.
20. Figure 1:24 in James A. Sauer, "Heshbon Pottery, 1971," *A Preliminary Report on the Pottery from the 1971 Excavations at Tell Hesbân.* (Berrien Springs, Mich.: Andrews University Press, 1973); Fig 9 and the accompanying photo in Pessah Bar-Adon, *The Cave of the Treasure the Finds from the Caves in Nahal Mishmar.* (Jerusalem: Israel Exploration Society, 1980), 209.
21. See Aharoni, "Expedition B—The Cave of Horror," Plate 31.
22. Mildenberg has conducted an extensive study on the Bar Kokhba coins particularly related to overstrikes. Leo Mildenberg and Patricia Erhart Mottahedeh, *The Coinage of the Bar Kokhba War*, ed. Patricia Erhart. Mottahedeh, trans. Patricia Erhart. Mottahedeh (Aarau: Sauerländer, 1984). Very interesting also is Mildenberg's observation on the importance of the year 68/69 CE as the sabbatical year that aids in determining the start of the second revolt in 132 CE on a subsequent sabbatical year. Leo Mildenberg, "Bar Kokhba Coins and Documents," *Harvard Studies in Classical Philology* 84 (1980): 318–19.

CHAPTER ELEVEN

Textiles, Threads AND Cordage FROM THE Cave OF Letters—2000–2001 Excavations

ORIT SHAMIR
Israel Antiquities Authority, Jerusalem, Israel

INTRODUCTION

The organic materials discussed in this chapter were discovered during excavations conducted in 2000 and 2001 at the Cave of Letters, under the direction of Dr. Richard Freund from the University of Hartford and Dr. Rami Arav from the University of Nebraska at Omaha.[1] The excavation yielded approximately 300 Roman period textile fragments. Twenty-two were cleaned and catalogued (Figure 11.1). Another 254 textiles were registered in a preliminary fashion (Figure 11.2) and others were not registered because of poor preservation. In addition, a small number of threads and cords were found. The artifacts were for the most part, small and deteriorated fragments. These fragments are some that were left in the so-called "Niche of Skulls" following the Yadin excavations in 1961.

LOCATION IN THE CAVE OF LETTERS

The textiles surveyed in this chapter came from two locations in the Cave of Letters. Most were found at locus 2y, which is Hall C, at the site known as the Niche of the Skulls (see Figure 11.3). This numbering on the map as locus 2y is related to Yadin's excavations in the cave ("y" indicating Yadin).[2] The second location is at locus 31 in the BB passage, which is across from 8y-Yadin's niche #8 (Figure 11.3).

THE TEXTILES

Twenty of the analyzed items (Figure 11.1) and 236 of those left un-catalogued are made of wool. All are S-spun with the exception of No. 2003–9021 (see Figure 11.4) whose wefts are Z-spun. Among them are found: variations of tabby weaves which include: tabby (10), weft-faced (6), balanced tabby (1) and extended tabby (2). The density of the warps ranges from 7 to 22 threads per cm, and 5 to 28 threads per cm for the wefts. Weaving mistakes are present in one piece (No. 2003–9011) where two warp threads cross each other. This indicates a faulty separation between warps and wefts or incorrect tying of the heddles.[3] Many textiles are dyed (Figure 11.5), but six only in the weft (e.g. No. 2003–9010; No. 2003–9012, see Figure 11.6) and only one in the warp. In 14 of the textiles the color of the warp is different from the weft (e.g. No. 2003–9007; No. 2003–9008; No. 2003–9016; No. 2003–9019).

Out of 258 wool textiles, 110 are dyed. Seven are dyed only in one direction, and six are decorated with bands (Figure 11.7). Of the six pieces that are decorated with bands in shades of red (Nos. 2003–9020; 2003–9021 (see Figure 11.4); 2003–9024), one is decorated with a black band (No. 2003–9017) and one is decorated with a green band (No. 2003–9014). In only one of these pieces (No. 2003–9021 (Figure 11.4)) is the band preserved to its whole width of 4 cm. The wefts of the bands are S-spun, with two exceptions (Nos. 2003–9017; and 2003–9024) which are Z-spun.

The number of wefts in the band is usually greater (20–34 per cm) than in the ground fabric, resulting in wefts that completely covers the variously colored warps. This effect was achieved by (1) crowding together the wefts in the band (e.g. No. 2003–9024) and (2) grouping two warps to work as one in a repeat pattern, which was found in four textiles (see Figure 11.7). Sample number 2003–9014 has a normal grouping pattern (without crossing) as 2,1,1,1,1,2,1,1,1,1,1,1,2. They are grouped at least 4.5 cm before the band. No. 2003–9017 has grouping pattern of 2,1,1,1,1,2,1,1,1,1,2. No. 2003–9021 (Figure 11.4) is grouped half cm before the band, usually 2,1,1,2,1,1,2 and the weft threads of the textile are more crowded near the band. In addition, No. 2003–9020 is also grouped and has thicker weft threads in the band. Three textiles have Z-spun threads in the band, contrasting the weft of the textile. This phenomenon makes the weft threads of the band more crowded.

After a few weft rows, the two warp threads which had been grouped in the band were separated and the weaving continued as usual. This method was also observed at Masada[4] as well as in the Crimea.[5] No crossed threads were found in the current excavation of the Cave of Letters, which is in contrast to Masada (with 3 examples), Muraba'at, Yadin's excavations at the Cave of Letters, 'En-Boqeq,[6] and Berenike.

There is only one case of grouped warp in a band (3/2/3/2/3/2 etc.) with warp crossing[7] and that is from Al-Tar.[8] In the Cave of Letters No. 2003–9019 has 2 red threads in the warp alternating with 2–3 un-dyed cream threads. The dyes from Yadin's excavations were analyzed by Abrahams and Edelstein, later on by Z. Koren and recently by N. Sukenik et. al. who found that the textiles from the Cave of Letters were dyed with three plant-based dyes: madder, plant-based indigotin, and weld.

Five reinforced selvedges have been preserved, with No. 2003–9007 having an additional red thread. The selvedge in this case served not only as reinforcement of the edge, but as decoration, as previously found in the Cave of Letters,[12] Wadi ed-Daliyeh,[13] and Al-Tar in Iraq.[14] Nos. 2003–9004 (see Figure 11.8), 2003–9007 and 2003–9011 has a reinforced selvedge which consists of three groups. It is impossible to know how many threads are in each group without unraveling the textile, and because of dirt imbedded in the fibers, except in the case of No. 2003–9011, which has four threads in each group. Additionally, No. 2003–9022 (see Figure 11.9) contains the remains of a closing cord, and No. 2003–9006 (see Figure 11.6) has the remains of a reddish-orange, S-spun sewing thread. The stitches are irregular, coarse and uneven in size. The irregular intervals between the stitches attest to an unskilled or careless hand, and this textile was possibly in secondary use. Of the additional 236 wool textile fragments that were registered in preliminary fashion (Figure 11.2), 16 of them are black.

LINEN TEXTILES

Two analyzed items and 18 of the un-catalogued items (Figure 11.2) are made of linen medium S-spun. One (No. 2003–9005 [see Figure 11.10]) is balanced tabby (11–12 threads per cm) and one (No. 2003–9015 [see Figure 11.11]) is extended tabby (4 × 2 threads per cm). Both analyzed items are un-dyed and cream in color. One is decorated with self-bands, consisting of 2 groups of 5–6 threads. The fact that both fragments were not bleached, is in contrast to other contemporaneous linen textiles found at sites from this same period (for example at En Rahel).[15]

DISCUSSION AND SUMMARY

Materials

The large quantity of woolen textiles is comparable to other arid-zone sites of the Roman period in Israel, as wool was the most common material for most clothing in the Roman Period in Israel.[16] In contrast, it was still of secondary importance in the Egyptian economy, but began to be more common in the Christian period from the sixth century CE onwards.[17] Linen was also very popular[18] and it became a major economic factor in Israel from the middle of the second century.[19] Similar to Yadin's excavations in the cave, no mixed wool and linen threads in the same textile (*sha'atnez*)[20] were found in the 2000–2001 excavations. In fact very few mixed wool and linen textiles have been found in Israel, even in non-Jewish sites.

Goat hair was not found in the Cave of Letters in any of the excavations, despite the fact that it was in regular use for items such as carry sacks.[21] Goat hair was used in Israel, but it was less common than in other locations outside of Israel.[22]

Spinning

The spin direction of wool and linen in the items collected in the cave is S, similar to other Roman-period sites in Israel. One item collected from the cave (No. 2003–9021) is S-spun warp and Z-spun weft (S/Z) within the same fabric. This phenomenon was found at a number of Roman period sites in Israel, Egypt and the Middle East. Small quantities were found at Murabba'at, previously in the Cave of Letters, and at Masada, where 11 of 122 such finds were of this type, and a large proportion of the finds from Dura-Europos and At-Tar were of this type as well. A. Sheffer and H. Granger-Taylor think that this type of technique originated as a Middle Eastern or Mesopotamian tradition that established itself in Judea, and they assume that these fabrics were produced on the two-beam upright loom.[23] S/Z is known to make a generally smoother fabric. It was suggested that the presence of two spin directions might be typical of sites on trade routes, influenced by local and outside traditions. Sheffer and Tidhar assume that in the case of fabrics from 'En Boqeq which derive from the Byzantine period had the influence of S-spun which is from Egypt and Z-spun which is derived from northern Mediterranean and Persian sources.[24]

Nevertheless it can be deliberate, in order to produce a denser weave, as well as visual effects. An alternative theory might be that it is characteristic of areas where the use of wool was highly developed. A number of Roman period textiles from northern Europe have combined spinning.[25] In textiles with warp threads twisted in the direction opposite to that of the weft threads, the weave is denser. Many of the woolen textiles from Dura-Europos are of this type.[26] Contrasting spin directions in the weft are found only in few cases. At Masada, for example, a few textiles have chevron patterning due to contrasting spin directions in the weft. First, weft yarn is S-spun; second weft yarn Z-spun.[27] In the Cave of Letters, the spinning is of a high standard, and the threads in the fabrics are of uniform thickness. Most of the textiles from the Roman period are S-spun, and few are Z-spun, and those are usually made of wool and imported.

Weaving

No weaving techniques more complex than variations of the plain weave were found in the 2000–2001 excavations at the cave, but in Yadin's various excavations[28] one twill 2/2 fabric was found at Wadi Murabba`at,[29] Masada,[30] Sha`ar Ramon, and Mo'a.[31]

Dyeing and Decoration

Most of the woolen fabrics found at the Cave of Letters were dyed or decorated with bands, more so than at other sites in Israel of this period. Linen fabrics are as a rule un-dyed, but are decorated with self-bands. Bands are common decoration in the Roman period. Roman weavers made good use of the contrasting decorative effect which can be achieved by beating up the weft tightly into bands or other reserved areas within a plain ground weave.

Self-bands were found only on linen textiles. Self-bands are an ancient decorative method common in local assemblages: the Cave of Letters,[32] Murabba'at,[33] Wadi ed-Daliyeh,[34] Qumran,[35] Masada,[36] and Avior Cave.[37] In this way it was possible to achieve a band typical to the Roman tunic, without using colored wool, and to avoid *sha'atnez*.[38] Outside of Israel, this style textile was found at Dura Europos[39]and Palmyra.[40] It seems that this decoration is typical from the Hellenistic period until the Byzantine period. Later periods, had more elaborate techniques of weaving and dyeing.

Use

Some of the bands on the woolen fabrics had ornamented sleeve ends, as at Dura Europos.[41] Usually, however, they were used for tunics, as demonstrated by two or more such vertical bands (*clavi*) flanking the neck opening.[42] The width of the clavi often indicated the rank of the wearer. In the Cave of Letters there are examples of *clavi* which indicate different ranks of the people in the Cave.[43] This is similar to finds at Dura Europos[44] and Nubia.[45] Additionally, the decorative bands on the textiles from the 2000–2001 excavations in the Cave of Letters are similar to those on textiles from other sites in the area including Wadi Murabba`at,[46] Yadin's finds in the Cave of Letters,[47] Masada,[48] and as previously mentioned, Dura Europos.[49] The mantle worn over a tunic consisted of a rectangular piece of cloth decorated in the corners with H-or gamma designs.[50] No samples of the

H or gamma were found in the fragments examined from the 2000–2001 Cave of Letters excavations.

Linen was used for sacks, as at, for example, Cave of Letters,[51] Wadi Daliyeh,[52] and Wadi Murabba`at.[53] In the Cave of Letters,[54] linen fabrics were used for garments, sacking, scarves, and to wrap scrolls.[55] Artifact no. 2003–9015 is made from a coarse basket-weave fabric usually used for sacking, and no. 2003–9005 is more delicate and probably used for clothing (see Figure 11.1).

Looms

The finds do not indicate what type of loom was employed in the making of the fabrics, and no loom-weights were found in the Cave of Letters. It has been suggested that the use of the warp-weighted loom ceased almost entirely in the country in the late first century CE and they were replaced by the upright loom.[56] Textiles which have grouping with crossed threads at the band[57] may indicate the use of a two beam upright loom. This phenomenon was observed at the Cave of Letters, and was noted at many sites of the Roman period. Weaving mistakes were found in one fabric collected in the Cave of Letters. Crowfoot[58] suggested that errors found in fabrics may have been due to the inexperience of the weavers during the transition from the warp-weighted loom to the upright loom. If the two-beam upright loom was used to produce the textiles found in the cave, it would have had only two heddle-rods. This assumption is based on the fact that textiles that required twill weaving or complex fabrics required at least three heddle rods, and only one textile of this type was recovered in the cave, and it is safe to assume that this piece was probably imported. I worked on the loom-weights from 'En Gedi[59] Level II dated to the Early Roman Period (until 70 CE), and this stratum produced samples of loom-weights. Unfortunately, at Level I at 'En Gedi (dated to the second century CE until the Byzantine period) loom-weights were not found.

Origin

For obvious reasons, most, if not all of the textiles found during the 2000–2001 excavations at the Cave of Letters were manufactured outside of the cave. The textiles were brought to the cave as clothing worn by the people or they were brought in bundles to the cave.[60] Some were probably manufactured at 'En Gedi or other nearby areas, except the one example of the twill weaving from Yadin's excavations, which probably was imported. Linens may well have been of local Galilean, Beth She'an or Jordan Valley production.[61]

Cordage and Threads

Seventeen cords and threads were analyzed, but it is worthwhile to mention only a few of those that were analyzed (Figure 11.12). Two threads are made of red wool fibers (Nos. 2003–9060; 2003–9061) and a few are un-dyed linen (e.g. Nos. 2003–9062), and four cords are made of date palm (Phoenix dactylifera) fibers. No. 571461 is made of S-plied from three Z-spun strings to create Z3S. The others are Z2S. One cord (No. 571647) is made of goat hair. Rope-making techniques have been described in a few different works[62] and these kinds of cords are very common.

Acknowledgements

I would like to express my appreciation to Richard Freund and Rami Arav for their support and encouragement, as well as their helpful comments in preparing this material. I also wish to thank Photographer, Christine Dalenta, and R. Vinizki of the Israel Antiquities Authority, for the treatment of textiles.

Locus	Basket	Additional No.	IAA No.	Mate-rial	Condi-tion	Size (cm)		Spinning Direction		Spinning Tightness		Threads per cm		Weaving Tech.
						WA	WE	WA	WE	WR	WE	WA	WE	
2	108	#10	2003–9009	W	M	4.7	12.3	S	S	T	L	10	20	WE.F
2	108	#12	2003–9010	W	M	17.5	9.2	S	S	T	L	10	16	T
2	108	#14	2003–9011	W	M	10	6.5	S	S	T	L	10	18	T
2	108	#15	2003–9012	W	M	6	6.2	S	S	T	L	8	16	T
2	108	#16	2003–9013	W	M	10	5.5	S	S	T	M	10	16	T
2	108	#17	2003–9014	W	M	5.2	4	S	S	T	L	7	28	WE.F
2	108	#7	2003–9005	L	M	21.5	4	S	S	M	M	12	11	BAL
2	108	#9a	2003–9006	W	M	2	3.5	S	S	T	L	13	16	T
2	108	#9b	2003–9007	W	P	3	1.8	S	S	T	L	11	18	T
2	108	#9c	2003–9008	W	M	3.5	3.2	S	S	T	L	10	18	T
2	108	#b	2003–9019	W	G	2	1.2	S	S	T	M	11	9	T
2	113	#11	2003–9020	W	M	6	7.5	S	S	T	M	10	24	WE.F
2	113	#1a	2003–9015	L	G	2	2	SS	SS	M	M	4x2	4x2	EX.T
2	113	#1b	2003–9016	W	G	1.5	2.5	S	S	T	M	9	5	T
2	113	#3	2003–9004	W	M	7	1	S	S	T	L			
2	113	#4	2003–9017	W	M	6.2	6.2	S	S	T	L	8	15x2	EX.T
2	113	#5	2003–9018	W	M	5.5	3.5	S	S	T	T	16	15	BAL
Survey	—	-	2003–9021	W	G	28	31	S	Z	T	L	9	16X2	EX.T
31	222	0034/1	2003–9022	W	M	6	12.5	S	S	T	L	22	26	WE.F
31	222	0034/2	2003–9023	W	M	11	8	S	S	T	L	12	15	T
31	222	0034/3	2003–9024	W	M	4.8	4.8	S	S	T	L	11	24	WE.F
31	222	0034/4	571460	W	M	8	11	S	S	T	L	12	22	WE.F

Locus	Basket	Additional No.	IAA No.	Colour Density			Edge	Decoration	Sewing	Other Fragments	Weaving Faults	Comments
					WR	WE						
2	108	#10	2003–9009	D	G	G						Not cleaned.
2	108	#12	2003–9010	D	UC	R						
2	108	#14	2003–9011	D	Y	R	S				+	Fault of two warp threads cross each other.
2	108	#15	2003–9012	D	UC	R						
2	108	#16	2003–9013	D	O	O						
2	108	#17	2003–9014	D	Y*	Y		B				*Probably faded from green.
2	108	#7	2003–9005	D	UC	UC		SB				
2	108	#9a	2003–9006	D	LY	LY			+			Remains of sewing thread R-O, S-spun.
2	108	#9b	2003–9007	D	DY	R-O	S*					*Reinforced with additional red thread.
2	108	#9c	2003–9008	D	DG	R						
2	108	#b	2003–9019	D	UC, R*	Black				1.4 x 1.2		*2 red threads alternating with 2–3 UC threads.
2	113	#11	2003–9020	D	R	DR		B		1.5 x 5		
2	113	#1a	2003–9015	D	UC	UC						
2	113	#1b	2003–9016	D	G	R						

Locus	Basket	Additional No.	IAA No.		Colour Density WR	WE	Edge	Decoration	Sewing	Other Fragments	Weaving Faults	Comments
2	113	#3	2003–9004	D	DG	DG	S*					*Only the selvedge was preserved.
2	113	#4	2003–9017	D	R-B	R-B		B				
2	113	#5	2003–9018	L	R-O	R-O						
Survey	—	-	2003–9021	D	Y	Y		B		16 x 17		
31	222	0034/1	2003–9022	D	Y	Y	CC					
31	222	0034/2	2003–9023	D	Y	Y	S					
31	222	0034/3	2003–9024	D	Y	Y		B				
31	222	0034/4	571460	D	Y	Y	S					

Explanation of the topology used for the columns in Figure 11.1

5. Material: L = Linen; GH = Goat hair; W = Wool.
6. Condition: G = Good; M = Medium; P = Poor.
7. WA = Size of warp (cm); WE = Size of weft (cm).
9. Spinning tightness: L = Loose; M = Medium; T = Tight.
10. Weaving Technique: ET = Extended Tabby; T = Tabby; WE.F = Weft-faced Tabby; BAL = Balanced tabby.
11. Density of weaving: D = Dense; M = Medium; L=Loose.
12. Color: B = Brown; C = Cream; D = Dark; G = Green; L = Light; LB = Light Brown; MC = Multi Color; O = Orange; R = Red; U = Un-dyed + Un-pigmented; Y = Yellow.
13. Edge: CC = Closing Cord; S = Selvedge.
14. Decoration: B = Band; SB = Self-bands.

Figure 11.1: Table of examined Roman period textile fragments.

Locus	Basket	IAA No.	Material	Color WR/WE	Comments
2	108	571652	W	R	
2	108	571650	L	UC	
2	108	571655	L	UC	
2	108	571656	L	UC	
2	108	571662	L	UC	
2	108	572281	L	UC	
2	112	572761	L	UC	
2	113	571621	L	UC	
2	113	571622	L	UC	
2	113	571640	L	UC	
2	113	571643	L	UC	
2	113	572304	L	UC	
2	113	572305	L	UC	
2	113	572306	L	UC	
2	113	572325	L	UC	
2	113	573299	L	UC	
2	117	572985	L	UC	
2	108	571648	W	UC	
2	108	571649	W	G	
2	108	571651	W	B	
2	108	571653	W	UC	
2	108	571654	W	B	
2	108	571657	W	R	Sewn with green thread
2	108	571658	W	Black	
2	108	571660	W	R	
2	108	571661	W	Y	
2	108	571663	W	Y	
2	108	571664	W	G	
2	108	571680	W	UC	
2	108	571700	W	UC	
2	108	571701	W	R	

Locus	Basket	IAA No.	Material	Color WR/WE	Comments
2	108	571702	W	UC	
2	108	571703	W	Y	
2	108	571703	W	Y	
2	108	571704	W	UC	
2	108	571705	W	UC	
2	108	571720	W	Black	
2	108	571721	W	O	
2	108	571740	W	Black	
2	108	571741	W	Black	
2	108	571761	W	UC/ UC,R,Black	
2	108	571800	W	UC	
2	108	572226	W	G	
2	108	572227	W	DR	
2	108	572230	W	G	
2	108	572230	W	G	
2	108	572231	W	B	
2	108	572280	W	UC	
2	108	2003–9064	W	R/O	Closing cord, selvedge
2	112	572607	W	G	
2	112	572608	W	Black	
2	112	572609	W	UC	
2	112	572621	W	R	
2	112	572640	W	UC	
2	112	572660	W	B	
2	112	572661	W	Black	
2	112	572662	W	UC	
2	112	572663	W	UC	
2	112	572664	W	G	
2	112	572665	W	G	
2	112	572680	W	UC	Double weft

Locus	Basket	IAA No.	Material	Color WR/WE	Comments
2	112	572681	W	UC	
2	112	572682	W	O	
2	112	572700	W	UC/DR	
2	112	572720	W	DR/R	
2	112	572721	W	O/R	
2	112	572740	W	Black	
2	112	572760	W	UC	
2	112	572762	W	Black	
2	112	572780	W	Black	
2	112	572781	W	LR	
2	112	572800	W	UC	
2	112	572801	W	UC	
2	112	572820	W	UC	
2	112	572821	W	UC	
2	112	572822	W	O	
2	112	572840	W	UC	
2	112	572860	W	R	
2	112	572880	W	R	
2	112	572901	W	UC	
2	112	572990	W	B	
2	113	571620	W	G	
2	113	571623	W	UC	
2	113	571624	W	UC	
2	113	571625	W	R	
2	113	571626	W	DR	
2	113	571628	W	R	
2	113	571629	W	UC	
2	113	571630	W	R	
2	113	571631	W	DR	
2	113	571632	W	R	
2	113	571641	W	R	

Locus	Basket	IAA No.	Material	Color WR/WE	Comments
2	113	571642	W	UC	
2	113	571644	W	G	
2	113	571645	W	R	
2	113	571646	W	R	
2	113	572302	W	G	
2	113	572303	W	B	Half basket weave, double weft
2	113	572307	W	DR	
2	113	572308	W	UC	Half basket weave, double warp
2	113	572309	W	UC	
2	113	572310	W	UC	Half basket weave, double weft
2	113	572311	W	Black	
2	113	572320	W	UC	
2	113	572321	W	UC	
2	113	572322	W	UC	Delicate. Open weave
2	113	572323	W	O/R	
2	113	572324	W	UC	
2	113	572326	W	UC	
2	113	572327	W	G	
2	113	572328	W	UC	
2	113	572329	W	LR	
2	113	572330	W	O	Half basket weave, double weft
2	113	572331	W	O/R	
2	113	572400	W	LR	
2	113	572460	W	LR	
2	113	572461	W	O	
2	113	572462	W	UC	
2	113	572463	W	UC	
2	113	572480	W	UC	
2	113	572481	W	UC	
2	113	572560	W	UC	
2	113	572581	W	UC	

Locus	Basket	IAA No.	Material	Color WR/WE	Comments
2	113	572582	W	O	
2	113	572583	W	UC	
2	113	572584	W	UC	
2	113	572600	W	UC	
2	113	572601	W	G	
2	113	572602	W	UC/B	
2	113	572603	W	UC	
2	113	572604	W	G	
2	113	572605	W	UC	
2	113	572606	W	UC	
2	113	573298	W	UC/G	
2	113	573300	W	UC	
2	113	573302	W	UC	
2	113	573324	W	R	
2	113	573325	W	R	
2	113	573326	W	O/R	
2	113	573327	W	R	
2	113	573328	W	B	
2	113	573329	W	UC	
2	113	573330	W	UC	
2	113	573331	W	UC	
2	113	573332	W	UC	
2	113	573380	W	UC	
2	113	2003–9026	W	Black	Decorated with one red warp thread
2	117	572920	W	Black	
2	117	572921	W	Black	
2	117	572922	W	LB	
2	117	572940	W	G	
2	117	572941	W	UC	
2	117	572942	W	UC	

Locus	Basket	IAA No.	Material	Color WR/WE	Comments
2	117	572943	W	R	Open weave
2	117	572960	W	UC	
2	117	572961	W	O	
2	117	572962	W	UC	
2	117	572963	W	UC	
2	117	572981	W	G/R,O	
2	117	572982	W	L Blue	
2	117	572983	W	R	
2	117	572984	W	Black	
2	117	572986	W	O	
2	117	572987	W	O	
2	117	572988	W	UC	
2	117	572989	W	UC	Selvedge
2	117	573000	W	O	
2	117	573001	W	UC	
2	117	573002	W	O	Selvedge
2	117	573003	W	UC	
2	117	573004	W	Black/B,UC	
2	127	571801	W	UC	
2	127	571802	W	UC	
2	127	571803	W	O	
2	127	571804	W	Black	
2	127	571820	W	G	
2	127	571821	W	G/UC	
2	127	571822	W	O	
2	127	571823	W	UC	
2	127	571824	W	UC	
2	127	571825	W	UC	
2	127	571826	W	UC	
2	127	571827	L	UC	
2	127	571828	W	UC	

Locus	Basket	IAA No.	Material	Color WR/WE	Comments
2	127	571840	W	R	
2	127	571841	W	R	
2	127	571842	W	R	
2	127	571843	W	UC	
2	127	571844	L	UC	
2	127	571845	W	Black	
2	127	571846	W	UC	
2	127	571861	W	R/DR	
2	127	571880	W	UC	
2	127	571881	W	DR	
2	127	571882	W	UC	
2	127	571883	W	R/DR	
2	127	571884	W	R	
2	127	571885	W	R	
31	22	474040	W	UC	
31	22	474041	W	UC	
31	22	474042	W	UC	
31	22	573726	W	UC	
31	22	573727	W	UC	
31	22	573728	W	UC	Closing cord
31	22	573729	W	UC	Closing cord
31	22	573730	W	UC	Reinforced Selvedge
31	22	573731	W	UC	
31	22	573732	W	UC	
31	22	573733	W	UC	
31	22	573982	W	UC	
31	22	573983	W	UC	
31	22	573984	W	UC	
31	22	573985	W	UC	
31	22	573986	W	UC	
31	22	573987	W	UC	

Locus	Basket	IAA No.	Material	Color WR/WE	Comments
31	22	573988	W	UC	
31	22	573989	W	UC	
31	22	573990	W	UC	
31	22	573991	W	UC	
31	22	573992	W	UC	
31	22	573993	W	UC	
31	22	573994	W	UC	
31	22	573995	W	UC	
31	22	573996	W	UC	
31	22	573997	W	UC	
31	22	573998	W	UC	
31	22	573999	W	UC	
31	22	574000	W	UC	
31	22	574001	W	UC	
31	22	574002	W	UC	
31	22	574003	W	UC	
31	22	574004	W	UC	
31	22	574005	W	UC	
31	22	574006	W	UC	
31	22	574007	W	UC	
31	22	574008	W	UC	
31	22	574009	W	UC	
31	22	574020	W	UC	
31	22	574021	W	UC	
31	22	574022	W	UC	
31	22	574023	W	UC	
31	22	574024	W	UC	
31	22	574025	W	UC	
31	22	574026	W	UC	
31	22	574027	W	UC	
31	22	574028	W	UC	

Locus	Basket	IAA No.	Material	Color WR/WE	Comments
31	22	574043	W	UC	
31	22	574044	W	UC	
31	22	574045	W	UC	
31	22	574046	W	UC	
31	22	574047	W	UC	
31	22	574048	W	UC	
31	22	574049	W	UC	
31	22	574050	W	UC	

Figure 11.2: Registry of Roman period textile fragments.

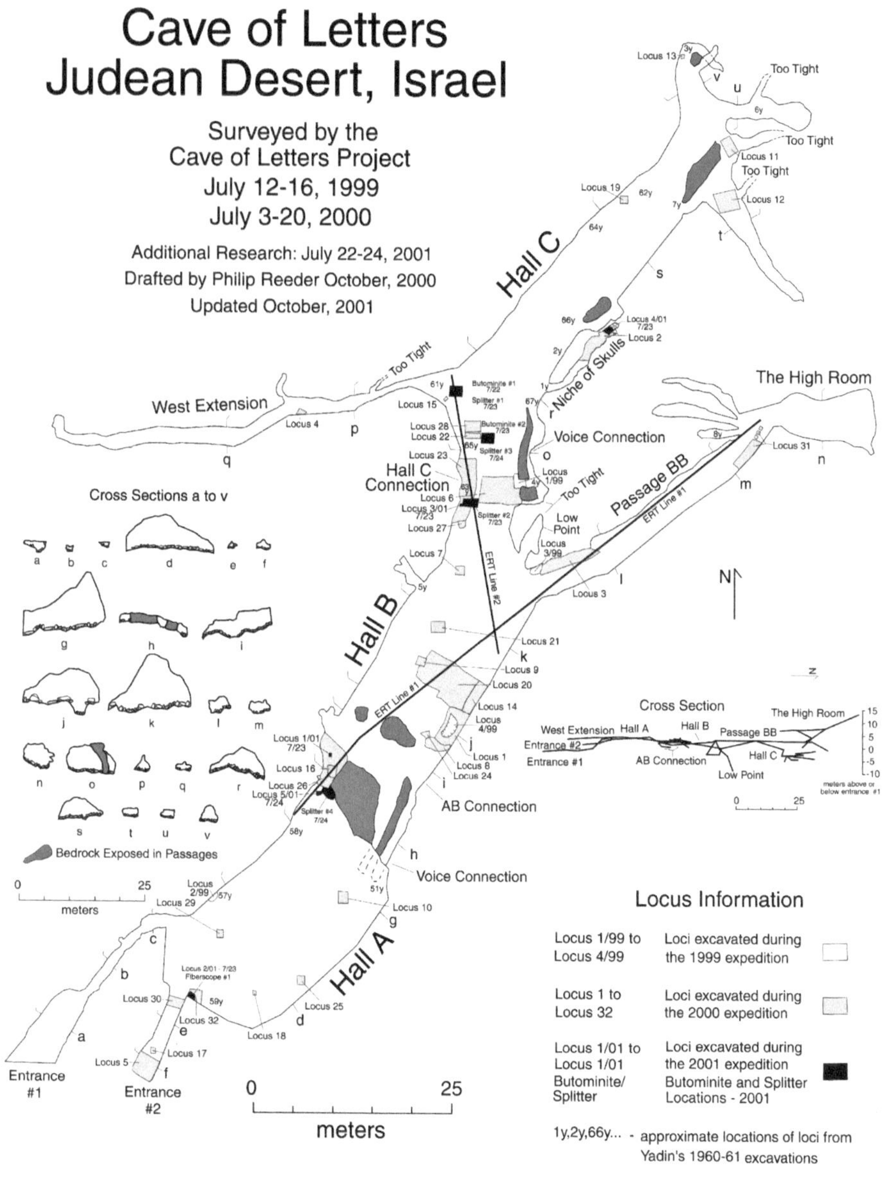

Figure 11.3: Map of Cave of Letters.

Figure 11.4: Wool textile, S-spun warp, Z-spun weft (No. 2003-9021).

Colors	Number of Textiles
Black	16
Blue	1
Brown	7
Green	18
Orange	14
Red (Different Shades)	40
Yellow	11
Orange (warp)/Red (weft)	5
Different Combination of Colors	15
Decorated with Bands	5

Figure 11.5: Table of colors appearing in textile fragments.

Figure 11.6: Textile dyed in the weft (No. 2003-9012).

IAA Number	Color WA	Color WE	Color Bands	Connection	Band Width (cm)	Threads Per cm in the weft	Threads Per cm in the band	Spinning Direction	Comments
2003–9005	UC	UC	UC	NOC	2	11	*	S	*2 bands, each consist of 2 groups of 5–6 threads. 1.1 cm between them.
2003–9014	Y	Y	G	GRO without crossing	0.8>	28		S	Only remains.
2003–9017	R-B	R-B	Black	GRO	5>	15x2	34	Z	Remains grouped usually 2,1,1,1,1,2,1,1,1,1,2.
2003–9020	R	DR	P-R	GRO	1.7>	24	20	S	Threads of the band are thicker from the weft threads. Because of poor preservation it was impossible to examine the grouping.
2003–9021	Y	Y	DR	GRO	4	16x2	20	S	Weft threads of the textile are more crowded near the band. Grouped half cm before the band, usually 2,1,1,2,1,1,2.
2003–9024	Y	Y	R-B	NOC	3.5>	24	30	Z	

Figure 11.7: Table of weaving pattern observations.

Figure 11.8: Reinforced selvedge (No. 2003-9004).

Figure 11.9: Reinforced selvedge and remains of a closing cord (No. 2003-9002).

Figure 11.10: Linen, balanced tabby (No. 2003-9005).

Figure 11.11: Linen, extended tabby (No. 2003-9015).

Figure 11.12: A red wool thread (No. 2003-9060).

Figure 11.13: A red wool thread (No. 2003-9061).

Locus	Basket	Additional Number	IAA Number	Material	Spin Direction	Length (cm)	Diam. (cm)	Twists per 10 cm	Color	Comments
2	113	#2	2003–9060	W	S2Z2S	3,2.7	0.2	4	R	
2	113	#8	2003–9061	W	S8Z2S	8	0.6		DR	Unraveled
2	118	#13	2003–9062	L	S20S	17	0.4		UY	
31	222	0034/5	571461	P	Z3S	15,10	0.9	11	ULB	
2	113		571627	L	S4Z2S	28	0.3		UY	
2	113		571647	GH	Z2S	4	0.3		Pig. B	
2	108		571760	L	S	4.5	0.5		UY	
2	127		571847	P	Z2S	4.5	0.3	19	UC	
2	127		571849	P	Z2S	4	0.4	28	UC	
2	127		571847	P	Z2S	4.7	0.3	22	UC	
2	127		571860	L	S	5,1.5	0.1		UC	
2	108/2		571886	L	S	5, 3.6	0.5		UY	
2	108		572200	L	S	Skein	0.2		UC	
2	108		572282	L	S	5	0.1		UY	
2	108		572283	P	Z2S	4.4	0.4		UC	
2	108		572284	P	Z2S	3.5	0.4		UC	
2	113		572301	P	Z2S	3.4	0.3		UC	

Figure 11.14: Table of cordage and threads.

NOTES

1. Rami Arav and Richard Freund, "Return to the Cave of Letters—What Still Lies Buried?", *Biblical Archaeology Review* 27 (2001): 25–38.
2. Yigael Yadin, *The Finds from the Bar Kokhba Period in the Cave of Letters.* (Judean Desert Studies; Jerusalem: Israel Exploration Society, 1963), 31–34.
3. J.P. Wild, *Textile Manufacture in the Northern Roman Provinces* (Cambridge Classical Studies; Cambridge, England: University Press, 1970). Rosalind Hall, *Egyptian Textiles* (Shire Egyptology; Princes Risborough, Aylesbury, Bucks, UK: Shire Publications, 1986).
4. A. Sheffer and H. Granger-Taylor, "Textiles from Masada: A Preliminary Selection," in *Masada IV: The Yigael Yadin Excavations 1963–1965* (ed. Y. Aviram, G. Foerster, and E. Netzer; Jerusalem: Israel Exploration Society/The Hebrew University of Jerusalem, 1994).
5. H. Granger-Taylor, I.D. Jenkins, and J.P. Wild, "From Rags to Riches:Two Textile Fragments from Cyprus," in *Cyprus and the East Mediterranean in the Iron Age: Proceeding of the Seventh British Museum Classical Colloquium.* (ed. G. Tatton-Brown; London, 1989), 150 Figure 4C; H. Granger-Taylor, "The Grouping of Warp Threads for Areas of Weft-Faced Decoration in Textiles of the Roman Period: A Means of Distinguishing Looms?" *Vlaamse Vereniging Voor Oud en Hedendaags Textiel Bulletin* (1992): 22.
6. Sheffer and Granger-Taylor, "Textiles from Masada: A Preliminary Selection," 193.
7. J.P. Wild and F.C. Wild, "Textiles," in *Berenike 1995: Preliminary Report of the 1995 Excavations at Berenike (Egyptian Red Sea Coast) and the Survey of the Eastern Desert* (ed. Steven E. Sidebotham and Willemina Z. Wendrich; CNWS Publications. Special Series 2; Leiden: Research School CNWS, School of Asian, African, and Amerindian studies, 1996), 249.
8. H. Fujii, K. Sakamoto, and M. Ichihashi, "Textiles from Al-Tar Caves, Part 1: Cave12," *Al-Rafidan (Journal of Western Asiatic Studies)* 10 (1989): 113.
9. D.H. Abrahams and S.M. Edelstein, "A Study of the Textiles from the Color Standpoint," in *The Finds from the Bar Kokhba Period in the Cave of Letters* (ed. Yigael Yadin; Judean Desert Studies; Jerusalem: Israel Exploration Society, 1963), 270–79.
10. Z.C. Koren, "Microscopic and Chromatographic Analyses of Decorative Band Colors on Nabatean 'En Rahel Textiles – Kermes and Shaded Bands," *'Atiqot* 28 (1999): 129–36.
11. N. Sukenik, A. Amar, A. Varvak, D. Iluz, "A Re-Evaluation of the Textile Dyes in the Cave of Letters." In *Textiles, Basketry and Dyes in the Ancient Mediterranean World. Proceedings of the Vth International Symposium on Textiles and Dyes in the Ancient Mediterranean World.* (ed. J. Ortiz, et. al.; Valencia: University of Valencia, 2016), 263–272.
12. Yadin, *The Finds from the Bar Kokhba Period in the Cave of Letters*, 205–206, 230.
13. E. Crowfoot, "Textiles," in *Discoveries in the Wâdi Ed-Dâliyeh* (ed. Paul W. Lapp, et al.; vol. 41 of *Discoveries in the Wadi Ed-Daliyeh*; Annual of the American Schools of Oriental Research; Cambridge, Mass.: American Schools of Oriental Research, 1974), 74.
14. Fujii, Sakamoto, and Ichihashi, "Textiles from Al-Tar Caves, Part 1: Cave12," 116, Plate 31.
15. Orit Shamir, "Textiles, Basketry, and Cordage from `En Rahel," *'Atiqot* 38 (1999): 91–124.
16. Orit Shamir, "Textiles in the Land of Israel from the Roman Period Till the Early Islamic Period in the Light of the Archaeological Finds," Supervised by Prof. Gideon Foerster and Dr. John Peter-Wild (Hebrew University of Jerusalem, 2005), 900 of 1635.
17. Sheffer and Granger-Taylor, "Textiles from Masada: A Preliminary Selection," 235–36.
18. Shamir, "Textiles in the Land of Israel from the Roman Period Till the Early Islamic Period in the Light of the Archaeological Finds," 662 of 1635.

19. Z. Amar, "Written Sources Regarding the Jaziret Fara'un (Coral Island) Textiles," *'Atiqot* 36 (1989a): 114.
20. Orit Shamir, "Two Special Traditions in Jewish Garments and the Rarity of Mixed Wool and Linen Threads in the Land of Israel," in *Ancient Near Eastern and Aegean Textiles and Dress: An Interdisciplinary Anthology* (ed. M. Harlow, C. Michel & M.-L. Nosch; Prehistoric, Ancient Textiles Series 18, Oxford: Oxbow Books, 2014), 298–308. Leviticus 19:19 and Deuteronomy 22:5, 22:9–11 prohibit the weaving together of woolen and linen threads.
21. Sheffer and Granger-Taylor, "Textiles from Masada: A Preliminary Selection," 177–81.
22. Shamir, "Textiles in the Land of Israel from the Roman Period Till the Early Islamic Period in the Light of the Archaeological Finds."
23. Sheffer and Granger-Taylor, "Textiles from Masada: A Preliminary Selection," 236.
24. A. Sheffer and A. Tidhar, "Textiles and Basketry at Kuntillat `Ajrud," *'Atiqot* 20 (1991): 1–26.
25. Sheffer and Tidhar, "Textiles and Basketry at Kuntillat `Ajrud."
26. Yadin, *The Finds from the Bar Kokhba Period in the Cave of Letters*, 182.
27. Sheffer and Granger-Taylor, "Textiles from Masada: A Preliminary Selection," 187.
28. Yadin, *The Finds from the Bar Kokhba Period in the Cave of Letters*, 198.
29. G.M. Crowfoot and E. Crowfoot, "The Textile and Basketry," in *Discoveries in the Judaean Desert II: Les Grottes De Murabba'ât* (ed. P. Benoît, J.T. Milik, and de Vaux; Oxford: Oxford University Press, 1961), 51–63.
30. Sheffer and Granger-Taylor, "Textiles from Masada: A Preliminary Selection."
31. Orit Shamir, "Textiles, Basketry, Cordage and Whorls from Mo'a," *'Atiqot* 50 (2006): 100.
32. Yadin, *The Finds from the Bar Kokhba Period in the Cave of Letters*, 198.
33. Crowfoot and Crowfoot, "The Textile and Basketry," 55–60, no. 70, 83–89.
34. E. Crowfoot, "Textiles," 69–70, no. 31.
35. G.M. Crowfoot, "The Linen Textiles," in *Judaean Desert I: Qumran Cave* (ed. D. Barthelemy and J.T. Milik; Oxford: Oxford University Press, 1955), no. 7 and 58.
36. Sheffer and Granger-Taylor, "Textiles from Masada: A Preliminary Selection," 174, 176–177.
37. A. Sheffer, "Bar Kokhba Period Textiles Fron Avior Cave," in *Refuge Caves of the Bar Kokhba Revolt* (ed. H. Eshel and D. Amit; Ramat Gan, Israel: Bar Ilan University, 1998), 172.
38. Ruth Precker, "Vegetable Dyes and the Dyeing Industry in the Eastern Mediterranean During the Hellenistic, Roman and Byzantine Period," MA Thesis (Ramat Gan, Israel: Bar-Ilan University, 1994), 153.
39. R. Pfister and Louisa Bellinger, *The Textiles* (The Excavations at Dura-Europos: Final Report; New Haven: Yale University Press, 1945), 25, no. 248.
40. R. Pfister, *Textiles de Palmyre Découverts par le Service Des Antiquités Du Haut-Commissariat de la République Française dans la Nécropole de Palmyre* (Paris: Les Éditions d'art et d'histoire, 1934), 14.
41. Pfister and Bellinger, *The Textiles*, Plate V.
42. Pfister and Bellinger, *The Textiles*, 14–15.
43. Yadin, *The Finds from the Bar Kokhba Period in the Cave of Letters*, 218–25.
44. Pfister and Bellinger, *The Textiles*, 12–15, Plates V–VII.
45. Ingrid Bergman, *Late Nubian Textiles* (with contributions by Hans-Åke Nordström and Torgny Säve-Söderbergh; Scandinavian Joint Expedition to Sudanese Nubia Publications 8 Scandinavian university books; Stockholm: Esselte studium, 1975).
46. Crowfoot and Crowfoot, "The Textile and Basketry," 52.
47. Yadin, *The Finds from the Bar Kokhba Period in the Cave of Letters*, 213.

48. Sheffer and Granger-Taylor, "Textiles from Masada: A Preliminary Selection," 188–95.
49. Pfister and Bellinger, *The Textiles*, 12.
50. Yadin, *The Finds from the Bar Kokhba Period in the Cave of Letters*, 210–15.
51. Yadin, *The Finds from the Bar Kokhba Period in the Cave of Letters*, 261, 264.
52. E. Crowfoot, "Textiles," 67.
53. Crowfoot and Crowfoot, "The Textile and Basketry," 59, no. 78.
54. Yadin, *The Finds from the Bar Kokhba Period in the Cave of Letters*, 259.
55. Yadin, *The Finds from the Bar Kokhba Period in the Cave of Letters*, 261, 264.
56. Orit Shamir, "Loomweights from Masada," in *Masada IV: The Yigael Yadin Excavations 1963–1965* (ed. Y. Aviram, G. Foerster, and E. Netzer; Jerusalem: Israel Exploration Society/The Hebrew University of Jerusalem, 1994), 277; Orit Shamir, "Loomweights and Whorls," in *Exacavations at the City of David 1978–85*, vol. 35 (ed. D.T. Ariel; Qedem; Jerusalem: Israel Exploration Society, 1996), 148; Sheffer and Granger-Taylor, "Textiles from Masada: A Preliminary Selection," 231.
57. Sheffer and Granger-Taylor, "Textiles from Masada: A Preliminary Selection," 193.
58. E. Crowfoot, "Textiles," 61.
59. Orit Shamir, "Loom Weights from En-Gedi," in *En-Gedi Excavations I: Final Report (1961–1965)* (conducted by B. Mazar, et al.; ed. Ephraim Stern; Jerusalem: Israel Exploration Society, Institute of Archaeology, Hebrew University of Jerusalem, 2007), 381–90.
60. Roussin, L.A., "Costume in Roman Palestine: Archaeological Remains and the Evidence from the Mishnah," in *The World of Roman Costume* (ed. Judith Lynn Sebesta and Larissa Bonfante; Wisconsin Studies in Classics; Madison, WI: University of Wisconsin Press, 1994), 182–90.
61. Ze'ev Safrai, *The Economy of Roman Palestine* (London; New York: Routledge, 1994), 155–57.
62. A. Baginski and O. Shamir, "Textiles, Basketry and Cordage from Nahal Omer," *'Atiqot* 26 (1995): 21–42 and Shamir, "Textiles, Basketry, and Cordage from `En Rahel," 106.

BIBLIOGRAPHY

Abrahams, D.H., and S.M. Edelstein. "A Study of the Textiles from the Color Standpoint." In *The Finds from the Bar Kokhba Period in the Cave of Letters*, edited by Yigael Yadin. Judean Desert Studies, 270–79. Jerusalem: Israel Exploration Society, 1963.

Amar, Z. "Written Sources Regarding the Jazıret Fara'un (Coral Island) Textiles." *'Atiqot* 36 (1989a): 114.

Arav, Rami, and Richard Freund. "Return to the Cave of Letters—What Still Lies Buried?" *Biblical Archaeology Review* 27 (2001): 25–38.

Baginski, A., and O. Shamir. "Textiles, Basketry and Cordage from Nahal Omer." *'Atiqot* 26 (1995): 21–42.

Bergman, Ingrid. *Late Nubian Textiles*. With contributions by Hans-Åke Nordström and Torgny Säve-Söderbergh. Scandinavian Joint Expedition to Sudanese Nubia Publications, vol. 8 Scandinavian university books. Stockholm: Esselte studium, 1975.

Crowfoot, E. "Textiles." In *Discoveries in the Wâdi Ed-Dâliyeh*. Vol. 41 of *Discoveries in the Wadi Ed-Daliyeh*, edited by Paul W. Lapp, Nancy L. Lapp, W. Lapp, and N.L. Lapp. Annual of the American Schools of Oriental Research, 60–77. Cambridge, MA: American Schools of Oriental Research, 1974.

Crowfoot, G.M. "The Linen Textiles." In *Judaean Desert I: Qumran Cave*, edited by D. Barthelemy and J.T. Milik, 18–38. Oxford: Oxford University Press, 1955.

Crowfoot, G.M., and E. Crowfoot. "The Textile and Basketry." In *Discoveries in the Judaean Desert II: Les Grottes De Murabba'ât*, edited by P. Benoît, J.T. Milik, and de Vaux, 51–63. Oxford: Oxford University Press, 1961.

Fujii, H., K. Sakamoto, and M. Ichihashi. "Textiles from Al-Tar Caves, Part 1: Cave12." *Al-Rafidan (Journal of Western Asiatic Studies)* 10 (1989): 109–65.

Granger-Taylor, H. "The Grouping of Warp Threads for Areas of Weft-Faced Decoration in Textiles of the Roman Period: A Means of Distinguishing Looms?" *Vlaamse Vereniging Voor Oud en Hedendaags Textiel Bulletin* 4:1 (1992), 18–28.

Granger-Taylor, H., I.D. Jenkins, and J.P. Wild. "From Rags to Riches: Two Textile Fragments from Cyprus." In *Cyprus and the East Mediterranean in the Iron Age: Proceeding of the Seventh British Museum Classical Colloquium*, edited by G. Tatton-Brown, 146–65. London, 1989.

Hall, Rosalind. *Egyptian Textiles*. Shire Egyptology. Princes Risborough, Aylesbury, Bucks, UK: Shire Publications, 1986.

Koren, Z.C. "Microscopic and Chromatographic Analyses of Decorative Band Colors on Nabatean 'En Rahel Textiles—Kermes and Shaded Bands." *'Atiqot* 28 (1999): 129–36.

Mannering, U. "Roman Garments from Mons Claudianus." In *Archéologie des Textiles des origins au Ve siècle Actes du colloque de Lattes, octobre 1999*, edited by D. Cardon and M. Feugère, 283–90. Montagnac, 2000.

Pfister, R. *Textiles de Palmyre Découverts par le Service Des Antiquités Du Haut-Commissariat de la République Française dans la Nécropole de Palmyre*. Paris: Les Éditions d'art et d'histoire, 1934.

Pfister, R., and Louisa Bellinger. *The Textiles*. The Excavations at Dura-Europos: Final Report. New Haven: Yale University Press, 1945.

Precker, Ruth. "Vegetable Dyes and the Dyeing Industry in the Eastern Mediterranean During the Hellenistic, Roman and Byzantine Period." MA Thesis, 153. Ramat Gan, Israel: Bar-Ilan University, 1994.

Roussin, L.A. "Costume in Roman Palestine: Archaeological Remains and the Evidence from the Mishnah." In *The World of Roman Costume*, edited by Judith Lynn Sebesta and Larissa Bonfante. Wisconsin Studies in Classics, 182–90. Madison, WI: University of Wisconsin Press, 1994.

Safrai, Ze'ev. *The Economy of Roman Palestine*. London; New York: Routledge, 1994.

Shamir, Orit. "Loom Weights from En-Gedi." In *En-Gedi Excavations I: Final Report (1961–1965)*, edited by Ephraim Stern, conducted by B. Mazar and I. Dunayevsky, with contributions by Y. Arbel, et al., 381–90. Jerusalem: Israel Exploration Society, Institute of Archaeology, Hebrew University of Jerusalem, 2007.

———. "Loomweights and Whorls." In *Exacavations at the City of David 1978–85*. Vol. 35, edited by D.T. Ariel. *Qedem*, 135–70. Jerusalem: Israel Exploration Society, 1996.

———. "Loomweights from Masada." In *Masada IV: The Yigael Yadin Excavations 1963–1965*, edited by Y. Aviram, G. Foerster, and E. Netzer. Jerusalem: Israel Exploration Society/The Hebrew University of Jerusalem, 1994.

———. "Textiles, Basketry, and Cordage from `En Rahel." *'Atiqot* 38 (1999): 91–124.

———. "Textiles, Basketry, Cordage and Whorls from Mo'a." *'Atiqot* 50 (2006): 100.

———. "Textiles in the Land of Israel from the Roman Period until the Early Islamic Period in the Light of the Archaeological Finds." Supervised by Prof. Gideon Foerster and Dr. John Peter-Wild, 208. Hebrew University of Jerusalem, 2005.

———. "Two Special Traditions in Jewish Garments and the Rarity of Mixed Wool and Linen Threads in the Land of Israel." In *Prehistoric, Ancient Near Eastern and Aegean Textiles and Dress:*

an Interdisciplinary Anthology, edited by M. Harlow, C. Michel & M.-L. Nosch. Ancient Textiles Series 18, 298–308. Oxford: Oxbow Books, 2014.

Sheffer, A. "Bar Kokhba Period Textiles Fron Avior Cave." In *Refuge Caves of the Bar Kokhba Revolt.*, edited by H. Eshel and D. Amit, 169–88 (Hebrew). Ramat Gan, Israel: Bar Ilan University, 1998.

Sheffer, A., and H. Granger-Taylor. "Textiles from Masada: A Preliminary Selection." In *Masada IV: The Yigael Yadin Excavations 1963–1965*, edited by Y. Aviram, G. Foerster, and E. Netzer. Jerusalem: Israel Exploration Society/The Hebrew University of Jerusalem, 1994.

Sheffer, A., and A. Tidhar. "Textiles and Basketry at Kuntillat `Ajrud." *'Atiqot* 20 (1991): 1–26.

Sukenik, N. and A. Amar, A. Varvak, D. Iluz. "A Re-Evaluation of the Textile Dyes in the Cave of Letters." In *Textiles, Basketry and Dyes in the Ancient Mediterranean World. Proceedings of the Vth International Symposium on Textiles and Dyes in the Ancient Mediterranean World*, edited by J. Ortiz, et. al., 263–272. Valencia: University of Valencia, 2016.

Wild, J.P. *Textile Manufacture in the Northern Roman Provinces.* Cambridge Classical Studies. Cambridge, England: University Press, 1970.

Wild, J.P., and F.C. Wild. "Textiles." In *Berenike 1995: Preliminary Report of the 1995 Excavations at Berenike (Egyptian Red Sea Coast) and the Survey of the Eastern Desert*, edited by Steven E. Sidebotham and Willemina Z. Wendrich. CNWS Publications. Special Series, no. 2, 245–56. Leiden: Research School CNWS, School of Asian, African, and Amerindian Studies, 1996.

Yadin, Yigael. *The Finds from the Bar Kokhba Period in the Cave of Letters.* Judean Desert Studes. Jerusalem: Israel Exploration Society, 1963.

CHAPTER TWELVE

THE Meaning OF Bar Kokhba Coins AND Their Distribution

FRED STRICKERT
Wartburg College, Waverly, IA 50677, USA

The coin finds from the Cave of Letters present a typical puzzle of numismatic research: Were they deposited by a single occupation at the site? Or were they left by a series of residents or visitors to the cave over a longer period of time? In addition, they raise particular questions concerning the geographic expanse of the Bar Kokhba revolt as well as its duration in time: How typical were these coin finds in the face of Bar Kokhba era research? How well do they fit the pattern of the dispersion of these numismatic artifacts? What do they show about the rebels who hid out in the cave?

Thus far eleven coins have been discovered at the Cave of Letters in Naḥal Ḥever.

Two Bar Kokhba coins were found by Yigal Yadin in his 1960–61 excavations.[1] These include:

- A medium Bar Kokhba bronze with palm tree on obverse and grape leaf on reverse. Inscriptions include "Shimon" and "for the freedom of Jerusalem." No date.
- A small Bar Kokhba bronze with palm tree on obverse and bunch of grapes on reverse. Inscriptions include "Shimon" and "for the freedom of Jerusalem." No date.

Excavations lead by Richard Freund in 1999 and 2000 produced another seven coins, published by Arieh Kindler in Hebrew in 2001.[2] These include another three Bar Kokhba coins:

- A small Bar Kokhba bronze with palm tree inside a wreath on the obverse and a four-stringed harp on the reverse. Inscriptions include "Shimon, president of Israel" and "for the redemption of Israel." Dated to year one of revolt (132 CE).
- A very small Bar Kokhba bronze with palm tree on obverse. Traces of inscriptions of "Shimon" and "for the freedom of Jerusalem." No date.
- A medium Bar Kokhba bronze with palm tree on obverse and grape leaf on reverse. Inscriptions include "for the freedom of Israel" and "Shimon." Dated to year two of the revolt (133 CE) (see Figure 12.2).

These excavations also produced four non-Bar Kokhba coins:

- A silver Trajan Arabian coin dated 113 CE.
- A bronze Nabatean coin of Rabbel dated from 70–106 CE.
- A silver coin of Vespasian from 70–71 CE (see Figure 12.3).
- A small bronze prutah from year two of the first Jewish revolt (68 CE).

In addition two Bar Kokhba coins were discovered by Richard Freund during a further survey in 2001. These are published here for the first time:

- A medium Bar Kokhba bronze (Figure 12.1: left side).[3]
  Obverse: Palm tree with seven branches and two bunches of dates.
  Inscription from right and left of the palm tree root: Sh M A(= Simon)
  Reverse: grape leaf. Inscription around from left to right: [SH] B LHR I [SRAL) (= the second year for the freedom of Israel) (= 133/34 AD).
- A medium Bar Kokhba bronze (Figure 12.1: right side).[4]
  Obverse: Palm tree with seven branches and two bunches of dates.
  Inscription from right and left of the palm tree trunk: Sh] M [AN] (= Simon)
  File marks from preparation for restriking. Countermark GAL[BA or KAI]
  Reverse: Five pointed grape leaf. Inscription around from left to right: [LHRO]T IRUSh [LM] (= For the liberation of Jerusalem).
  Image of Nero from original use.

THE YEARS OF CAVE OCCUPATION

The presence of seven Bar Kokhba coins, discovered at various locations throughout the cave, provides a large enough sample to conclude without a doubt that Bar Kokhba sympathizers occupied the cave during the war which lasted from 132 to 136 CE. These seven coins provide evidence spanning the entire period, including one coin from the first year[5] of the revolt, two dated to the second year, and four with no date. Through the careful analysis of Leo Mildenberg tracing marks from cracks in the dies, it is now established that the undated coins all come from the latter stages of the revolt.[6] Thus it is clear that Bar Kokhba rebels occupied the cave at least during that latter period.

Coin evidence for earlier years is more tendencial since those fleeing to the caves could easily have carried along these earlier coins. Analysis of economic documents (deeds, bills of sale, and contracts) suggests that life did continue with some normalcy in the region of Ein Gedi through the years 133 and 134, but desperation sales began in the winter and spring of 135.[7] Following the fall of Betar and the death of Bar Kokhba in the summer of 135, rebels would have fled to the caves of refuge in Naḥal Ḥever at least by the fall of 135, holding out until the end during the winter months of 136 CE. The presence of large amounts of firewood in the cave may point to more extensive preparation and perhaps some earlier occupation, but main occupation of the Cave of Letters can be dated to these last nine or ten months.

Similarly one is faced with the presence of four coins predating the revolt. The small prutah from the second year of the first Jewish revolt (68 CE) may well signal occupation during that period. Similar first revolt coins have been found in other underground complexes in Judaea.[8] However, for the other coins, the likelihood is that these coins belonged to the coffers of the later rebels. The Bar Kokhba coins had limited economic value, offering purchasing power primarily within the areas controlled by the rebels. Other coinage was needed for the purchase of supplies from less sympathetic parties and from sources reaching further afield. One can easily assume that the earlier coins were brought to the cave by these same Bar Kokhba rebels.

BAR KOKHBA OVERSTRIKES

There was a second purpose for Roman coins among the Bar Kokhba hoards. They also served as potential flans to be restruck as Bar Kokhba coins.[9] As a rebel outfit, Bar Kokhba did not have the luxury of an established permanent mint where metal could be melted and struck with dies by professionals. Rather coins were minted in a somewhat crude fashion.[10] Old coins were hammered flat and in some cases filed to remove the previous inscriptions and images. Examples of these filed coins have been included in catalogues by Mildenberg and Meshorer. In other cases, it is possible to see the earlier inscriptions and images beneath the Bar Kokhba characters due to careless workmanship.

One of the coins from the 1999 excavations at the Cave of Letters, a medium-sized bronze Bar Kokhba coin from year two of the revolt (Figure 12.2: left side and center), shows evidence of a restrike. Beneath the secondary image of a palm tree emerges the still-present face of the emperor Vespasian from a coin of the 70s CE.

The resemblance can be seen when comparing the Vespasian silver coin discovered in the year 2000 (Figure 12.3: right side). Ironically, many of the Vespasian coins display on the reverse the personified Judaea figure sitting underneath a barren palm tree with the inscription "Judaea Capta," marking the failure of the first Jewish Revolt. For Bar Kokhba, the goal was simple: a reversal of fate with a restoration of Jewish rule where residents enjoy the fruits of their labors. The messianic overtones are unmistakable: each will have his own vine and fig tree—in this case date palms dropping plenteous nourishing dates. Prominently displayed underneath the tree is the name Shimon, representing the leader Shimon Bar Kosiba (Kokhba). Thus it is not surprising that most of the Cave of Letters coins display this imagery of palm tree and vine leaf. Endurance of such current hardships—holding out in a barren cave—is tolerable only with such an idyllic vision of the future when the oppression of Rome is stamped out.

A second example of an overstrike is an undated bronze found in 2001. In this case, the filing marks are clear on both sides (Figure 12.4: left side and center). Yet the crude workmanship has not produced the same clear letters of the name Shimon, as is found on the above example, and only three letters of the inscription on the reverse. On the obverse, there remains evidence of an earlier countermark with two clear Greek letters and part of a third GAL. The complete countermark was likely either GALBA (Galba) or GALKAI (Galba Caesar) for the Emperor Galba who ruled briefly in 69 CE. Similar countermarks are found on coins of Nero that were often given during his reign to soldiers serving in the eastern provinces. On the reverse of this coin, one can readily see the image of Nero (Figure 12.4: right side). On the far left of the coin there remains a ribbon tying a wreath worn over his full head of hair.

THE COIN HOARDS

Among artifacts of ancient Jewish history, Bar Kokhba coins are most prolific. The reason for this is that these coins minted in the context of a major revolt were stashed away in caves and underground hiding places for safe keeping from occupying Roman troops. While a significant number of coins have surfaced from controlled excavations as at the Cave of Letters, by far the majority of these coins have been discovered by treasure hunters in hoards numbering in the hundreds and even the thousands.[11]

This process of discovery began already in 1889 with a major cache of three to four hundred Bar Kokhba coins originating from Dahariyeh near Hebron.[12] From 1965–1982, following the initial excavations of the Dead Sea area caves, no less than a dozen hoards were reported from locations like El Fawar, Idna, and Duma, all located on a Herodium-Hebron axis south of Jerusalem in the Judean hills. Subsequently Leo Mildenberg catalogued 6,793 Bar Kokhba coins.[13] Through further discovery, the number of Bar Kokhba coins today approaches 10,000.

These coin hoards, however, did not merely contain Bar Kokhba minted coins, but also dozens of Roman coins that were available for purchasing goods, as well as for minting re-struck Bar Kokhba coins. For example, the Beit Mirsim discoveries of 1973 included three hundred Roman silver coins, in addition to one-hundred Bar Kokhba silver coins, and other Bar Kokhba bronzes.[14] This leaves open the possibility that such collectors altered the composition of these hoards. However, as we shall see, more recent controlled excavations have demonstrated a similar pattern. Therefore presence of the Vespasian and Trajan silver coins in the Cave of Letters should not be surprising.

THE DEAD SEA CAVES

The early efforts of Yadin at the Cave of Letters were part of a program in 1960–1961 to explore caves overlooking the Dead Sea. Prior to this time finds had been made at:

- Qumran—six bronze Bar Kokhba coins (1949–52);
- 'En Feshkha—three bronze Bar Kokhba coins
- Wadi Murabba'at—nine bronze Bar Kokhba coins (1952).

Then with the push to excavation in 1960–1961, coins turned up at several sites:

- The Cave of Letters (1960–61)—two bronze Bar Kokhba coins;
- The Cave of Horrors across from the Cave of Letters in Naḥal Ḥever (1961)—four bronze Bar Kokhba coins;
- The Cave of the Pool in Nahal David (1961)—one bronze Bar Kokhba coin;
- The Cave of the Reservoir in Nahal Hardof (1960)—one bronze Bar Kokhba coin;

The nine additional coins excavated at the Cave of Letters from 1999–2001 add significantly to this record.

Further exploration has continued to turn up individual Bar Kokhba coins in still other caves near the Dead Sea, including:

- The Cave of the Spear—two bronze Bar Kokhba coins (2001);
- The Cave of the Denarius—one silver Bar Kokhba coin (2001);
- 'En Gedi settlements—one bronze Bar Kokhba coin (2000).[15]
- The Cave of the Tetradrachm, Upper Naḥal Ḥever—one silver Bar Kokhba Tetradrachma.[16]

Perhaps the most significant aspect of these finds near the Dead Sea is the large number of undated Bar Kokhba coins. By way of comparison that large hoard of 822 Bar Kokhba coins discovered at the Herodium includes a majority from the first two years of the revolt.[17] The large majority of coins from the third and fourth year of the war—four out of seven bronze Bar Kokhba coins at the Cave of Letters—among the Dead Sea area finds points to occupation during the latter phase of the war.

Also significant is the virtual absence of Bar Kokhba silver coins in this area. The 1993 discovery of a tetradrachma is a definite anomaly. This cave must be considered in relation to the Cave of Letters because it is identified with Naḥal Ḥever. However, the cave is located twelve kilometers west of the Cave of Letters in the upper regions of the valley, half way to Hebron. It seems that silver coinage had been hidden near the permanent homes of the refugees rather than carried along to the caves.

Of course, this leaves unanswered particular questions concerning refugees from Ein Gedi and regions to the east of the Dead Sea. No coins have been discovered east of the Dead Sea where Jews had been involved in struggles in years shortly before the outbreak of the Bar Kokhba revolt.[18] The Babatha correspondence found by Yadin in the Cave of Letters suggests that they were still living in Mehoza in August, 132 CE and that the departure from Arabia may have been some months later when rebel success in Judaea made vulnerable a Jewish presence in Arabia and led them to move to Ein Gedi.[19] From there, the family of Babatha made their way to the Cave of Letters. Thus the Trajan Arabia coin in the Cave of Letters is a significant find.

THE GEOGRAPHICAL EXTENT OF THE REVOLT

Several decades ago Dan Barag wrote an article noting that Bar Kokhba coins have been discovered only in a limited area, almost entirely south of Jerusalem.[20] This has lead to revision concerning traditional views about the revolt. The report of Dio Cassius gives the impression that rebel-controlled territory had been rather extensive:

> Fifty of their most important outposts and nine hundred and eighty-five of their most famous villages were razed to the ground. Five hundred and eighty thousand men were slain in the various raids and battles, and the number of those that perished by famine, disease and fire was past finding out. Thus nearly the whole of Judaea was made desolate, a result of which the people had had forewarning before the war.
> (Dio Cassius, Roman History 14:1–2)

Following Dio Cassius, a number of modern scholars have argued that battles also took place in the Galilee[21] and that Jerusalem itself had been captured by Bar Kokhba's men.

Yet, as Barag noted, the distribution of coin finds seemed to provide a corrective to Dio Cassius' seemingly exaggerated report. Only rebel sympathizers would have risked possession of these coins in face of a hostile Roman army. Their mere existence challenged Roman rule[22] and their message was pure anti-Roman propaganda as inscriptions on Cave of Letters coins demonstrate: "For the Freedom of Israel" and "For the Freedom of Jerusalem." The first-year coin displaying the words "Shimon the President"[23] declared a challenge that could not go unchecked. So the discovery of such coins would mark the extent of Bar Kokhba held territory.

As noted above, by far the bulk of coin finds have been along the Herodium-Hebron axis, the Judaean countryside that had remained in the hands of Jews following their expulsion from Jerusalem by the Romans in 70 CE. By contrast, the Gaza mint continued throughout the course of the revolt, and no Bar Kokhba coins have been discovered along the Mediterranean Coast. Barag found also that no coins had surfaced in the Galilee or ancient Samaria.

Barag concluded that the northern extremity was likely Wadi ed-Daliyeh, northeast of Jerusalem, where a single coin was found.[24] Mildenberg correctly showed skepticism over extending the northeast limits of Bar Kokhba controlled territory on the basis of a single coin find.[25] At the same time, he seemed to neglect non-coin evidence such as pottery, keys, bronze vessels, and a wooden comb dating from this period. Further exploration of this same Wadi ed-Daliyeh cave in 1980 yielded a small hoard of silver coins from Trajan and Hadrian, including one dated to 129 CE shortly before the beginning of hostilities.[26] This clearly fits the pattern of rebel finds. More recently, however, Hanan Eshel's exploration of other caves northeast of Jerusalem has yielded further Bar Kokhba coinage. In 1987, he reported another Bar Kokhba coin find in Wadi el-Mackuk between Michmash and Jericho.[27] In 1998, he made further coin discoveries in the el-Jai Cave in Wadi Suweinit (Nahal Mickmash). This hoard of sixteen coins included four Bar Kokhba coins, three of which were dated to the second year of the revolt.[28] The fourth coin, an undated, bronze, points to activity in the later stages of the revolt. There is now no question that Bar Kokhba rebels were active northeast of Jerusalem.

To the northwest of Jerusalem, the extent of Bar Kokhba activity is likewise being rethought. In the same volume that published Barag's geographical distribution proposal, Emmanuel Damati had reported on the discovery of four coins found at Khirbet El-'Aqd (Emmaus-Nicopolis).[29] Barag noted this as the first such discovery on that side of Jerusalem. Ten years later, M. Gihon and M. Fisher discovered another ten Bar Kokhba coins at this same site.[30] The Khirbet El-'Aqd discovery turned out to be rather significant. Coins from both seasons of excavations included undated bronzes that pointed to activity late in the war. The site has now been described as "a stronghold of the forces of Bar Kokhba,"[31] Shortly thereafter, another eleven kilometers to the north in the Samaritan foothills; a 1990s survey of hiding places at Al-Midya Ar-Ras uncovered two silver Bar Kokhba coins.[32] Even more recent finds at Qiryat Sefer, Hurvat Tittora, and Hurvat Burnat—the latter now marking the limit to the north—all point to activity of Bar Kokhba troops northwest of Jerusalem extending into the last period of the war.[33]

The Khirbet el-'Aqd discoveries led to a comprehensive study of underground hiding places in Judaea. Amos Kloner and Boaz Zissu have now recorded 125 sites containing hiding complexes.[34] Many of these were carved out shortly before the revolt erupted and yield artifacts from the Bar Kokhba period, including coins from Khirbet Hamadan, Kibbutz Gat, Khallet Qeis, and Deir Nakhkhas.[35] Thirteen sites with twenty underground hiding places have now been uncovered in the area of western Benjamin and the Samaritan foothills.[36] Likewise, the work of Kloner and Zissu has confirmed similar complexes on the eastern side of Benjamin, many noted above, with evidence of activity throughout the revolt.[37] The area located in between offers a different picture. They explain this in simple terms: "The absence of hiding complexes in central Benjamin (the area west of Ramallah and north of the Jerusalem-Emmaus line) is quite likely due to a geological factor: the region contains strata of hard limestone that is difficult to cut through."[38] Thus the territory where Bar Kokhba rebels were active—limited to some degree by geological issues—is more extensive than Barag had proposed and more in line with Dio Cassius' description.

Further analysis of the Khirbet el-'Aqd coins by Arieh Kindler found that one of these medium bronzes fits the pattern of "irregular" coins, that is, coins that did not come from the main dies traced by Mildenberg.[39] Therefore, in a recent work, Dan Barag has proposed the existence of two separate mints, the main one at Herodium and a second that produced these "irregulars," possibly north of Jerusalem.[40] Kindler suggested that this may have been a "mobile mint." Numerous "irregular" coins came from southern locations including the hoard at al-Fawar. However, there is some evidence for a northern connection. Among the coins discovered by Eshel at the el-Jai Cave, one of the medium bronzes was also identified as "irregular." The implications of this may be significant if a link can be shown between rebels on the northeast of Jerusalem and those to the northwest. Thus far no excavations have turned up Bar Kokhba coins in the area directly to the north of Jerusalem. However, J. Ch. Kauffman has recently published a number of "irregulars," two of which he identifies as coming from Beit Ur (biblical Bethoron) near Ramallah, north of Jerusalem.[41] This would serve to link Al-Midya Ar-Ras and Khirbet El-'Aqd to the northwest with the caves of Wadi ed-Daliyeh, Wadi el-Mackuk, and Wadi Suweinit to the northeast.

THE QUESTION OF JERUSALEM

The question of Jerusalem itself has long been a matter of debate. Emil Schürer argued that "Jerusalem also was certainly beset by rebels."[42] The assumption was that the successes of the early years of the war extended to Jerusalem and that possibly efforts to rebuild the temple had been initiated. Yet at this point, only three Bar Kokhba coins have turned up in Jerusalem excavations[43]—comprising only .00006 of all Jerusalem coin finds. Meshorer concludes that "these coins were brought there by Roman soldiers" and that this is "evidence that it was a sort of enclave beyond his [Bar Kokhba's] influence."[44] Likewise, Barag suggests that they were "possibly brought there as souvenirs by Roman soldiers."[45] While it is likely that Bar Kokhba never wrested control of the city from Roman forces, these comments seem to misunderstand the nature of guerilla warfare. With a picture emerging of rebel activity north of Jerusalem, as well as extensive strength in the south, one can easily imagine periodic forays into the city and non-conventional attacks against Roman strongholds, especially considering that Jerusalem was not a walled city during this period and still in a state of partial ruin.[46]

One should take note that two of these three coins were discovered near the temple mount and not in the center of the Roman city of that period. One cannot discount the symbolic importance of the temple to the rebel cause. The prominence of temple images on Bar Kokhba silver coins provided a constant reminder that the ultimate goal was the establishment of the third temple. Thus the early inscriptions "For the redemption of Israel" and "Eleazar the priest" identified this as a religious quest. It is true that no Bar Kokhba silver issues (with temple imagery) were found at the Cave of Letters. However, in 1991, Eshel and D. Amit found at Wadi Hedrah in the same Naḥal Ḥever a Bar Kokhba silver Tetradrachma with the Jerusalem Temple on one side and the four species of lulav, myrtle, willow, and etrog on the other.[47]

According to both Eusebius[48] and the Mishnah,[49] the founding of Jerusalem as a Roman city took place after the final battle of the war at Betar as punishment for Jewish rebellion. In contrast Dio Cassius connects the beginning of the revolt with the renaming of Jerusalem and the establishment of a pagan temple:

> At Jerusalem, he [Hadrian] founded a city in place of the one which had been razed to the ground, renaming it Aelia Capitolina, and on the site of the temple of the god he raised a new temple to Jupiter.[50]

From a logical point of view, this seems more likely. Hadrian's visit to Gaza in 130 CE is well documented. From that year, city coins of Gaza included a double date which confirms the time of Hadrian's visit. One Gaza coin celebrates Hadrian's visit with the inscription "Adventvi Avg Ivdaeae" and the personification of a female Judaea and two children holding palm branches. A number of sources had suggested false hopes that the Emperor would respond more favorably to the Jews. As Dio Cassius would suggest, the Emperor instead turned his back on the Jews, and war erupted.

Dio Cassius also notes the early success of the rebels, so much that Hadrian sent an important general Julius Severus and three legions to squelch the revolt. Coin evidence supports such a picture since the main inscription of first year coins "for the redemption of Israel" was modified on many later coins to "for the liberation of Jerusalem." The reason, as the common argument goes, was that the successes against Roman troops in the Judaea hill country, led Bar Kokhba to instill confidence that even Jerusalem could be taken.

Recent coin finds suggest that a revision may be necessary. The coin discoveries by Eshel and Zissu in the el-Jai Cave in 1998 included two Aelia Capitolina coins, along with a Hadrian coin for Gaza, and four Bar Kokhba coins.[51] Previously, most of the coin hoards from south of Jerusalem did not include non-Bar Kokhba coins dated after the outbreak of the revolt in 132 CE. An exception to this was a coin hoard brought to Meshorer's attention in 1967 where one Aelia Capitolina coin was mixed in with Bar Kokhba coins.[52] Justifiably Meshorer was suspicious that the hoard had been silted. If the coins had been hidden together prior to the end of the war then this would be evidence that Eusebius' report was wrong: Aelia Capitolina had been founded prior to the end of the war. This is the problem with black market coin hoards and the value of controlled excavations like those at the Cave of Letters.

The controlled dig at el-Jai provides the kind of evidence that favors Dio Cassius' report over Eusebius. One coin in the cave was a Hadrian city coin of Gaza with a double date equivalent to 133/134 CE, demonstrating that at least some Bar Kokhba forces had contact with the external economy after the beginning of the revolt. One of the Aelia Capitolina coins shows Hadrian ploughing with a bull and a cow and the inscription "COL AEL KAPIT" and "[CO]ND," Latin for founder. The other included an image of Hadrian on the obverse and his wife Sabina on the reverse. Both coins included an unusual Greek abbreviation KAP rather than the typical Latin CAP for Capitolina, suggesting that they were minted simultaneously.[53] However, both Aelia Capitolina coins were undated, so the exact date of the city's founding was not so certain.

As a result of this find, Kindler states that this is "positive proof that Aelia Capitolina was indeed founded before the Bar Kokhba War."[54] However, the matter may not be so simple. Is there a third option that places the establishment of the city somewhere between the visit of Hadrian to Gaza in 130 CE and the end of the war in 136 CE? The revolt did not erupt until 132 CE, nearly two years after Hadrian's visit. While Hadrian likely announced his plan in 130 CE, the catalyst for the beginning of the revolt was likely the initiation of construction. This would make most sense if such construction focused on the Temple of Jupiter, an inflammatory act from the Jewish perspective. Thus from the beginning of the revolt, Bar Kokhba's coins displayed images of the Jewish Temple and the religious declaration "For the Redemption of Israel." Coin imagery of a Jewish Temple would serve well to counter pagan temple construction in Bar Kokhba's propaganda campaign. This coincided with a period of relative success for the Bar Kokhba rebels.

Such success is recognized with the "year two" designation as the revolt continued past the summer of 133 CE. What is interesting with the second year coins is that many continue with the same inscriptions from the year-one coinage. Yet at some point in year two, two new inscriptions were introduced: in place of "redemption" the term "liberation" was used and instead of "Israel," one finds Jerusalem—"for the liberation of Jerusalem." What was the catalyst for this alteration? This would seem to be the natural point to date the initiation of the Roman mint in Jerusalem and the first Aelia Capitolina coins: 133/134 CE.

It has often been recognized that Bar Kokhba made use of coins for propaganda purposes, instilling confidence in the rebels for the purposes of the revolt. The coinage of Hadrian would likewise have been minted for similar propaganda value. In this case, the initial minting of Aelia Capitolina coins would thus correspond to a critical moment in the war, perhaps the turning point announcing the arrival of the general Julius Severus in 134 CE and Roman Legions.[55] While it is true that the one Aelia Capitolina coin features the imagery of Hadrian himself ploughing the ground to mark the boundaries of the new Roman city, one cannot overlook the presence of a legionary standard in the background. This would demonstrate Hadrian's determination to continue with his already announced plans, especially significant if rebels had indeed made periodic attacks upon the city as the three Jerusalem Bar Kokhba coin finds suggest. The presence of images of both Hadrian and his wife Sabina on the second Aelia Capitolina coin would thus symbolize the normalization of city life and the return to family activities with the suppression of the revolt.[56]

Such iconic symbolism with the inscription "AEL CAPIT" would only further inflame Bar Kokhba and his rebel troops. It is understandable that Bar Kokhba would have countered with equal determination and the new coin inscription "For the liberation of Jerusalem." The rest is history. Julius Severus provided the leadership and the additional legions provided the strength to turn the tide of the revolt. By the summer of 135 CE, the significant battle of Betar brought a decisive defeat to the rebel troops, death to Bar Kokhba, and the ultimate demise of the rebel cause. At this point, the hold-outs sought safety in various places of refuge including the Cave of Letters in Naḥal Ḥever.

Figure 12.1: Two medium Bar Kokhba Bronzes discovered in 2001 at Cave of Letters.

Figure 12.2: A medium-sized bronze Bar Kokhba coin, found during the 1999 excavations, which shows evidence of a re-strike (left side) with other coins for comparison.

Figure 12.3: A second example of an overstrike is this undated bronze found in 2001 (left).

Figure 12.4: Reverse side of same coin presented in Figure 12.3 (center), with the obverse and reverse of an additional coin of Nero for comparison.

NOTES

1. Yigal Yadin, Bar Kokhba, *The Rediscovery of the Legendary Hero of the Second Jewish Revolt Against Rome* (New York: Random House, 1971), 86–87.
2. Arieh Kindler, "Coins from the Cave of Letters (1999–2000)" in H. Eshel and B. Zissu (eds.), New Studies on the Bar Kokhba War (Ramat Gan: 2001), pp. 14–15 [Hebrew].
3. Ya'akov Meshorer, *A Treasury of Jewish Coins: From the Persian Period to Bar Kokhba* (Nyack, New York: Amphora, 2001), # 261.
4. Meshorer, # 289. This example is larger than most coins classified as medium bronzes. During years one and two large bronzes were minted with amphora on one side and wreath on the other. However, the large bronzes are not found in the later years of the war.
5. As well as literary evidence, the first year of the revolt is established by a Bar Kokhba medium bronze that was overstruck upon a coin from Gaza dated 131/132 CE. Leo Mildenberg, "Numismatische Evidenz zur Chronologie der Bar Kochba-Erhebung," *Schweizerische Numismatische Rundschau,* 34 (1948/49), 19–27.
6. Leo Mildenberg, *The Coinage of the Bar Kokhba War* (Frankfurt: Verlag Sauerländer, 1984), 18–20.
7. Hanan Eshel, "The Dates used during the Bar Kokhba Revolt," in Peter Schäfer, ed. *The Bar Kokhba War Reconsidered: New Perspectives on the Second Jewish Revolt against Rome* (Tübingen: Mohr Siebeck, 2003), 93–105.

8. That is Ain 'Arrub, H. 'Ethri, Susya, and Kh. Zeita. See Amos Kloner and Boaz Zissu, "Hiding Complexes in Judaea: An Archaeological and Geographical Update on the Area of the Bar Kokhba Revolt," in *The Bar Kokhba War Reconsidered*, 188.
9. Mildenberg, 24–27.
10. Dan Barag suggests that the main mint was based at Herodium and another northwest of Jerusalem. Dan Barag, "The Two Mints of the Bar Kokhba War," *Israel Numismatic Journal*, 14 (2002), 153–156. Kindler referred to the latter as a "mobile mint" in Arieh Kindler, "Coin Remains from a Mobile Mint of Bar-Kokhva at Khirbet el-'Aqd," *INJ* (1986–7), 46–50.
11. Mildenberg, 49–57.
12. Mildenberg, 49.
13. This number includes 1,177 silver tetradrachma, 3,586 silver denarii, and 2,030 bronze coins of three sizes. Mildenberg, 119–348.
14. Mildenberg, 56.
15. Zissu and Eshel, "The Geographical Distribution of Coins of the Bar Kokhba War," *INJ*, 14 (2002), 157–167, especially 163.
16. D. Amit and H. Eshel, "A Tetradrachm of Bar Kokhba from a Cave in Naḥal Ḥever," *INJ*, 11 (1993), 33–35.
17. Zissu and Eshel, "Distribution," 164.
18. There is documentary evidence that Jews of Peraea participated in the revolt as also Nabataeans. Hannah M. Cotton, "The Bar Kokhba Revolt and the Documents from the Judaean Desert: Nabataean Participation in the Revolt (P. Yadin 52)," in *The Bar Kokhba Revolt Reconsidered*, 133–152.
19. Eck, 80–81. Menachem Mor, "The Geographical Scope of the Bar-Kokhba Revolt," in *The Bar Kokhba Revolt Reconsidered*, 107–132, especially 124–127.
20. Dan Barag, "A Note on the Geographical Distribution of Bar Kokhba Coins," *INJ*, 4 (1980), 30–33.
21. Werner Eck, "The Bar Kokhba Revolt: The Roman Point of View," *JRS*, 89 (1999), 76–89.
22. The minting of coins implied statehood. The minting of silver tetradrachmas and denarii was even more of a challenge since Rome limited the minting of silver to a few choice sites.
23. The term Nasi, translated as president or prince, was often used in Messianic terms. It corresponded to one of the documents discovered by Yadin in the Cave of Letters.
24. P.E. Lapp and N.L. Lapp, "Discoveries in the Wadi ed-Daliyeh," *AASOR*, 41 (1974), 102.
25. Mildenberg, note 117, page 50.
26. Emmanuel Damati and Zeev Erlich, "A Hoard of Denarii and a Tridrachm from Wadi ed-Daliyeh," *INJ*, 5 (1981), 33–37.
27. H. Eshel, "A Coin of Bar Kokhba from a Cave in Wadi el-Mackuck," *INJ*, 9 (1987), 51–52. B. Zissu and H. Eshel, "Distribution," 157–167.
28. H. Eshel and B. Zissu, "Coins from the el-Jai Cave in Nahal Mikmash (Wadi Suweinit)," *INJ*, 14 (2000), 168–175.
29. E. Damati, "Four Bar Kockba Coins from Khirbet el-'Aqd," *INJ*, 4 (1980), 27–29.
30. Arieh Kindler, "Coin Remains from a Mobile Mint of Bar-Kokhva at Khirbet el-'Aqd," 46–50.
31. Jacob Janai, "A Find of Bar Kokhba Coins from the Al Miya Ar-Ras Area," *INJ*, 13 (1994), 78.
32. Janai, "A Find of Bar Kokhba Coins from the Al Miya Ar-Ras Area," 78–82.
33. These sites are yet to be published, but are noted in Zizzu and Eshel, "Distribution," 157–158.

34. Amos Kloner and Boaz Zissu, "Hiding Complexes in Judaea: An Archaeological and Geographical Update on the Area of the Bar Kokhba Revolt," in *The Bar Kokhba War Reconsidered*, 181–216, especially 182.
35. Zissu and Eshel, "Distribution," 159.
36. Kloner and Zissu, 192.
37. Kloner and Zissu, 193–196.
38. Kloner and Zissu, 195.
39. Meshorer refers to these irregulars as "crude coins" created by less skilled craftsmen. Meshorer, 161. However, they may merely represent an alternate mint. Arieh Kindler, "Coin Remains from a Mobile Mint of Bar-Kokhva at Khirbet el-'Aqd," *INJ* (1986–7), 46–50.
40. Dan Barag, "The Two Mints," 154.
41. Numbers 88 and 89. J.Ch. Kaufman, "Additions to the Corpus of Leo Mildenberg's Coinage of the Bar Kokhba War," *INJ*, 14 (2002), 129–152, especially 152.
42. Emil Schürer, *The History of the Jewish People in the Age of Jesus Christ*, vol. I, revised edition, Geza Vermes and F Millar (Edinburgh: T. T. & Clark, 1973), 545.
43. B. Mazar, *The Excavation in the Old City of Jerusalem near the Temple Mount. Preliminary Report of the Second and Third Seasons, 1969–1970* (Jerusalem: 1971), 8. D.T. Ariel, "A Survey of the Coin Finds in Jerusalem (Until the end of the Byzantine Period)," *Liber Annuus*, 32 (1982), 293–294.
44. Ya'akov Meshorer, 164. This view is repeated by Zissu and Eshel, "Distribution," 160.
45. Barag (1980), 33.
46. Yaron Z. Eliav, "The Urban Layout of Aelia Capitolina: A New View from the Perspective of the Temple Mount," in *The Bar Kokhba War Reconsidered*, 241–77.
47. D. Amit and H. Eshel, "A Tetradrachm of Bar Kokhba from a Cave in Naḥal Ḥever," *INJ*, 11 (1991), 33–35.
48. Eusebius, *Ecclesiastical History* IV, 6.
49. Mishnah *Ta'anit* IV, 6.
50. Dio Cassius, 69.10.
51. Sixteen coins were discovered in all. Eshel and Zissu, "Coins from the el-Jalai cave," 172.
52. Hanan Eshel, "The Date of the Founding of Aelia Capitolina" in L.H. Shiffman, E. Tov, and J.V. VanderKam, eds. *The Dead Sea Scrolls Fifty Years after their Discovery* (Jerusalem: Congress Proceedings, 2000), 637–643, especially 638.
53. Kindler assumes the minting of these coins was prior to the outbreak of the revolt, 179.
54. Arieh Kindler, "Was Aelia Capitolina Founded before or after the Outbreak of the Bar Kokhba War? Numismatic Evidence," *INJ*, 14 (2000), 176–179. Quotation from 176. Yoram Tsafir is less convinced, based on the assumption that more than one person could have left the various coins in the large el-Jai cave. For him proof positive would be an Aelia Capitolina coin overstruck with a Bar Kokhba coin. Yoram Tsafrir, "Numismatics and the Foundation of Aelia Capitolina—A Critical Review," *The Bar Kokhba War Reconsidered*, 31–36.
55. Shimon Applebaum, "Tineius Rufus and Julius Severus," in *Judaea in Hellenistic and Roman Times: Historical and Archaeological Essays* (Leiden: E.J. Brill, 1988), 120–123.
56. Her title "Augusta" clearly predates the end of the war since her death occurred shortly thereafter in 136 CE.

CHAPTER THIRTEEN

THE Date OF THE Psalms Scroll FROM THE Cave OF Letters

WALTER C. BOUZARD
Wartburg College, Waverly, IA 50677, USA

As we climbed, as stealthily as possible, up the four-hundred-foot cliff above the Naḥal Ḥever with temperatures that soared above one hundred and twenty degrees Fahrenheit, I again wondered how, exactly, I had come to be in this remote part of Israel. For nearly three weeks, I had been with a group of scholars re-exploring the Cave of Letters. We had spent day after day in the darkness of the cave, accompanied by bats and certain parasitical ticks, moving rocks and sifting dirt in spaces that were often cramped, frequently dangerous and always uncomfortable. The expedition was a challenging one, not the least because each day began and ended in a long trek across the trackless Judean wilderness and an arduous climb down the cliff to the cave and then back up again. But our mission was clear: we sought better to understand the history of the occupation of this cave that, years ago, had yielded many important artifacts. But if the purpose of our mission was clear, my existential location was not. How in the world, I wondered, did I come to a place where, in my late forties, I was dangling by a carabineer from a cliff in Israel?

It occurred to me then that it was Michael's fault. Michael Stadtmueller had been my student in 1996. Together with about forty other students, we had traveled together during our college's "May Term" in order to tour the biblical sites of Israel and to dig as volunteers at Bethsaida, just north of the Sea of Galilee. I could shepherd that many students only because my colleague and friend, Dr. Fred Strickert, was willing to surrender part of his sabbatical to accompany me for the month of May. I could not have known then that, four years later, Strickert and I would be together in the Cave of Letters project.

Even with so many students, Michael was exceptional. Not only was he cheerful and hardworking, Stadtmueller was a driven man. Day after day, he was the first up the tel in the mornings, the last to come to breakfast, the first to head back to the dig after breakfast, and the last off the tel as the bus loaded in the early afternoon. Michael connived not to take his turn washing pottery or preparing breakfast so he could dig. He was determined, he told me, to uncover "something cool."

And he did.

On one of our last mornings of the season, Michael encountered what he at first thought was yet another brass casing from modern Syrian bullet. On closer inspection, however, he realized it was something quite unique. Michael soon unearthed the handle and then the body of a rare Roman period bronze incense shovel of the sort known to be associated with Roman temples. Although Michael is not identified, his picture appeared in the early 1997 "Dig" issue of Biblical Archaeology Review as he unearthed what came to be identified as that magazine's "Prize Find" for that year.

And it was that incense shovel, a near twin of one of the shovels found by Yigal Yadin in the Cave of Letters in the 1960's, that gave rise to a new hypothesis about many of the previously known artifacts Yadin discovered. Professor Richard Freund, co-director of the "2000 Return to the Cave of Letters" expedition was inspired, initially on the basis of the incense shovel, to propose that, pace Yadin, not all of the artifacts recovered in previous expeditions and associated by Yadin with Bar-Kokhba are properly linked with that second-century revolt. Drawing upon evidence from the Copper Scroll, numismatic materials from the Cave of Letters, and an array of other evidence, Freund proposed that

> The objects that appear to date to the first century C.E.—the incense shovels and rest of the bronze hoard, the stone ware, the Psalms fragment, the First Revolt era coins—not only appear to date to the first century C.E., they really do. Perhaps they were taken to the cave by Jerusalem priests fleeing the city's destruction by the Romans. The second-century C.E. remains—Babatha's archive, the Bar-Kokhba letters—pose no dating difficulty (they are clearly second century C.E.), but they should not preclude our dating of the cave's other artifacts to the first century C.E.[1]

The purpose of the Cave of Letters expedition, therefore, was to recover evidence that would confirm or deny Freud's hypothesis. Freund invited me, in part, because of my training in biblical studies. Freund was cautiously hopeful that one of the new discoveries in the Cave of Letters would include a new scroll fragment. That hope was frustrated. Although Strickert and I did uncover—among other artifacts—a small scroll fragment, it was blank.

Nevertheless, enough new artifacts were recovered in this latest expedition so as to encourage a reassessment of Yadin's supposition that all of the artifacts found in the Cave of Letters were properly associated with the second century Jewish revolt. I propose here, therefore, to reconsider one artifact mentioned by Professor Freund above, namely the famous Psalms fragment known as 5/6ḤevPs, in the light of this new hypothesis. Any difficulties associated with such an investigation have been eased considerably by the very recent publication of this text, among others, by Peter W. Flint in the Discoveries in the Judean Desert series.[2] My own contribution will be to bring the fruits of that and other investigations to bear on the Freund hypothesis, articulating what can and cannot be claimed relative to the other artifacts discovered in the Cave of Letters on the basis of the Psalms fragment.

The details of Yadin's discovery of the Psalms fragment are sufficiently well known that they need not be rehearsed in detail. On April 3, 1960, Yadin's team, investigating Naḥal Ḥever, discovered a small Psalms fragment containing Psalm 15:1–16:1 near the western wall of the first chamber of Cave 5/6,[3] soon to be known as the Cave of Letters.[4] The following March saw the discovery of a second small biblical fragment, this one from the book of Numbers, found in the eastern entry of the same cave.[5] Another biblical text discovered in Naḥal Ḥever by Yadin, fragments of the Greek Minor Prophets Scroll, came from Cave 8, the so-called "Cave of Horror."[6] Yadin did, of course, also find numerous non-biblical scrolls in the course of his twin expeditions to the Cave of Letters, most notably the Bar-Kokhba letters and the Babatha archives. Indeed, it was primarily on the basis of these readily dated documents that Yadin concluded that all of the artifacts from the Cave of Letters are properly associated with the Second Revolt.

The documents discovered by Yadin's team and the subsequent study of those materials also confirmed a rumor relative to the discovery of other scrolls that had been sold by Bedouin to the Rockefeller Museum and other authorities in the early 1950's, namely, that these scrolls had been found in Israeli territory. Indeed, it is now certain that the Bedouin entered the caves of Israeli-occupied Naḥal Ḥever and removed documents from there rather than having discovered them, as they claimed, in Jordanian controlled Naḥal Ṣe'elim (Wadi Seiyâl). For example, corresponding fragments of the following documents, all putatively discovered near Naḥal Ṣe'elim, were discovered in the caves of Naḥal Ḥever: a Nabatean text published already in 1954,[7] an Aramaic sales deed known as the Kefar Bebayu conveyance,[8] and the first published fragments of the Greek Minor Prophets scroll.[9] These few examples are by no means exhaustive: Jonas C. Greenfield[10] and Hannah Cotton with Ada Yardeni[11] have amassed further instances of texts claimed to be from Naḥal Ṣe'elim but which, in fact, were removed from Naḥal Ḥever. In his consideration of these and numerous other materials, Jonas C. Greenfield concludes, simply, "There can therefore be little doubt that Naḥal Ḥever is the source of the so called Ṣe'elim texts."[12]

There can likewise be little doubt that Naḥal Ḥever served as the source for at least two of the longer biblical scrolls in the Ṣe'elim collection. After his own comparative examination of the biblical fragments from the respective sites, Peter

Flint reports that "For both the Numbers and the Psalms scrolls, similarities in the script and in the texture and appearance of the leather confirm that the two fragments found by Yadin at Naḥal Ḥever are indeed part of the larger scrolls discovered earlier by the Bedouin."[13] Indeed, Flint is so confident of the identification of the two collections that he employs the siglum 5/6ḤevPs for both the fragment found by Yadin and the "Seiyâl" portion[14]—a practice I will hereafter follow.

Flint reasons that the tiny size of Yadin's fragments eliminate the possibility that the Bedouin had divided the scrolls intentionally and taken them to two locations. It appears far more likely that small portions of the first Numbers scroll and the Psalms scroll were accidently left behind when they were removed from Naḥal Ḥever's Cave 5/6. It may also be true that the second Numbers scroll and the Deuteronomy scroll of the Seiyâl collection were likewise found in Cave 5/6. However, since no fragments of these documents have been discovered in that cave, that likely supposition cannot be proven.[15]

As it is the case that Yadin's fragment, Psalm 15:1–16:1, is a portion of a larger scroll known to be from the Cave the Letters, not just the fragment but the scroll as a whole—or at least what has survived of it—becomes pertinent for this discussion.[16] The scroll is especially important as it is the only Psalms scroll from the Judaean desert that preserves material from Psalms 7, 8, 10, 11, 14, 15, 23, 24, and 29. It is also, significantly, one of the only three Psalms scrolls not found at Qumran, the others being MasPs[a] and MasPs[b].[17] Eleven columns are preserved, and each column was originally contained twenty-eight lines.[18] The surviving text represents portions of eighteen Psalms ranging from Psalm 7 to Psalm 31. The pristine state of this manuscript is described as having been "fine and delicate" with both the recto and verso smooth and the writing surface well prepared.[19] The format is stichometric, normally including two cola per line. Unless the Psalm begins at a top of a column, new Psalms consistently begin after a full one-line interval. A. Steudel, with Flint, believes that "the surviving fragments represent cols. III–XVI of a much larger scroll which probably contained all 150 Psalms in some seventy-five to eighty columns, and was between c.6.75 and 7.20 m in length, excluding the handle sheets."[20] Flint does, however, allow the for the possibility that the original scroll contained only a Davidic Psalter of Psalms 1–41 arranged in 21 columns and measuring about 1.9 meters, excluding the handle sheets.[21]

A crucial factor in the consideration of the significance of this scroll for the history of the Cave of Letters is its age, determinable only on palaeographical grounds. Although originally thought by Yardeni to be a Herodian book hand of the early first century C.E.,[22] F.M. Cross believes the script to be post-Herodian and locates the script in last quarter of the first century C.E. As recently as 1998, the latter scholar published his view that the scroll represents a highly evolved formal script that has its closest parallel in very late Herodian type represented by 4QPs[b], but is one step beyond it, i.e., it is "post-Herodian."[23] Peter Flint, the editor of this scroll, previously adopted Cross's dating.[24] Upon fuller personal investigation of the documents, however, he now asserts that a late Herodian date (c. 50–58) is appropriate not only for 5/6ḤevPs but for XḤevNum[a], XḤev/SeNum[b], and XḤev/SeDeut as well.[25] It should be noted that Flint uses Cross's own palaeographic criteria and descriptions to revise the date of this and the other scrolls in the Naḥal Ḥever/Seiyâl collection. Conveniently, several letters exhibiting telltale characteristics of the late Herodian script appear clearly on that portion of the scroll discovered by Yadin, including the following[26]: (i) The alep appears in lines 1, 4, and 5. The characteristic Herodian *keraia* appears on the top right stroke and the oblique axis while the left leg is penned as an inverted "V". (ii) Bet, appearing clearly in lines 1 and 4, is written with a baseline drawn from left to right breaking through the right down stroke. (iii) The tick, or *keraia*, of the Herodian *gimel* can be seen in line 1. (iv) The he, surviving in lines 1, 3, and 5, exemplifies that Herodian style whereby the right leg, penned upwards, moves first into a triangular motion with the horizontal and then in another triangular motion to the down stroke. (v) The *yod* and *waw*, virtually indistinguishable in the early Herodian period, are again distinguishable in the late Herodian era. On the Yadin fragment, this development is discernible on lines 1 through 5. (vi) The triangular *keraia*, typical of the Herodian *ḥet*, can be seen on lines 3 and 6. (vii) The *keraia* of the Hasmonean lamed appears on lines 1, 3, 5, 6, and 8. (viii) Line 4 witnesses to the Herodian practice of forming a tick, or *keraia*, on the top of the medial nun. (ix) The closed, somewhat triangular *samekh* appears in lines 4 and 5. (x) The habit of rotating the *'ayin* clockwise, evident in lines 2 through 6, show a late Herodian development. (xi) The Herodian tendency to curl the heads under in the *pe* in both medial and final forms can be seen in line 2. (xii) Line 2 also includes a *ṣade* with a *keraia* appearing on both the right and left arms. (xiii) The long tail of the Herodian *qoph*, lengthened already in the Hasmonean period, appears in line 2. (xiv) The slowly evolved *keraia* of the *sin*, while absent in line 3, is clear in line 6. Many more examples could be adduced from other columns of 5/6ḤevPs, but this is sufficient to demonstrate that the script of this scroll is Herodian and not post-Herodian as asserted by Cross.

Finally, it is well to note that the surviving text contains very few variant readings against the Codex Leningradus as in the BHS; most variants are against the minor Masoretic manuscripts.[27] Two of the most interesting of these variants, however, occur in the fragment discovered by Yadin and are worthy of mention.

First is the omission of verse 15:3a. That is, 5/6ḤevPs presents two colas for verse 3, as opposed to the three of the MT and LXX. This variant is otherwise unattested, and one suspects that we have here to do with a simple case of parablepsis triggered by homoioarkton: the sequence of *waw* (here representing a third masculine singular suffix) preceding the negative word *lō'o* occurs at beginning of this colon and the one that follows it. Even though the phrase is difficult to translate in the MT (the *qal* form of the verb *rāgal* is otherwise unattested in the MT or at

Qumran,[28] and the exact meaning of the verb in that *binyān* is difficult to assess[29]), the MT here seems to have greater claim to originality. This claim is made not only on the obvious text critical grounds, but also because the trilogy of "heart/tongue/eyes" (verses 2b, 3a, and 4a), here used positively to describe the righteous, appears in the same sequence elsewhere in the Psalms, albeit negatively, to describe the deeds of the wicked.[30]

A more telling variant is the Davidic superscription that appears in the MT and LXX but appears to be lacking in 5/6ḤevPs. The Yadin fragment was originally at the top of the seventh column of text, a column that once included all of Psalms 15:1–17:4ab. Although the bottom of column VI is not preserved, the absence of the superscription seems certain since a) the alignment and spacing precludes it having been originally placed before the preserved text in line 1, and b) as occurs elsewhere in this scroll, were the superscription originally present in column VI, the superscription would have been preceded by a blank line, yet there is no room for two additional lines (i.e., a blank line followed by the superscription) in that twenty-eight line column.[31]

The absence of the Davidic superscription here is puzzling since, as Flint has elsewhere convincingly argued, the association of superscriptions with particular Psalms was a phenomena that is not only attested early in the Dead Sea Psalms scrolls, but seems to have been a constitutive component of the early canonical shaping of what is now the Psalter's Books I to III (Psalms 1–89).[32] To be more specific on this point: the surviving thirty-nine Judaean Psalms scrolls show only two instances of deviation from the Masoretic ordering of Psalms 1–89: Psalm 33 follows Psalm 31 in 4QPs[a] and 4QPs[q] and Psalm 71 follows Psalm 38 in 4QPs[a]. Flint claims of this

> Both deviations coincide with the rare absence of superscriptions in Books I to III of the Masoretic Psalter, which signals a correlation between stabilization and the presence of titles for Psalms 1–89. The absence of superscriptions in Books I–III seems to identify those Psalms whose positions were not finalized as late as the first century CE (cf. 4QPs[q]), thus rendering them prone to varying combinations with other Psalms.[33]

Based on these observations, as well as corroborative evidence from his study of the relative age of the Judaean Psalms manuscripts and the sequencing of Psalms found in them, Flint concludes that the stabilization of the Psalter took place in two distinctive stages, with Psalms 1–89 (or thereabouts) having reached its canonical shape prior to the first century BCE and the balance of the present Psalter towards the end of the first century CE: "Although the manuscript evidence is not complete, the scrolls strongly suggest that during the entire Qumran period Psalms 1–89 were virtually finalized as a collection, while Psalms 90 and beyond remained much more fluid."[34]

Flint's careful study has obvious and important implications for the ongoing conversation about the canonical shape and shaping of the Psalms. His observations are important for this present discussion as well. As will be noted presently, the omission of the Davidic superscription on a scroll containing Psalms—the structure and order of which had already long achieved stability, and that in correspondence to the presence of superscriptions in the copied texts—provides a clue as to the questions that surround the Naḥal Ḥever Psalms scroll. Before turning to matter, however, it will be helpful to summarize what has been discerned about 5/6ḤevPs.

We have noted that 5/6ḤevPs is: (i) one of three Psalms scrolls discovered in the Judaean wilderness that are not associated with Qumran, the other two having been discovered in the ruins of Masada. (ii) from the late Herodian period, c. 50–58 CE, and is one of two (or perhaps four) biblical scrolls discovered in Naḥal Ḥever, all copied in the same period. (iii) a scroll that was of exceptionally high quality in its pristine state. (iv) represents a Psalms collection that is largely consistent with the MT in terms of the content and order of the preserved Psalms, although the apparent absence of the superscription before Psalm 15 represents a curious variant.

From this, the question naturally emerges, "Where was this scroll produced?" In spite of the fact that not a few of the Qumran Psalms scrolls date from the same period as does 5/6ḤevPs,[35] it is highly unlikely that this Psalms scroll was produced by the scribes of Qumran, especially given the absence of the Davidic superscription for Psalm 15. This claim is based on the following considerations: (i) The Psalms were considered as "scripture" at Qumran. Besides the fact that there are more surviving Psalms scrolls than any other biblical book,[36] the scrolls themselves signal the scriptural status of the Psalms. For example, the Halakhic Letter 4QMMT enjoins the study of three scriptural groupings, namely "the book of Moses," "the Prophets," and "David,"[37] the latter referring to the nascent third division of the Hebrew scriptures by means of signaling out the foremost Davidic book, the Psalms.[38] (ii) The Psalms were viewed by the Qumran scribes as prophetic.[39] 11QPs[a] 27.11[40] articulates the belief that 4,050 musical compositions (including those preserved in the MT) were given to David from the Most High, and that explicitly, by prophecy.[41] Something of that same understanding of the prophetic and revelatory character of the Psalms also appears to stand behind the statement included in the daily prayers of 4Q503: "you have made known to us the psalms of your glory."[42] Moreover, pesharim—a genre mostly confined to prophetic texts—were written on the Psalms at Qumran including 4Q171, 4Q173 and, perhaps, 4Q172.[43] (iii) The Davidic authorship of the Psalter was strongly asserted in Qumran. The already mentioned 11QPs[a] 27 ("David's Compositions") "is directed rather patently to those who might doubt the Davidic authorship of the Psalter, of whatever size, so that those who would limit the Psalter might be persuaded otherwise."[44] Flint, expanding upon the earlier work of Wilson,[45] has shown that the emphasis on David is a major organizing element of 11QPs[a] since the redactor of that scroll has dispersed Davidic Psalms in such a way as to highlight the Davidic character of the entire collection.[46] This same emphasis is transparent in the concatenation formed by Psalm 154 (col. 18), the Sirach 51 canticle (col. 21), and the "Hymn to the Creator" (col. 26) preceding "David's Compositions" (col. 27): the four compositions serve to stress the role of Wisdom in David's life and musical gifts.[47]

None of the above considerations, of course, eliminates the possibility that a Qumran scribe copied this text and, for whatever reason, omitted the superscription. Nevertheless, given the relative stability of the first half of the Psalter throughout the Qumran period, the prophetic value that was placed on the Psalms, and the Davidic emphasis that features so prominently in that community, it would be an especially curious omission at Qumran. Couple those considerations with the following observation regarding the Psalm scrolls collected at Qumran: Of the six Qumran Psalms scrolls that are generally classified as "Herodian" (1QPs[c], 2 QPs, 4QPs[h], 4QPs[m], 4QPs[p], and 4QPs[r48]), five are written in prose format. Only 4QPs[h], which evidently included no more than Psalm 119:10–21, was written stichometrically by full line. However, the scribal penchant for copying Psalm 119 stichometrically manifested itself early in the tradition. 1QPs[a], which dates from the middle of the first century BCE, is arranged in a prose format except for Psalm 119 which is presented with two hemistichs per line.[49] Similarly, 5QPs (loosely assigned to the first century CE) and 4QPs[g] (mid-first century CE) present Psalm 119 with two hemistichs per line.[50]

The remaining Qumran scrolls that have been located in the first century CE include the following: 4QPs[s], 4QPs[t], 4QPs[q], 4QPs[j], 4QPs[e], 1QPs[b], 11QPs[a], 11QPs[b], 11QPs[c], 11QPs[d], 11QPsAp[a], 3QPs, 8QPs[a], 4QPs[c], 5/6ḤevPs and MasPs[a].[51] The scrolls from Naḥal Ḥever and Masada will be discussed separately below. Of the remaining scrolls, eleven present the psalms in a prose format, with 3QPs, 8QPs[a], and 4QPs[c] serving as apparent exceptions. Thus, and discounting those scrolls that preserve only Psalm 119, of the nineteen Qumran psalms scrolls classified as Herodian periods and including the first century CE, sixteen, or 84.2%, are written in a prose format. Moreover, the evidence surrounding the remaining texts is mixed. 3QPs is a small fragment that includes only a few words of Psalm 2:6–7 which "as far as may be ascertained from the remaining text" is stichometrically arranged, a surmise based on the relative positions of a few remaining letters in verses six and seven which begin successive lines.[52] Obviously, the evidence remains sparse. 4QPs[c], placed between 50 and 68 CE, is mostly—but not completely—written stichometrically, usually with two hemistichs per line without any spaces between them. The scroll also includes some Psalms written in prose format.[53] Finally, there is 8QPs, putatively written in bi-cola lines and, according to that Psalm scroll's editor, M. Baillet, is to be loosely located in the first century CE.[54] The fourteen fragments preserve portions of Psalms 17 (frgs. 8–13) and Psalm 18:6–13 (frgs. 8–13). Baillet reconstructs this poorly preserved scroll as representing bi-cola lines with the cola separated by one or more intervals of widely varying length. In light of subsequent discoveries and publications, however, it may be time to reconsider Baillet's reconstruction of this Psalm scroll. One notes, for example, that except for the lines representing Psalm 17:8 and 18:11, the intervals between the cola are all supplied by Baillet himself. Of the two surviving lines which do seem to attest to a space between cola, the first (Psalm 17:8, fragment 5) does not appear to represent an extraordinarily large gap between the words *ʿyn* and *bṣēl*, and about the second (Psalm 18:11, fragment 13) Baillet himself remarks, "Après *wyʿp*, espace en blanc attesté, ***mais assez réduit***" (emphasis added).[55]

The just mentioned scrolls (and the 3QPs fragment) may be exceptions for another reason: neither 4QPs[c] nor 8QPs is copied in what Emanuel Tov has dubbed the "Qumran system"[56] and both were certainly among a number of scrolls that were imported to the Qumran settlement rather than copied there.[57] If so, it appears to be the case that, all of the Psalms scrolls identified as "Herodian" in script and the vast majority—and perhaps all—of the Qumran Psalms scrolls dated to the early or mid-first century CE are written in a prose format except the scrolls that, apparently, originally included only Psalm 119.

One must, of course grant that the significance of stichometric arrangements remains unclear, as does their place in their history of transmission.[58] For example, it is difficult to know how to evaluate those few psalms that were presented stichometrically from the first century CE—if there were, in fact, any full representatives of this type—or how to judge the several first century BCE scrolls arranged stichometrically.[59] A lack of clarity on the meaning and purpose of stichometric arrangements notwithstanding, it is nevertheless startling to realize how unique two stichometric Psalms scrolls appear in contrast to the scrolls copied in the same period and, moreover, how similar they are to each other. 5/6ḤevPs and MasPs[a60] are both copied in late Herodian script on, respectively, twenty-eight and twenty-nine(?) line columns, including blank lines separating the Psalms.[61] They both are written in a stichometric format with two cola per line with space between each stich.[62] In this respect these texts are distinct from the stichometric structure of 4QPs[c]. Also noteworthy are the similarities in the blank spaces separating the psalms in these two scrolls. Column II of MasPs[a] features a 1.5 centimeter blank that divides Psalms 82 and 83. Talmon has noted that "a similar separation of two consecutive psalms by a blank line" can be observed in 5/6ḤevPs between Psalms 15 and 16.[63]

Further similarities and differences between these scrolls need to be investigated. The argument at this point, however, is as follows: First, these two Psalms scrolls are similar in format and that in contradistinction to other Psalms scrolls from the Judaean desert dating from the first century CE. In this connection, it is worthy to note that MasPs[b] is written in what Talmon says might be a Herodian script from the first decades of that era. Although older than MasPs[a] and exhibiting a format that is somewhat different, the script type and the stichometric arrangement for both of these scrolls—and especially MasPs[a]—suggests a link of some description with 5/6Ḥever/Ṣe'elim as well as a link to the later MT tradition. Second, it seems unlikely that any of these scrolls originated at Qumran. Quite apart from the evidence adduced above, Tov has argued that none of the biblical texts discovered at Masada originated at Qumran. This is so, he avers, because the Masada biblical texts are closer to the medieval Masoretic texts than are the Qumran texts that also belong to the Masoretic family.[64] Indeed, this same observation pertains to other biblical texts deriving from the time of the Bar Kokhba revolt,[65] all of which point to "groups which adhered to one single biblical texts, probably the Jerusalem central text."[66] With respect to Masada, Tov believes that the Zealots were the most likely group to have brought the biblical texts to that fortress.[67] This assumption appears probable since the terminus ad quem of the deposit of MasPs[a] is certain: discovered at the base of casemate wall 1039, south of the fortress synagogue, the scroll was left by the rebels of the First Revolt—or dropped their by their Roman conquerors—in 74 CE. Tov does not, of course, transfer the same assumption to the Hebrew biblical scrolls of Naḥal Ḥever/Ṣe'elim because those texts are generally understood as deriving from the time of Bar Kokhba. In principle, however, there is no reason not to assume that Zealots of the first century CE—or sympathetic priests from Jerusalem—rather than the Bar Kokhba rebels of the second century CE, were responsible for depositing biblical texts in the Cave of Letters. The notion that these scrolls were left by Bar Kokhba's confederates is driven primarily by the fact that indisputable Bar Kokhba artifacts were found in the same cave. Against this conclusion, however, are the following considerations: i) The Cave of Letters, like other caves of the Judaean wilderness were known throughout antiquity and frequently visited by various groups. For example, sundry Chalcolithic artifacts were discovered in the Cave of Letters,[68] and at a small cave next to the Cave of Letters finds of both the Chalcolithic and Bar Kokhba periods were found intermingled.[69] ii) The fragments of biblical scrolls obviously were not in their original location when Yadin's team discovered them and, consequently, nothing of their original stratigraphic location can be known. In any event, stratigraphy in the Cave of Letters is extraordinarily difficult to discern owing to the nature of the cave itself and ceiling collapse over the centuries. Even had the larger scrolls been found in situ, the picture would not necessarily be make clearer. iii) Nothing about the surviving scrolls themselves—including the Psalms scroll under discussion—testifies to the conclusion that the Bar Kokhba rebels deposited the scrolls. To the contrary, given the obvious general similarities between 5/6ḤevPs and MasPs[a] (e.g. palaeography, format, and proximity to the MT) the possibility that the Zealots were responsible for secreting the scrolls in the Cave of Letters appears at least as likely as the traditional interpretation. In other words, had the Bedouin removed the second century CE Bar Kokhba letters and Babahta archives rather than the two (or four?) biblical scrolls, quite another conclusion about the deposit of those scrolls would likely have been reached. In short, the similarities between that scroll and the one from the Cave of Letters, while not providing proof of the terminus ad quem for the deposit of the latter, does make conceivable a scenario whereby the nearby cave served as refuge for first-century Jewish rebels, just as it did for the comrades of Bar-Kokhba.

In this connection, the caves of Wadi Murabba'ât provide a telling parallel. The Hebrew literary remains discovered at Murabba'ât include a palimpsest from the eighth century BCE, ostraca and (likely) a papyrus from the first century BCE, a recognition of debt dated 55/56 CE along with other literary fragments deposited during the First Revolt, and various documents associated with the Bar Kokhba period.[70] In contrast to the Cave of Letters, however, the biblical materials (including fragments of Genesis, Numbers, Deuteronomy, and Isaiah) and the phylactery found at Murabba'ât have all been dated, on palaeographic grounds, to the late first or early second century CE.[71] That is to say, these scrolls—which were doubtless deposited by the participants of the Second Revolt under Bar Kokhba—were all copied subsequent to the Herodian era. They are post-Herodian. This observation underscores the curious fact that the 5/6Ḥever/Ṣe'elim biblical scrolls originate in the late Herodian period and antedate the First Revolt. Is it accidental that a cache of biblical scrolls, putatively left in the Cave of Letters by the allies of Bar Kokhba, all date from the late Herodian period while a second collection, found at Murabba'ât, were all copied at a later time? Such an accident of history remains, of course, possible. And yet, the facts might more easily be explained if the scrolls of the Cave of Letters were deposited not by Bar Kokhba's confederates, but rather by those who allied themselves with the Zealots of Masada. Precisely this sort of scenario is demonstrable at Murabba'ât, albeit there the First Revolt documents are not biblical texts. On the other hand, at Naḥal Ḥever evidence of Bar Kokhba occupation appears intermingled with artifacts from the Chalcolithic era. Why should it be assumed that other artifacts—artifacts from the First Revolt—would not be intermingled with the Bar Kokhba material as well? This is, of course, Professor Freund's hypothesis.

Returning to that hypothesis: Freund's claim is that artifacts from the Cave of Letters that are datable to the First Revolt period were, in fact, deposited in the cave during that time. Moreover, he asserts, the more precious and valuable of these items, such as the bronze hoard, came from the precincts of the Second Temple. With respect to the Psalms scroll, is Freund's hypothesis possible? The palaeographic evidence and the links between this Psalms scroll and MasPs[a] answer that question in the affirmative. The scroll is old enough and, not incidentally, of sufficiently high quality in its pristine state for that scenario to be credible. The late Herodian fragment of the larger 5/6ḤevPs scroll (as well as XḤevNum[a], XḤev/SeNum[b], and XḤev/SeDeut) provides, if not proof, then at least one more link in the chain of circumstantial evidence adduced by Professor Freund. But could it not also be possible that the Bar-Kokhba rebels had this and other eighty-year-old scrolls in their possession and that they deposited them in the Cave of Letters? While this scenario also remains possible and cannot be ruled out on the evidence of the Psalms scroll alone, the Murabba'ât materials, at the very least, cast suspicion on the traditional interpretation of the Psalms fragment and the other scrolls originally deposited in the Cave of Letters. Recognizing that the state of the evidence to date does not allow for an irrefutable conclusion, the circumstantial evidence here adduced, coupled with evidence drawn from other first-century artifacts found in the Cave of Letters and elsewhere in the Judaean desert, nevertheless tilts the balance in favor of Freund's hypothesis. Perhaps a definitive answer will be had only with another expedition to the Cave of Letters. Until such time as that difficult and expensive expedition can be undertaken, however, the above remarks point to the importance of re-evaluating widely accepted conclusions about the archaeological data associated with the Cave of Letters as well as the other caves of the Judaean wilderness.

NOTES

1. R. Freund and R. Arav, "Return to the Cave of the Letters. What Still Lies Buried?" *BAR* 27/1 (January/February 2001) 39.
2. P. Flint, "5/6ḤevPsalms," *Miscellaneous Texts from the Judaean Desert* (DJD 38; Oxford: Clarendon Press, 2000) 141–66. In the same volume, Flint has published other important biblical texts from Naḥal Ḥever including 5/6Numbers[a], XḤev/SeNumbers[b], and XḤev/SeDeuteronomy.
3. Cave 5/6 was so designated as the cave has both an eastern and a western entrance. See Y. Yadin, "The Expedition to the Judaean Desert, 1960, Expedition D," *IEJ* 11 (1961) 40.
4. See Y. Yadin, "The Expedition to the Judaean Desert, 1960, Expedition D," (esp. 40 and plate XXD) as well as the popular account found in his *Bar-Kokhba. The Rediscovery of the Legendary Hero of the Last Jewish Revolt Against Imperial Rome* (London: Weidenfeld and Nicolson, 1971) 113–14.
5. Y. Yadin, "The Expedition to the Judaean Desert, 1961, Expedition D," *IEJ* 12 (1962) 229 and plate D.
6. B. Lifshitz, "The Greek Documents from the Cave of Horror," *IEJ* 12 (1962) 201–7; E. Tov, *The Greek Minor Prophets Scroll from Naḥal Ḥever [*8ḤevXII Gr*]* (DJD 8; Oxford: Clarendon, 1990) 1.
7. J. Starcky, "Un contrat nabatéen sur Papyrus," *RB* 61 (1954) 161–81.
8. J. Milik, "Un contrat juif de l'an 134 après J.-C," *RB* 61 (1954) 182–90.
9. D. Barthélemy, "Redécouverte d'un chaînon manquant de l'histoire de la Septante," *RB* 60 (1953) 18–29.
10. J. C. Greenfield, "The Texts from Naḥal Ṣe'elim (Wadi Seiyal)," *The Madrid Qumran Congress. Proceedings of the International Congress on the Dead Sea Scrolls Madrid 18–21 March, 1991* (*STDJ* 11,2; eds. J. T. Barrerra and L. V. Montaner; Leiden: E.J. Brill, 1992) 661–5.
11. H. M. Cotton and A. Yardeni, *Aramaic, Hebrew, and Greek Documentary Texts from Naḥal Ḥever and Other Sites* (DJD 27; Oxford: Clarendon, 1997) 1–6.

12. Greenfield, "The Texts from Naḥal Ṣe'elim (Wadi Seiyal)," 662.
13. P. Flint, "Biblical Scrolls from Naḥal Ḥever and 'Wadi Seiyal': Introduction," *Miscellaneous Texts from the Judean Desert* (DJD 38; Oxford: Clarendon Press, 2000) 134; *cf.* P. W. Flint, *The Dead Sea Psalms Scrolls and the Book of Psalms* (STDJ 17; Leiden: Brill, 1997) 43–4; Greenfield, "The Texts from Naḥal Ṣe'elim [Wadi Seiyal]," 662–3.
14. P. Flint, "5/6ḤevPsalms," 134–5.
15. P. Flint, "Biblical Scrolls from Naḥal Ḥever and 'Wadi Seiyal': Introduction," 134.
16. For the following see P. Flint, "5/6ḤevPsalms," 141–5.
17. See P. W. Flint, *The Dead Sea Psalms Scrolls and the Book of Psalms*, 43–5. On MasPs[a], see below. Shemaryahu Talmon, editor of MasPs[b], judges that manuscript to date to the last half of the first century BCE. *Cf.* S. Talmon, "Fragments of a Psalms Scroll from Masada, MPs[b] (Masada 1103–1742)," *MinḤah le-NaḤum: Biblical and Other Studies Presented to Nahum M. Sarna in Honour of His 70th Birthday* (JSOTSup 154; ed. M. Brettler and M. Fishbane; Sheffield: *JSOT*, 1993) 318–27 + pl.
18. Not preserved are columns I–II with their estimated contents of Psalms 1:1–7:5, column X with its estimated content of Psalm 18:44–22:3, and columns XIII–XIV with their estimated contents of Psalms 25:7c-28:9. See P. Flint, "5/6ḤevPsalms," 141, chart 1.
19. P. Flint, "5/6ḤevPsalms," 142.
20. *Ibid.*
21. *Ibid.*
22. As cited by Greenfield, "The Texts from Naḥal Ṣe'elim (Wadi Seiyal)," 663.
23. F. M. Cross, "Palaeography and the Dead Sea Scrolls," *The Dead Sea Scrolls After Fifty Years. A Comprehensive Assessment* (eds. P. W. Flint and J. C. VanderKam; Leiden: Brill, 1998) 388–9 and plate 10. In his discussion of the evolution of the Herodian script, Cross neglects to explain how 5/6ḤevPs is to be distinguished from other examples of the late Herodian formal script. For an early assessment of a late first-century C.E. date which perhaps has unduly influenced Cross (pace tanti viri) see Y. Yadin, "The Expedition to the Judaean Desert, 1960, Expedition D," 40.
24. See P. W. Flint, *The Dead Sea Psalms Scrolls and the Book of Psalms*, 43, 253; "Psalms Scrolls from the Judean Desert," *The Dead Sea Scrolls: Hebrew, Aramaic, and Greek Texts with English Translation 4a* (eds. J. H. Charlesworth, H. W. Rietz and P. W. Flint; Tübingen: Mohr Siebeck, 1997) 289.
25. P. Flint, "5/6ḤevPsalms," 143.
26. For the characteristics of the Herodian and late Herodian script described here, see P. Flint, "5/6ḤevPsalms," 143, and Cross, "Palaeography and the Dead Sea Scrolls," 388–92. Line numbers 1 through 5 below refer to Flint's publication of the Yadin fragment ("5/6ḤevPsalms," 153–4); lines 7 and 8 refer to the blank space and fragmentary remains of Psalm 16:1, respectively.
27. P. Flint, "5/6HevPsalms," 144. All of the variants are listed in his publication of the individual portions of 5/6ḤevPs, pp. 146–66. For a list of the major variants see, conveniently, Flint's *The Dead Sea Psalms Scrolls and the Book of Psalms*, 82–3.
28. The only attestation of **rgl* at Qumran is a Dp participle in 1 QM 5.13, referring to the lower foot of a sword (J. Duhaime, "War Scroll [1QM; 1Q33; 4Q491-496 = 4Qm1-6; 4Q497]," *The Dead Sea Scrolls: Hebrew, Aramaic, and Greek Texts with English Translations 2* (eds. J. H. Charlesworth, H. W. Rietz and P. W. Flint; Tübingen: J.C.B. Mohr [Paul Siebeck], 1993, 108) and a curious astrological text (4Q186 2 1.2, see J. Allegro, *Qumrân Cave 4 [4Q158—4Q186]* (DJD 5; Oxford: Clarendon, 1968, 88–91) the elusive meaning of which seems not pertinent.
29. See, however, the attractive suggestion of M. Dahood (*Psalms I: 1–50* [*Anchor Bible 16*; Garden City: Doubleday & Company, 1966], 83–4) who understands *rāgal* to represent a denominative

verb formed from the name of a body part, here "foot." He thus translates the phrase, "He who does not trip (*rāgal*) over his tongue."

30. *Cf.* Psalm 10:6–8.
31. P. Flint, "5/6ḤevPsalms," 155.
32. See P. W. Flint, *The Dead Sea Psalms Scrolls and the Book of Psalms*, 117–49, esp. 148–49.
33. *Ibid.*, 146.
34. *Ibid.*, 148. Flint continues: "For Psalms 1–89, the order of contiguous Psalms almost always corresponds with that found in the MT-150 Psalter (92% supportive, 8% contradictory), with the only two exceptions involving Psalms that lack superscriptions in the Received Text." See also M. C. Dailey, "The Psalter of the Dead Sea Scrolls: Arguing for an Early Formulation of the Psalms," MA thesis (Ohio State University, 1996) 24–44.
35. Besides 5/6ḤevPs, Psalms scrolls dated from the early to middle first century CE include 4QPs[e], 4QPs[g], 4QPs[j], 4QPs[q], 4QPs[t], 11QPs[a], 11QPs[c], 11QPs[d], and from Masada, MasPs[a]. Only marginally later are 4QPs[c] (50–68 CE), 4QPs[s] (50 CE or later), and 11QPsAp[a] (50–70 CE). *Cf.* P. W. Flint, "Psalms Scrolls from the Judean Desert," 288–9; *The Dead Sea Psalms Scrolls and the Book of Psalms*, 49, notes 150 and 151.
36. Thirty-six Psalms manuscripts have been recovered from Qumran. Other well represented biblical books include Deuteronomy (27) and Isaiah (24). These numbers are in contrast to the two manuscripts representing Qoheleth (4QQoh[a, b]) and one each for Ezra (4QEzra) and Chronicles (4QChron). *Cf.* P. W. Flint, "Psalms Scrolls from the Judean Desert," 221. The special role of the Psalms for the Qumranic community, as well as the importance of books of Deuteronomy and Isaiah, is likewise evidenced by the fact that these books are the most frequently quoted in the sectarian writings of the Qumran sect. *Cf.* E. Tov, "Hebrew Biblical Manuscripts from the Judaean Desert: Their Contribution to Textual Criticism," Jewish Civilization in the Hellenistic-Roman Period (ed. S. Talmon; Philadelphia: Trinity, 1991) 122 (previously published in JSS 39 (1988): 5–37); J. Carmignac, "Les citations de l'ancien testament dans 'La guerre des fils de lumiére les fils de ténébres'," *RB* 63 (1956) 375–90; P. Wernberg-Møller, "The Contribution of the Hodayot to Biblical Textual Criticism," *Textus* 4 (1964) 173–5.
37. 4Q397 14–21 C, lines 9–10: "And we have also written to you that you should examine the book of Moses, and the books of the Prophets and David…" in E. Qimron and J. Strugnell, *Qumran Cave 4: V. Miqṣat Ma'aśe Ha-Torah* (DJD 10; Oxford: Clarendon, 1994) 58–9.
38. P. W. Flint, *The Dead Sea Psalms Scrolls and the Book of Psalms*, 23.
39. E. Urlich, "Pluriformity and the Biblical Text," *The Madrid Qumran Congress. Proceedings of the International Congress on the Dead Sea Scrolls Madrid 18–21 March, 1991* (STDJ 11,1; eds. J. T. Barrerra and L. V. Montaner; Leiden: E.J. Brill, 1992) 34; P. W. Flint, *The Dead Sea Psalms Scrolls and the Book of Psalms*, 218–9.
40. 11QPs[a] = 11Q5.
41. See J. A. Sanders, J. Charlesworth and H. Rietz, "Non-Masoretic Psalms [4Q88 = 4QPs[f], 11Q5 = 11QPs[a], 11Q6 = 11QPs[b]]," *The Dead Sea Scrolls: Hebrew, Aramaic, and Greek Texts with English Translation 4a* (eds. J. H. Charlesworth, H. W. Rietz and P. W. Flint; Tübingen: Mohr Siebeck, 1997) 214–5. The full line reads: *kwl 'lh dbr bnbw'h 'šr ntn lw mlpny h'lywn*. Flint (*The Dead Sea Psalms Scrolls and the Book of Psalms*, 219) sagely cautions against identifying Psalms with the later collection of Nebi'im in the MT, especially since 4QMMT distinguishes between the "prophets" and "David." He concludes, however, by stating, "It is nevertheless clear that the Psalms were among those Scriptures which contained the revealed truth that was to be interpreted, and served as the basis for ordering of the community."

42. 4Q503 frgs. 51–55, 9. See D. T. Olson, "Daily Prayers (4Q503 = 4QPrQuot)," *The Dead Sea Scrolls: Hebrew, Aramaic, and Greek Texts with English Translation 4a* (eds. J. H. Charlesworth, H. W. Rietz and P. W. Flint; Tübingen: Mohr Siebeck, 1997) 263.
43. Urlich, "Pluriformity and the Biblical Text," 34; P. W. Flint, *The Dead Sea Psalms Scrolls and the Book of Psalms*, 218–9.
44. Sanders, Charlesworth and Rietz, "Non-Masoretic Psalms (4Q88 = 4QPs^{f}, 11Q5 = 11QPs^{a}, 11Q6 = 11QPs^{b})," 213.
45. G. Wilson, *The Editing of the Hebrew Psalter* (SBLDS 76; Chico, CA: Scholars, 1985).
46. P. W. Flint, *The Dead Sea Psalms Scrolls and the Book of Psalms*, 193–5, 224.
47. Sanders, Charlesworth and Rietz, "Non-Masoretic Psalms (4Q88 = 4QPs^{f}, 11Q5 = 11QPs^{a}, 11Q6 = 11QPs^{b})," 213.
48. Flint (*The Dead Sea Psalms Scrolls and the Book of Psalms*, 49, note 148 and page 45), on paleographical grounds, also includes a seventh text of uncertain provenance among the Herodian Psalms scrolls. The fragment in question was discovered in Yadin's study and has been variously catalogued as Mas1g(?), MasPs^{c}(?) and now, by Flint, as 4QPs^{v}.
49. Emanuel Tov, "Special Layout of Poetical Units in the Texts from the Judean Desert," in *Give Ear to My Word: Psalms and Other Poetry in and Around the Hebrew Bible. Essays in Honour of Professor N.A. van Uchelen* (eds. J. Dyk, et al.; Amsterdam: Societas Hebraica Amstelodamensis, 1996) 118–19. A second century BCE scroll, 4QPs^{a}, presents the psalms as a running text except for the preserved portion of Psalm 119 which is copied with two hemistichs per line.
50. Tov, "Special Layout," 120–121. 5QPs has two hemistichs per line separated by a space while 4QPs^{g} has no such divider between the stichs.
51. Flint, *The Dead Sea Psalms Scrolls and the Book of Psalms*, 31–37, 252–253.
52. *Ibid.*, 32 and note 23.
53. *Ibid.*, 34.
54. *DJD* 3: 148–9 and plate xxxi.
55. *Ibid.*, 149.
56. On the "Qumran system" see E. Tov, "Hebrew Biblical Manuscripts from the Judaean Desert: Their Contribution to Textual Criticism," 114–9. Tov divides the Qumran scrolls between those copied elsewhere and imported into the settlement and those scrolls characterized by the "Qumran system." The features of the latter include peculiarities of orthography, language, and scribal markings, the placement of initial-medial letters in final positions, the generally higher quality of the physical scroll itself, and the writing of the divine names ʾl(hym) and yhwh in palaeo-Hebrew letters in texts written in the Assyrian script. In addition, he notes, "The scrolls written in Qumran orthography and language display a tendency to modernize biblical Hebrew spelling and languange" (119). See, however, Eugene Ulrich [*The Dead Sea Scrolls and the Origins of the Bible* (Grand Rapids: William B. Eerdmanns, 1999) 110–120] who expresses reservation as to whether or not the idea of a "Qumran practice" remains a useful concept. He suggests instead that variations such as described by Tov are "generally typical of contemporary Palestinian copyists rather than specific to the copyists at Qumran" (112).
57. E. Tov, "Hebrew Biblical Manuscripts from the Judaean Desert: Their Contribution to Textual Criticism," 120, especially note 1.
58. Tov, "Special Layout," 126 and passim. See, however, the earlier suggestion by Joseph M. Oesch who proposes a functional origin for sundry copying patterns ["Textgliederung im alten Testament und in den Qumranhandschriften," Henoch 5 (1983) 317]. Specifically, Oesch asserts "dass die Wahl der Gliederungsmethode und des dadurch geschaffenen Gliederungssystems von Zweck

abhängt, für welchen eine HS angefertigt wurde." Evidence for this proposal is indirect only and is drawn by analogy from findings at the Cairo Geniza.

59. There are eleven scrolls dated to the first century BCE by means of palaeographic analysis. Of these, five are copied in a prose format (4QPs[f], 4QPs[k], 4QPs[n], 4QPs[o], and 4QPs522), and one (4QPs[u], including only Psalm 99:1!) is too fragmentary for stichometric analysis. The remaining scrolls, 1QPs[a], 4QPs[b], 4QPs[d], 4QPs[l] and MasPs[b] present their psalms variously. 4QPs[b] and 4QPs[l] have one hemistich per line, 1QPs[a] has two hemistichs on each line with an intervening space, while MasPs[b] displays what the masoretes would later call the as "a half-brick atop a half-brick" layout [see Tov, "Special Layout," 121; Talmon, *Masada VI*, 92]. It is interesting to note that two of these scrolls either omit psalms included in the MT (4QPs[b]) or present the psalms in a different order (4QPs[d]) albeit the orthography of both—as well as that of 4QPs[l] and MasPs[b]—is close to that of the MT [see Flint, *The Dead Sea Psalms Scrolls and the Book of Psalms*, 31, 33–38].

60. MasPs[a] appears in S. Talmon, *Masada VI: Yigael Yadin Excavations, 1963–1965: Final Reports: Hebrew Fragments from Masada* (Masada Reports; Jerusalem: Israel Exploration Society, 1999) 76–90. See also two photos of the scroll published in J. K. McDonald, "Treasures from the Holy Land. Ancient Art from the Israel Museum," *BA* 43 (1986) 163 and E. Netzer, "Masada," *The New Encyclopaedia of Archaeological Excavations* (ed. E. Stern; Jerusalem: Israel Exploration Society and Carta, 1993) 982as well as the earlier helpful edition of this text (based on the McDonald photo) by G. Wilhelm Nebe, "Die Masada-Psalmen-Handschrift M1039-160 nach einem jüngst veröffentlichen Photo mit Text von Psalm 81.2–85.6," *RevQ* 53 (1989) 89–97.

61. The best preserved columns of MasPs[a] include 27 lines of text plus blanks of various widths. See Talmon, Masada VI, 78; *cf.* G.W. Nebe, "Die Masada-Psalmen-Handschrift," 91.

62. Tov, "Special Layout," 121.

63. Talmon, *Masada VI*, 78 and note 11.

64. E. Tov, "A Qumran Origin for the Masada Non-Biblical Texts?" *DSD* 7/1 (2000) 61.

65. *Ibid.*, 61. Tov lists, as examples, the texts discovered at Murabba'at (MurGen, MurExod, MurNum, MurDeut, MurIsa, MurXII), Naḥal Ḥever/Ṣe'elim (5/6ḤevNum[a], 5/6ḤevPs; XḤev/SeNum[b], XḤev/SeDeut, 34SeNum) and Sdeir.

66. *Ibid.*, 61.

67. *Ibid.*, 61, 67.

68. Y. Yadin, *The Finds from the Bar-Kokhba Period in the Cave of the Letters* (Jerusalem: Israel Exploration Society, 1963) 27.

69. *Ibid.*, 9.

70. P. Benoit, J. Milik and R. d. Vaux, *Les Grottes de Murabba'ât* (DJD 2; Oxford: Clarendon, 1961) 70, noting the author's reservation about dating the ostraca. See also R. d. Vaux, "Grottes de Murabba'ât et leurs documents," *RB* 60 (April 1953) 246–67; G. W. L. Harding, "Khi*RB*et Qumran and Wady Muraba'at," *PEQ* 84 (May-October 1952) 104–9; F. M. Cross, "Manuscripts of the Dead Sea Caves," *BA* 17 (Feb 1954) 7–12; W. Baumgartner, "Neues von den palästinischen Handschriftenfunden," *TZ* 9 (July-August 1953) 316; W. Baumgartner, "Wiederum die palästinischen Handschriftenfunde," *TZ* 9 (November-December 1953) 469–70.

71. Benoit, Milik and Vaux, *Les Grottes de Murabba'ât*, 70,77–81; Cross, "Manuscripts of the Dead Sea Caves," 11.

CHAPTER FOURTEEN

Archaeology, Museology, and Identity

The Display of the Cave of Letters Materials in the Shrine of the Book (1965–2004)*

ADOLFO D. ROITMAN
The Israel Museum, Jerusalem

INTRODUCTION: THE ORIGINAL PROGRAM OF THE SHRINE OF THE BOOK

The concept behind the world-famous building known as the Shrine of the Book originates with the discovery of the first seven scrolls. These were found by Bedouin between the end of 1946 and the summer of 1947, on the northwestern shore of the Dead Sea, in a cave (afterwards called cave 1), near Khirbet Qumran. In July 1947, four of these ancient manuscripts were acquired by the Syrian Orthodox Metropolitan of the Saint Mark monastery in East Jerusalem, Mar Athanasius Samuel. Another three were acquired by Prof. Eliezer L. Sukenik, of the Institute of Archaeology, Hebrew University, in November-December of the same year.

In 1948, following suggestions by American scholars at the American School of Oriental Research (known today as the Albright Institute) in East Jerusalem, Mar Samuel smuggled his scrolls from Palestine, through Beirut, to the States. Between 1949–51, three of the four scrolls in his possession were exhibited in art galleries and academic institutions (i.e., Library of Congress, the Institute of Oriental Studies [University of Chicago], Duke University, etc.), with the object of selling them. Finally, in 1954, Samuel's Scrolls were acquired by Yigael Yadin (Sukenik's son), who had responded to an advertisement in The Wall Street Journal dated June 1st. They were returned to the Land of Israel at the beginning of 1955.[1]

The idea of exhibiting these recently discovered scrolls was expressed in a Government decision, February 13, 1955, which established the Shrine of the Book Foundation. This decision spelled out the responsibilities of the Foundation. It would act as trustee for the Scrolls and collect manuscripts, documents, and other remains connected to the Bible. In addition, the Foundation "...would build a special hall in the National and University Library in Jerusalem which would be a repository and museum for preserving and displaying these items and any other material connected to the Bible."[2]

From the beginning, then, the Directors of the Foundation emphasized the ties between the Scrolls and the Bible, attempting to identify the Shrine of the Book as the Shrine of The Book. During the annual meeting of the Foundation's Directors, May 6, 1956, when discussing the proposition that the Shrine of the Book be part of the section of the National and University Library, the then Minister of Education, Zalman Aran, suggested taking advantage of the opportunity to create, in the same hall, an area "...for a permanent exhibition of the Bible in Jerusalem."[3]

The ancient documents were first exhibited in Israel during Passover, 1955, in the Senate Hall of the Hebrew University, located in those days in the Terra Sancta Building in West Jerusalem. In July 1957, they were put on permanent display in a small room in the Hebrew University's administration building in the campus at Giv'at Ram. The curator of this exhibition was Y. Yadin, then professor of archaeology at the Hebrew University.[4]

In the meantime, in 1957 two American-Jewish architects, Armand Bartos and Frederik Kiesler, were nominated to design the hall to house the Dead Sea Scrolls. Between 1957–59, several proposals were advanced. None were accepted since the architects wanted an independent building, rather than just an exhibition hall as part of the National Library project, as originally suggested by the 1955 Government decision.

In 1959, the original scheme was cancelled by the University authorities and transferred to the Israel Museum project. This was thanks to the efforts of Teddy Kollek, in those days General Director of the Prime Minister's office, and Yigael Yadin, on behalf of the Hebrew University's steering committee for the archeological section of the future National Museum. In October 1959, a final decision was made. The Shrine of the Book would be built on one of the hills of Neve Sha'anan, near the University and overlooking the Valley of the Cross. After meticulous negotiations, the agreement was finally signed on May 11, 1960.[5]

On that day, the museological program of the Shrine of the Book was very clear to all. The plan was to show the Scrolls along with materials related to the Bible. However, on that very night, May 11, 1960, something would happen to introduce a dramatic change in the original agenda.

THE FINDINGS FROM THE CAVE OF LETTERS

That night a meeting was held at the residence of the then President of the State of Israel, Yitzhak Ben-Zvi. The meeting was held in presence of the President, the Prime Minister David Ben-Gurion, Cabinet ministers, the Chief-of-Staff, the Speaker and Members of the Knesset (Israeli Parliament), and other distinguished personalities. The four archaeologists, who had supervised the expeditions dispatched to the Judean desert last April, Prof. Yohannan Aharoni, Prof. Nahman Avigad, Mr. Pesach Bar-Adon and Prof. Yigael Yadin, gave an account and presented results of their investigations.

Three archaeologists summarized their findings, and then came Yadin's turn to give his report. What happened then and afterwards is told by Yadin himself, as befitting a gifted storyteller:

> A screen had been erected at Mr Ben Zvi's house, and when my turn came to report, I projected on to the screen through a film slide the coloured photograph of part of a document and read out aloud the first line of the writing upon it: 'Shimeon Bar Kosiba, President over Israel'. And turning to our Head of State, I said: 'Your Excellency, I am honored to be able to tell you that we have discovered fifteen dispatches written, or dictated, by the last President of ancient Israel, 1,800 years ago.'
>
> For a moment, the audience seemed to be struck dumb. Then the silence was shattered with spontaneous cries of astonishment and joy. That evening the national radio interrupted its scheduled programme to broadcast news of the discovery. Next day, newspapers came out with banner headlines over the announcement.[6]

Why was a whole nation overwhelmed by the discovery of these fragmentary papyri? According to Yadin, "the answer lies in the magic of the name, a name treasured in folklore but almost lost to authenticated history, and the realization at this meeting that after nearly two thousand years the desert had given up factual links with the man who led the last attempt of his people to overthrow their Roman masters."[7]

However, as was brilliantly shown by Prof. Yael Zerubavel in her paper, "Bar Kokhba's Image in Modern Israeli Culture" the answer is much more sophisticated.[8] In fact, the memory of Bar Kokhba had been manipulated by Hebrew intellectuals, writers, poets and educators since the end of the 19th century to support the ideology of national revival and liberation. In contrast to ancient Jewish sources, which had presented his image as a false Messiah and arrogant hero,[9] Zionist memory showed a highly selective attitude toward traditional Jewish memories. It now focused on the ancient portrayal of Bar Kokhba as a charismatic leader and daring hero, playing down or ignoring the highly negative stories about him.[10]

In light of the changes and developments in Bar Kohba's image in modern Israeli culture, the answer to the question asked above is clear, "Why was a whole nation overwhelmed by the discovery of the fragmentary papyri?" As stated by Zerubavel, "[t]he archaeological findings from the Bar Kokhba period provided material proof of the historicity of the ancient leader whom Hebrew national culture had celebrated as a legendary figure in its pantheon of Jewish heroes."[11] Therefore, hearing news about such an exceptional discovery, Israelis naturally felt that a national myth had become a reality and, as stated by one of the newspapers of that day, Yedioth Aharonot: "Bar Kokhba is still alive—in his letters."

THE FINDINGS FROM THE CAVE OF LETTERS AT THE SHRINE OF THE BOOK

As a consequence of the dramatic announcement, and taking into account the cultural significance of Bar Kokhba's image for the Israelis in the 1960's, and also the fact that Yadin himself was closely involved in the Shrine of the Book project,[12] it came as no surprise when it was decided to introduce a strategic change in the original museological program of the Shrine of the Book. The original scrolls from Qumran would be displayed together with the papyri of Bar Kokhba and Babatha and the other exceptional artifacts discovered between the years 1960–61, in "The Cave of Letters."[13]

The Shrine of the Book, built between 1961–64, is composed of two distinct areas. One, the outside area, where the White Dome and the Black Wall are standing symbolizing a central theological motif in the scrolls, i.e. the struggle between the "Sons of Light" and "the Sons of Darkness". The other area is underground and is divided into three spaces: an orientation room, a corridor built in the form of a cave, reflecting the fact that the Scrolls (and Bar Kohba's findings) were discovered in caves, the round main hall and a further exhibition space down a curving staircase. The museum was built, not merely as a functional building for the preservation and display of archaeological materials, but as a symbolic structure. As it name suggests, the designers conceived of a temple-like structure, relating to the idea of a State renewing its links with the past. The guiding concept was that the architectural design should convey certain messages, and offer the visitor a spiritual experience combining architectural and symbolic design.[14]

The new curator M. Broshi, appointed by Yadin at the end of 1963, started officially serving in January 1964, and during that year developed his strategy for the exhibition. The Scrolls from Qumran were exhibited in the main hall, in accordance with the original idea, while the Bar Kokhba and Babatha's archives were displayed in the corridor. The bronze jugs and the exceptional glass plate were shown at the lower level, together with the shovels, patera, the keys and the fabrics.

The Shrine of the Book was officially inaugurated on April 20, 1965 in the presence of the President of the State and diplomats from all over the world. From that day on, thousands of visitors came to the Shrine, queuing for hours, compelled with an almost religious fervor, to see the exceptional findings on display.[15]

Probably, some readers would query the idea of "religious fervor," seeing in these words a kind of exaggerated metaphor. However, this museum was consciously (or unconsciously) conceived to serve as the secular temple of the new-born State of Israel.[16] Archaeology, in those days, functioned as a kind of pseudo-religion in Israeli society,[17] and Yadin played the role of mythmaker and a kind of cultural "high-priest."[18] If the Shrine, as argued, was a kind of secular temple of this pseudo-religion, then the Scrolls and the treasures from the Cave of Letters (and for a while, some few biblical and non-canonical manuscripts from Masada, discovered also by Yadin between 1963–65) were, without exaggeration, kinds of "sacred objects."[19]

According to Zerubavel's interpretation, the exhibition of Bar Kokhba's findings in the Shrine of the Book might also have had a clear nationalistic significance for the people of that generation: "the display of artifacts associated with his revolt at the Israel Museum was presented as an act of historical justice and a proof of the ancient rebels' ultimate victory over the Romans."[20] As stated very clearly by Yadin himself:

> Archaeologists are also human beings, and as human beings, they are often emotionally attached to the history of their own people.
>
> Descending daily over the precipice, crossing the dangerous ledge to the caves, working all day long in the stench of the bats, confronted from time to time with the tragic remains of those besieged and trapped—we found that our emotions were a mixture of tension and awe, yet astonishment and pride at being part of the reborn State of Israel after a Diaspora of 1,800 years. Here were we, living in tents erected by the Israel Defence Forces, walking every day through the ruins of a Roman camp which caused the death of our forefathers. Nothing remains here today of the Romans save a heap of stones on the face of the desert, but here the descendants of the besieged were returning to salvage their ancestors' precious belongings.[21]

Yadin played a key role in promoting the significance of Bar Kokhba for modern Israelis through public lectures, scholarly and popular publications, and personal guiding at the Shrine itself. It must be recalled the fact that he himself had been the second Chief-of-Staff of the Israeli army and served during the 1948 Independence War. Therefore, it is understandable that Yadin identified himself with Bar Kokhba the hero and rebel against the Roman forces, as he led the Israeli forces in the war against the Arab nations.

Noteworthy, is the conversation held between the then Minister of Education, Z. Aran, and Yadin, at the meeting of the Board of Trustees of the Shrine of the Book Foundation on April 14, 1966, that is a year after the inauguration of the museum. According to the minutes of that meeting, Aran asked the participants a question: "What can we do to impart special value to young people as they visit the Shrine?" And Yadin replied:

> Regarding our young people, I am not concerned: they have the appropriate background, they have heard about Bar Kokhba, and when they see a letter from Bar Kokhba, it deepens their consciousness, and then they return home and read about it again. [...] [A]s for the foreign visitor, I think that the Shrine of the Book does an excellent job of public relations for the State of Israel and for Zionism in general. When visitors who have never heard of Bar Kokhba are told the story of the rebellion against the Romans and shown documents testifying to the historicity of the event, they become better acquainted with the Jewish past.[22]

After its inauguration, no significant changes were introduced in the Shrine of the Book's permanent exhibition approach until 1997. The original display had been in place for nearly thirty years, apart from a display in 1984 of twelve Cairo Genizah manuscripts, which temporarily replaced Bar Kokhba's and Babatha's material.[23]

THE REPLACEMENT OF BAR KOKHBA'S AND BABATHA'S MATERIAL

Since the inauguration of the museum in 1965, the Bar Kokhba's and Babatha's papyri had been installed in the corridor before the Scrolls shown in the main hall, and the artifacts from the Cave of Letters had been put on display downstairs, leaving the unconnected story of the Scrolls in the middle. A fact which produced confusion among visitors, and countless problems for tourist guides.

In July 1997, against a background of the international conference held at the Israel Museum to commemorate the fiftieth anniversary of the discovery of the Dead Sea Scrolls, a dramatic and far-reaching change was introduced to the old-fashioned conception. Bar Kokhba's and Babatha's papyri and related materials were taken out of the Shrine of the Book. In their place, a new permanent exhibition, called A Day at Qumran, was installed by the recently appointed curator (the writer of the present paper). Its purpose was to present everyday life of the people who lived in Qumran two thousand years ago. This showed their life from sunrise to sunset on the basis of information in Josephus' and Philo's works on the Essenes, in the Scrolls themselves, and in the information provided by the archaeological finds discovered in Khirbet Qumran and the caves.[24]

The reasons for this change were practical, museological and archaeological. As apparent when installing the Cairo Genizah show some years ago, it was clear there was no other suitable place in the building to display this exhibition apart from the corridor. Also, the exhibition, A Day at Qumran, was to serve as an introduction to the manuscripts on show in the main hall, providing a sociological and historical perspective lacking previously. And therefore, logically the exhibition had to be installed in the area located before the main hall. And finally, although the Scrolls and the Cave of Letters materials were both discovered in the Judean desert and belonged to the history of the Jewish people in Second Temple times, no real and essential relationship can be established between the discoveries. Strictly speaking, they are basically two different and non-connected stories, and therefore there was not any archaeological justification to display both discoveries at the same place.[25]

Some remnants of Bar Kokhba's finds from the Cave of Letters located downstairs were left untouched. However, a few years later, during the renewal of the building in the years 2003–04, the artifacts from the Cave of Letters were totally removed from the Shrine. In the space vacated, a new exhibition was installed, called A Wandering Bible: The Aleppo Codex, devoted to the history of the Hebrew Bible, and the development of this sacred book from Qumran through to modern times.[26]

In this way, after forty years, the original program of the Shrine of the Book was finally restored, giving real content to the name of the institution. The Shrine of the Book now became a real Shrine of The Book—i.e., the Shrine of the Bible, and not, as had happened in the past, the Shrine of the Scrolls.[27]

THE SIGNIFICANCE OF THE REPLACEMENT OF BAR KOKHBA'S MATERIAL

Finally, a crucial question must be raised: Why is it that no objections were raised to these new and radical changes by the Israel Museum's authorities? What has happened to Bar Kokhba's significance for the modern Israeli?

Since the '70s, heroic myths of the early years of the State, as those of Masada and Tel Hai, had become the subject of criticism and revisionism in Israeli political discourse, literature, art, and popular culture.[28] Thus, the Bar Kokhba revolt also became the subject of a heated debate during the early '80s, initiated by Yeoshafat Harkavi, a professor of international relations at Hebrew University. He argued that Bar Kokhba's revolt had been a catastrophic political mistake that brought about a massive destruction of Jewish life, and in contradiction to the traditional Zionist interpretation of the revolt, he rejected its positive and inspiring significance for contemporary Israel.[29]

In spite of the counter-criticism made by ideologists (as Israel Eldad, a Revisionist-Zionist scholar), historians (as Prof. David Rokeah) and archaeologists (as Yadin and Broshi), "Harkabi's views," as argued by Zerubavel, "contributed to the public awareness of the problematic aspects of the revolt and highlighted the recent, constructed character of its commemoration in modern Israeli culture."[30] Since then, the weakening of Bar Kokhba's image in Israeli culture has only deepened, being transformed by alternative ideological and interpretative frameworks, as evidenced in educational narratives and public rituals.[31]

As concrete proof of the dramatic change of Bar Kokhba's image in recent years in the Israeli public, a personal anecdote comes to mind. Having started working on this paper in 2002, I asked my then 13 year-old son Nethanel about the identity of Bar Kokhba. After a few moments, he replied: "Daddy, I do not know!." Supposing that he was making fun, I asked him a further question about Bar Kokhba's era. And his answer was hesitatingly: "In Jesus' times?" And finally, when I asked him about the connections of Bar Kokhba with any Jewish festival, his answer was again hesitatingly: "Hannuka?"[32]

All this shows, then, that the decision to tell the story of the Qumran community and the Aleppo Codex at the Shrine of the Book, instead of the findings from the Cave of Letters, was probably not only the consequence of a rationale process intended to add a museological coherence to the narrative on display, and to return to the original program of the institution. It might have been also a sublimate and unconscious expression of the dramatic change happening to modern Israeli society. From being a one-sided Zionist and militaristic-orientated society, it had become by those years pluralistic and peace-orientated. In this new cultural atmosphere, the once mythical significance of Bar Kokhba and the display of the Cave of Letters materials in the Shrine of the Book had become anachronistic.[33]

NOTES

* A first draft of this article was read at the Conference held at New York University (February 3, 2002) on "The Dead Sea Scrolls and the Bar Kokhba Documents." I would like to thank Mr. Cyrill Hirsch for his editing assistance.

1. For the history of the discovery and acquisition of the first seven scrolls, see Y. Yadin, *The Temple Scroll.* Tel Aviv: Sifriyat Maariv, 1990, chap. 2 (Heb.).
2. Quoted from the Shrine of the Book constitution (5.15.1957).
3. The General Archive, the Hebrew University of Jerusalem, 1956, File no. 2252; from the minutes of the meeting of the Board of Directors of the Shrine of the Book Foundation, May 6, 1956, p. 7.
4. For more details, see A. Roitman, "Exhibiting the Dead Sea Scrolls: Some Historical and Theoretical Considerations," in: N. A. Silberman and E. S. Frerichs, eds. *Archaeology and Society in the 21st Century. The Dead Sea Scrolls and Other Case Studies.* Jerusalem: Israel Exploration Society/The Dorot Foundation, 2001, 51–52.
5. On this issue, see A. Roitman, "From a Functional Room to an Ideological Building," in: L. Mazor, ed. *On a Scroll of a Book. Articles on the Dead Sea Scrolls.* Jerusalem: The Magnes Press, 1997, 98–109 (Heb.); *Idem*, "Exhibiting the Dead Sea Scrolls" (see note 4 above), 51–57.
6. Y. Yadin, Bar Kokhba, *The Rediscovery of the Legendary Hero of the Last Jewish Revolt Against Imperial Rome.* Tel Aviv/Haifa: Steimatzky's Agency Ltd, 1971, repr. 1978, 15.
7. *Idem.*

8. Y. Zerubavel, "Bar Kokhba's Image in Modern Israeli Culture," in: P. Shaeffer, ed. *The Bar Kokhba War Reconsidered. New Perspectives on the Second Jewish Revolt against Rome.* Mohr Siebeck, 2003, 279–297.
9. On the perceptions of Bar Kokhba's revolt in ancient sources, see A. Reinhartz, "Rabbinic Perceptions of Simeon Bar Kosiba," *Journal for the Study of Judaism* XX/2 (1989): 171–194; A. Oppenhaimer, s. v. "Bar Kokhba, Shim'on," in: L. H. Schiffman and J. C. VanderKam, eds. *Encyclopedia of the Dead Sea Scrolls.* Oxford: Oxford University Press, 2000, I, 78–80.
10. Zeruvabel, "Bar Kokhba's Image" (see note 8 above), 280–289. See recently, S. Turel, "Bar Kokhba: Construction of a Myth," in: *Bar Kokhba. Historical Memory and the Myth of Heroism.* Tel Aviv: Musa Eretz Israel Museum Tel-Aviv, 2016, 8e–17e.
11. *Idem*, 290.
12. On this issue, see A. Roitman, "Yigael Yadin and the Shrine of the Book," *The Israel Museum Journal XIX* (2001): 40–41.
13. For a survey of the findings from the Cave of Letters, see Y. Yadin, s. v. "The Cave of Letters," in: E. Stern, ed. *The New Encyclopedia of Archaeological Excavations in the Holy Land.* Rishon Le-Zion: Israel Exploration Society/The Ministry of Defence/Karta, 1992, III, 907–911 (Heb.). About the discussions held concerning the needs for a change in the original museological program, see the minutes of the meeting of the Trustees of the Shrine of the Book Foundation, August 1st, 1961.
14. On the architecture and symbolism of the building, see M. Sgan-Cohen, "La Sanctuaire du Livre et l'art de transformer le superflu en necessaire," in: Ch. Béret, ed. Frederick Kiesler; *Artiste-architecte*, Paris, 1996, 229–239; A. Roitman, "The History, Architecture, and Symbolism of the Shrine of the Book," *The Israel Museum Journal XV* (1997): 15–34; *Idem*, The Bible in the Shrine of the Book. Jerusalem: The Israel Museum, 2006, 68–92.
15. As described by M. Perelman, "Even for the layman, seeing these writings—direct communications from the ancient world to the present—is a moving encounter. For the Hebrew-speaking Israeli, able to recognize, read and follow the ancient Hebrew text, there is an added dimension of excitement and significance." (*The Dead Sea Scrolls in the Shrine of the Book.* Jerusalem: The Israel Museum, 1988, fifth edition 1993, 94).
16. See Roitman, "Exhibiting the Dead Sea Scrolls" (see note 2 above), 59–60; *Idem, Envisioning the Temple. Scrolls, Stones, and Symbols.* Jerusalem: The Israel Museum, 2003, 143–145. As argued by A. Elon: "Consciously or not, Kiesler absorbed to a remarkable degree local feelings towards the Scrolls, in all their complexity. No building in Israel is as clearly based on the exploitation of anatomical shapes and erotic symbols. In the Shrine of the Book, archaeology and nationalism are mated as an ancient rejuvenation and fertility rite." (*The Israelis: Founders and Sons.* New York: Holt, Rinehart and Winston, 1971, 284).
17. On Israeli archaeology and politics, see A. Elon, "Politics and Archaeology," in: N.A. Silberman and D. Small, eds. *The Archaeology of Israel. Constructing the Past, Interpreting the Present.* Sheffield: Sheffield Academic Press, 1997, 34–47; Y. Shavit, "Archaeology, Political Culture, and Culture in Israel," in: Silberman-Small, *op. cit.*, 48–61; M. Broshi, "Religion, Ideology, and Politics and Their Impact on Palestinian Archaeology," *The Israel Museum Journal* 6 (1987): 17–32, especially, 26–30.
18. See N. A. Silberman, *A Prophet from Amongst You. The Life of Yigael Yadin: Soldier, Scholar, and Mythmaker of Modern Israel.* Reading, MA: Addison-Wesley Publishing Company, 1993.
19. That archaeology was, in those days, imbued with almost "mystical" significance for the Israelis and Jews comes out very clearly from Yadin's words on the significance of the Dead Sea Scrolls: "It was a tremendously exciting experience, difficult to convey in words, to see the original

scrolls and to study them, knowing that some of the biblical manuscripts were copied only a few hundred years after their composition, and that these very scrolls were read and studied by our forefathers in the period of the Second Temple. They constitute a vital link—long lost and now regained—between those ancient times, so rich in civilized thought, and the present day. And just as a Christian reader must be moved by the knowledge that here he has a manuscript of a sect whom the early Christians may have known and by whom they may have been influenced, so an Israeli and a Jew can find nothing more deeply moving than the study of manuscripts written by the People of the Book in the Land of the Book more than two thousand years ago." (*The Message of the Scrolls*. New York: Simon and Schuster, 1969, 14).

20. Zeruvabel, "Bar Kakhba's Image" (see note 8 above), 290. However, the display of the artifacts of the Cave of Letters seems to have had also a very practical and non-ideological reason. In a private conversation, the former curator M. Broshi confessed to me that the display of the Bar Kokhba's materials in the Shrine of the Book resolved a built-in problem, i.e., that the museum had more showcases than the findings to be displayed (the Jordanian scrolls and the other archaeological materials from Qumran would arrive in only after the Six-Day War in 1967). And, therefore, the decision to display the Cave of Letters materials was made in order "to fill in the showcases."
21. Yadin, *Bar Kokhba* (see note 6 above), 253.
22. From the minutes of the meeting of the Trustees of the Shrine of the Book Foundation, April 14, 1966, 20 (Shrine of the Book Archive).
23. The twelve manuscripts were brought from the library of the Jewish Theological Seminary in New York. Some of the documents were related to texts found in the Qumran scrolls, such as the Second Book of Chronicles (Bible), the Wisdom of Ben-Sira (Apocrypha), and the Vision of Daniel (apocalyptic literature). For details, see *Twelve Pages from the Cairo Genizah. Exhibition Catalog*. Jerusalem: The Israel Museum, 1984, 3rd ed. 1997.
24. See A. Roitman, ed. *A Day at Qumran. The Dead Sea Sect and Its Scrolls*. Jerusalem: The Israel Museum, 1997.
25. As thought by the majority of scholars, the Dead Sea Scrolls were part of the library of the radical, apocalyptic, ascetic and segregated (Essene?) community, which lived in Qumran and its surroundings between the second-first century B.C.E. and 68 C.E. On this issue, see M. Broshi, s. v. "Qumran: Archaeology," in: L. H. Schiffman and J. C. VanderKam, eds. *Encyclopedia of the Dead Sea Scrolls*. Oxford: Oxford University Press, 2000, II, 733–739; D. Dimant, s. v. "Qumran: Written Material," in: *Idem*, 739–736. On the other hand, the findings from the Cave of Letters were related to Bar Kokhba's rebels, who supported by some Pharisaic sages (the best known was Rabbi Akiva), fought against the Romans between 132–135 C.E. On this issue, see A. Oppenheimer, s. v. "Bar Kokhba Revolt," in: L. H. Schiffman and J. C. VanderKam, eds. *Encyclopedia of the Dead Sea Scrolls*. Oxford: Oxford University Press, 2000, I, 80–83. However, it must be stressed the fact that a brief occupation of Khirbet Qumran by Bar Kokhba's forces happened during the Second Revolt (132–135 C.E.), as identified by the archaeologists.
26. Roitman, *The Bible in the Shrine of the Boo*k (see note 14 above), 65–66.
27. By the way, as a consequence of this strategic change, also the Aleppo Codex finally found its logical place in the Shrine of the Book's narrative among other late biblical manuscripts, and not as it had happened since its arrival in the early '90s, that the manuscript had been shown in the round hall upstairs along with the Dead Sea Scrolls, confusing the visitors and creating the feeling that the Aleppo Codex was originally part of the sectarian library of Qumran. For a critical evaluation of the present Shrine of the Book exhibitions, see the article published by Y. Elitzur

and Y. Offer, "The Shrine and the Book—Impressions of the Visit at 'The Shrine of the Book' after Its Reopening," *Al Atar* 13/14 (2006): 201–220 (Heb.).

28. On this issue, see Y. Zerubavel, *Recovered Roots: Collective Memory and the Making of Israeli National Tradition.* Chicago: University of Chicago Press, 1995, 147–213; N. Ben-Yehuda, The Masada Myth: Collective Memory and Mythmaking in Israel. Madison: University of Wisconsin Press, 1995.
29. See Y. Harkabi, *The Bar Kokhba Syndrome: Risk and Realism in International Relations.* New York, NY: Rossel Books, 1982.
30. See Zerubavel, "Bar Kokhba's Image" (see note 8 above), 292.
31. See *Idem*, 294–296.
32. The festival of Hannuka is related to the Maccabean revolt against the Greeks in 167 B.C.E., and not to Bar Kohkba! By the way, since then I have asked many young people the same question, and in all the cases the answers showed a clear-cut ignorance about Bar Kokhba and his revolt. In this sense, it should be added that the Jewish tradition had accustomed to link the figure of Bar Kokhba with the festival of Lag Ba'omer. However, today this same celebration is principally associated with the figure of Rabbi Shimon Bar Yohai (II Century CE) and the pilgrimage to his tomb at Meron in the Upper Galilee. On this issue, Y. Schwartz, "Hillula of Rabbi Shimon Bar Yohai," in: R. Gonen (ed.), *To the Tombs of the Righteous: Pilgrimage in Contemporary Israel.* Jerusalem: The Israel Museum, 1999, 47–62.
33. After finishing the renewal project of the Bronfmann archaeological wing in the Israel Museum in 2010, the Cave of Letters materials were put on display in the framework of Second Temple Judaism and the Roman period gallery. This development confirmed, Zeruvabel's prognosis advanced in 2003: "...if the trajectory of decline continues, memory will submit to history and the Bar Kokhba revolt will recede from the more volatile domain of collective memory into the relative calm and obscurity of scholarly books, conferences, and museum displays" ("Bar Kokhba's image" [see note 8 above], 297).

Addendum 1

Dendroarchaeological Investigations at the Cave of Letters

NILI LIPHSCHITZ
Institute of Archaeology, The Botanical Laboratories,
Tel Aviv University, Tel Aviv-Yafo, Israel

INTRODUCTION

The Cave of Letters is located on the northern cliff of Naḥal Ḥever, ca. 100 m below a Roman siege camp and ca. 200 m above the wadi bed, and in close proximity to Nahal Arugot, 'En Gedi and the Dead Sea. According to Yigal Yadin,[1] the cave is about 150 m long, has two entrances—a northern and a southern, at a distance of 7 m apart, leading to three natural halls; A, B and C. Most archaeological remains were stored, presumably on purpose, by the refugees of the Bar Kokhba revolt. The finds were discovered beneath stones and in crevices at various locations in the cave.

During spring 1960 and 1961 Yadin completed excavations in the cave. Letters sent by Bar Kokhba (Simon Bar Kosiba) to the leaders of the revolt were found in Locus 7 (see Figure A1.1). It was these letters that gave the cave its name. Numerous wooden objects were found in a basket buried in the cave. This basket also contained 35 letters dating to 93 AD–132 AD, with this collection of artifacts named after the owner of the treasure—Babtha the daughter of Simon.

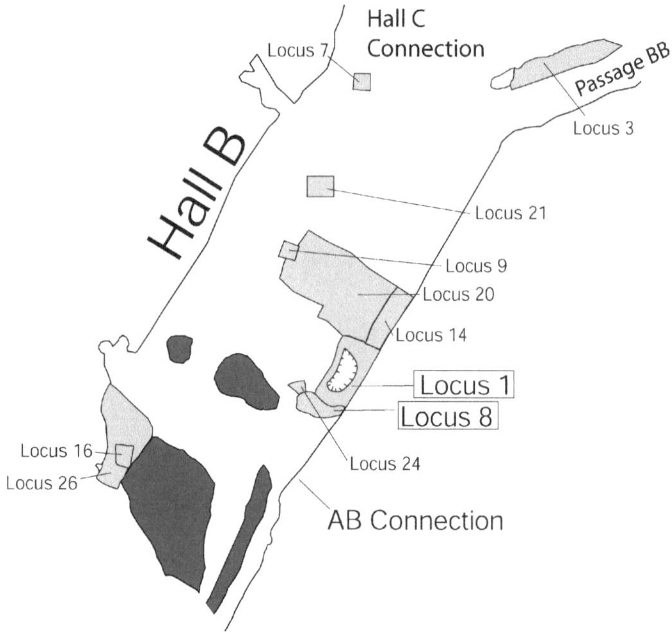

Figure A1.1: A section of the Cave of Letters Locus Map depicting Hall "B", with Locus 1 and Locus 8 (the locations where woods samples were collected) highlighted.

Some of the wooden objects dating to the Bar Kokhba period include:

A box made of *Tilia* (Lime), a cylindrical box and a cover made of *Arbutus* (Strawberry), two mirror handles made of *Tilia*, four vessels of *Acacia* (Acacia), remnants of four other vessels, two spoons, a comb made of *Buxus sempervirens* (Boxwood), a handle and three spindles. Many of the wooden finds were charred, and most probably were used as firewood for heating and light while in the cave.

MATERIALS AND METHODS

During two seasons of excavations in 1999–2000 numerous pieces of firewood from 0.5–2 cm in diameter were collected in the Cave of Letters. Many of the branches were not charred and some retained their bark. The branches were found in Hall B, distributed around an area that was interpreted to be a hearth. Altogether, 83 branches were collected in the cave. A single branch was found on the floor of what was interpreted to be the kitchen area in front of the hearth, which dates to the Bar Kokhba period. This single object was labeled as found in Locus 8, and all other 82 branches (pieces of firewood) were labeled as found in Locus 1 (see Figure A1.1).

Samples that were 0.5–1 cubic cm in size were taken from each branch for botanical identification. Cross, longitudinal, tangential and radial sections were prepared for each sample, stained with Safranin and immersed in Glycerol. Identification, up to the species level, based on the three dimensional structure of the wood, was made microscopically from these sections. Comparison was made with reference sections prepared from systematically identified, living trees and shrubs, and with comparison anatomical atlases.

RESULTS AND DISCUSSION

The branches collected in the cave were determined to be four woody species (Table 1): (1) *Ziziphus spina christi* (Christ thorn; Jujube), (2) *Retama raetam* (White broom), (3) *Acacia raddiana* (Acacia), and a member of the Chaenopodiaceae family—most probably *Anabasis sp.* The majority of the branches were from two species: 34 (41%) were of *Ziziphus spina christi*, and 45 (54%) were of *Retama raetam*. One branch (1.2%) was of *Acacia raddiana* and three branches (3.6%) were made of the Chaenopodiaceae shrub—most probably *Anabasis sp. Ziziphus spina christi* trees inhabit the wadi beds and oases in the Negev desert, the Jordan Valley and Dead Sea area.[2] *Retama raetam* is one of the most common desert shrubs. It grows mainly in the desert on sand, rocky and gravelly ground in the East, Central and South Negev, in the Judean Desert and in the Arava Valley. *Acacia raddiana* trees are native to hot desert wadis, depressions and oases in the West, North and Central Negev, Lower Jordan Valley, Dead Sea area and Arava Valley. Both species of *Anabasis*: *Anabasis articulata* and *Anabasis setifera* are shrubs native to stony, gravelly and sandy deserts in the Judean Desert and the Negev.[3]

All the branches found in the Cave of Letters during the 1999–2000 seasons were therefore brought into the cave from the local area. Only radiocarbon dating of each of the branches will enable a determination of their age. Two branches from the 1999–2000 assemblage were sent to the Radiocarbon Laboratory at the Weizmann Institute, in Israel, for dating. Branch No. 62—*Ziziphus spina christi* was dated to 40 BC–60 CE (calibrated age) and branch No. 64—*Retama raetam* was dated to 45 BC–55 CE (calibrated age). The period of Bar Kokhba was ca. 135 CE. Three other samples collected during these two seasons, which were not identified botanically, were sent to the USA for radiocarbon dating. These three samples yielded in 2-sigma calibrated age the following dates: 45–430 CE (sample A), 95–405 CE (sample B) and 260–610 CE (sample C). Branch 62 and branch 64 are therefore older by ca. 80–150 years than the period of Bar Kokhba, and the other three branches represent later periods than the time of Bar Kokhba.

Editors' Note

The five dates obtained could be interpreted a number of ways: (1) the samples represent wood being brought into the cave prior to and after the Bar Kokhba period, or (2) the first two samples represent wood that was previously stored, or preserved by the naturally dry conditions associated with the natural landscape, but this seems unlikely. Hence, the first condition seems more plausible.

Figure A1.2: Table of woody species found in the Cave of Letters during 1999–2000 seasons.

Species	Locus Number	Number of Branches
Retama raetam	1	44
Retama raetam	8	1
Ziziphus spina christi	1	34
Acacia raddiana	1	1
Chaenopodiaceae Anabasis sp.?	1	3

NOTES

1. Yigael Yadin, *The Finds from the Bar Kokhba Period in the Caves of Letters.* (Jerusalem: Israel Exploration Society, 1975) 126–140 (Hebrew).
2. Zohary, M. *Flora Palaestina Vol. II.* (Jerusalem: The Israel Academy of Sciences and Humanities, Jerusalem, 1972).
3. Zohary, M. *Flora Palaestina Vol. I.* (Jerusalem: The Israel Academy of Sciences and Humanities, 1966).

Addendum 2

Provenance Determination of Some Selected Pottery Shards from the Cave of Letters by Instrumental Neutron Activation Analysis

JAN GUNNEWEG
Hebrew University, Jerusalem, Israel

MARTA BALLA
Budapest University of Technology and Economics, Budapest, Hungary

INTRODUCTION

In mid-March 2002, Richard Freund approached us to solve one of the most interesting questions he and his research team had encountered in connection with the Cave of Letters. He asked: Is the content of the cave that of 130–135 AD on the basis of the letters of Bar Koziba found in the cave, or was the cave also occupied earlier by pre-70 AD sectarians who had fled Jericho, Jerusalem or Qumran?

Our answer was that the neutron activation technique we employ was not designed to provide a date to artifacts. However, we let him know that if we would be able to trace the cave's pottery to a manufacturing site located either at Jerusalem or in Qumran; we would indirectly obtain a date since both sites were entirely destroyed by the Roman general Titus in 70 AD. Pottery in the Cave of Letters being traced to either one or both of these places would provide a date that antedates their destruction after which no local Judaean pottery could have been manufactured. As a result, some of the pottery in the Cave of Letters might be from pre-70 AD. At the end of March 2002, a pottery provenance study was started that would serve as a pilot study for further research, if needed.

SAMPLING

Pottery was sampled from Loci 1 (Hall B), Locus 2 (Hall C), Locus 6 (BC Passage), Locus 8 (kiln lining), Locus 16 (AB Passage), Locus 29 (Hall A) and Locus 34 (pottery found at Entrance No. 1 of the Cave). See base map of the Cave of Letters for location of the pottery sampling locations (see Plate 3).

J. Clark initially considered the fifteen samples that were to be chosen as being representative of the pottery that has been found in the Cave of Letters, and from these samples we proceeded. See Table 1, for the associated laboratory number and associated short description of each pottery sample tested.

TECHNIQUE

Instrumental Neutron activation analysis (henceforth INAA) is a nuclear technique that during past 35 years has successfully been employed to establish an accurate chemical fingerprint of a ceramic. A piece of pottery is removed from a vessel, ground and its powder placed into a vial that is irradiated in a nuclear reactor. We used the nuclear facility on the premises of the Technical University at Budapest, Hungary. Irradiation induces nuclear reactions, producing radioisotopes of the elements in the sample. The decay of these radioactive nuclides is accompanied by the emission of gamma-quanta of characteristic energies, and the radiation can be used both to identify and accurately quantify the elements of the sample. Measurements result in 20–30 concentration values per sample that can be used as a "chemical fingerprint" of an artifact. Typically, each clay source on Earth has its own specific fingerprint and thus ceramic can be traced to the location where it was manufactured.

By combing through various INAA databanks from Jerusalem, California-Berkeley, and Bonn, Germany, which one of us (Gunneweg) was familiar with, it was thought possible to find a statistical match between chemical compositions of pottery found in the Cave of Letters and those of other pottery, which would indicate the place of origin for the pottery. This would establish possible trade relationships, and in the case of the Cave of Letters, any relationship between its inhabitants and nearby or more remote locations.

No potter's kiln has been found in the cave that would indicate a local production of the pottery found within the cave. It was, therefore, assumed that the cave's pottery came from elsewhere. In that case, it would be sufficient to trace the chemical compositions of the pottery to those of pottery that was manufactured elsewhere.

RESULTS

In our data bases, are stored INAA data for pottery that has been manufactured in such places as Masada, Jericho, Jerusalem, Tell en-Nasbeh, Qumran and Arad, as well as sites near Petra and Tawilan in Jordan. If the newly obtained INAA data from the Cave of Letters can be traced to one of these sites, then the location of where the pottery in the cave was imported from will be proven.

THE RED WARES

The Red Wares (pottery with a red/brown fabric) were all chemically similar to one another with minor variations. These minor variations point to minute differences in the potter's practices.

COL 1 analyzed as two balsam-like juglets (QUM 295–6) found in Tell ez-Zara on the eastern shore of the Dead Sea. The latter were established by INAA as being juglets made in the immediate Qumran region.[1] COL 2 analyzed as a two-handled ovoid jar (QUM133) found in Qumran that also was locally made at Qumran as INAA has shown.[2]

COL 8 also analyzed the same as COL 1 and 2, as did COL 12, 13 and 14, which indicates that all the red wares match local Qumran ceramics. There are minor differences in chemical abundances of elements in various artifacts, but the general chemical pattern is similar to Qumran. This is, however, not the end of the story, because new evidence has shown that clay and marl was brought into Qumran by a torrential flood that originated in the Judaean hilly country. The flood deposited a clay layer that has the same chemical composition as a rain puddle sample, and reference material from Qumran, as found in the INAA data base. Brill Publishing has published the latter in the Proceedings of the workshop entitled "Holistic Qumran" which was held in the Physics Department of the Lorentz Center at the University of Leiden.[3] It is very likely that similar torrential rain-deposited clay layers have been deposited elsewhere along the Dead Sea's western shores. Furthermore, a stylistic comparison between the Cave of Letters and the aforementioned Qumran-made artifacts must be taken into consideration before the chemical outcome attested by the INAA results can be finalized.

WHITE "LIMESTONE-LIKE" WARES

In previous studies centered on Qumran and Roman En Hazeva in the Arava, White Wares, or "limestone-like" wares as mentioned here, have been analyzed by

INAA. We have run the results against the data from various data bases and we have not been able to trace them to a specific site.

COL 4,5,6,7 and 11 are five chemical compositions of ceramic White Ware vessels for which we did not find any match with earlier analyzed pottery in our data banks. There are minute differences between these chemical compositions, which may suggest that the "limestone-like" ware has a provenance at different places, but it is not yet known where these vessels are from.

OVEN LININGS

COL 9 and 10, taken from the inner lining material of an oven, are made of the same material that has been found in the vicinity of the Dead Sea which is known as Dead Sea Mud. This clayish substance can be found in lenses deposited everywhere around the Dead Sea area. Dead Sea mud (QUM226) was previously analyzed for the Qumran project.[4]

COOKING WARE

Cooking ware has always been difficult to trace to its manufacture center. A cooking pot demands certain requirements that probably have to do with the heat-shock treatment. Such a pot has to withstand great differences in externally applied temperatures when placed over or into a fire. Special ingredients, such as additional quartz, coupled with firing the pot at higher elevated temperatures than those used in making other pottery makes it equipped to come in direct contact with fire in a hearth or bonfire without breaking.

COL 3 has a chemical composition that was encountered in Michniewicz and Krzysko's contribution.[5] Michniewicz analyzed a dish of clay that was found to have originated from clay sampled from the Upper Aqueduct at the cliff above Qumran [Qum98]. The water arrives through Wadi Qumran and thus this clay deposit could well have been used everywhere along Wadi Qumran from Hebron to the Dead Sea. Nonetheless, we have not found a statistical match with ceramic analyzed by INAA in our own data banks.

CHALCOLITHIC

Finally, sample COL 15, a Chalcolithic shard that points to an early cave habitation of 4500 BC, has a chemical composition that is similar to that of Motsa clay. This clay was probably found in a deposit in a wadi that ran off from the Motsa Clay Formation. However, we have to emphasize that there is no direct statistical match with Motsa clay like the one found in Jerusalem; the Chalcolithic chemical fingerprint is just Motsa-alike.

CONCLUSION

The Red Wares from the Cave of Letters are similar to those analyzed at Qumran, which, in turn, are local to Qumran clay, deposited by torrential floods in that area. In the case that the Cave of Letters' pottery was made in Qumran, it could well be from before 70 AD. The caveat, however, is whether this pottery is also found in Qumran itself. If not, the pottery could have been made in a site that used similar clay for its ceramics, coming from the same inundation as that of Qumran. In the latter, however, one is not allowed to give the Cave of Letters pottery the same date as that found in Qumran, because it could have been made from other clay/marl deposits along the Dead Sea.

Joan Taylor mentioned at the Qumran Conference at Brown University, that Period III at Qumran lasted about 25 years instead of the five years mentioned by R. de Vaux. The main problem, she sees is the lack of typological pottery studies from 70 to 132 AD. She added that the occupants of Qumran Period III did not make pottery because the kiln at Qumran was turned into a chalk-storage place. Taylor did not provide any scientific proof, though.

On the other hand, the analyzed samples from the Cave of Letters chemically have nothing in common with pottery found at Masada, Jericho and Jerusalem that has been analyzed by INAA and that is stored in the data bases.

We, therefore, are inclined to conclude that more research and analyses have to be performed on local clays from the environment of the cave itself, as well as of pottery unearthed in Engedi and/or Masada before one can claim with some certainty that the Cave of Letters' ceramics are from before 70 AD.

Figure A2.1: Table of pottery samples numbers with brief description of the type of pottery. (The abbreviation COL stands for the laboratory code for the Cave of Letters.)

Laboratory Sample #	Pottery Description
COL 1	Red ware body shard
COL 2	Red ware body shard
COL 3	Blackened cooking pot
COL 4	Limestone imitation ceramic
COL 5	Limestone imitation ceramic
COL 6	Limestone imitation ceramic
COL 7	Limestone imitation ceramic

Laboratory Sample #	Pottery Description
COL 8	Red Ware
COL 9	Oven lining
COL 10	Oven lining
COL 11	Storage jar with incised lines
COL 12	Storage jar with incised lines
COL 13	Red ware body shard
COL 14	Red ware body shard
COL 15	Chalcolithic shard

NOTES

1. Gunneweg, J., and Balla, M. "Neutron Activation Analysis: Scroll jars and Common ware," in *Qumran and 'Ain Feshkha, II,* Humbert, J.-B. and Gunneweg, J. (eds). 3–57. (Goettingen: Academic Press Fribourg, Vandenhoeck & Ruprecht, 2003); see page 20 & Table 7.
2. Gunneweg, J., and Balla, M. "Neutron Activation Analysis: Scroll jars and Common ware," see page 11.
3. Gunneweg, J., and Balla, M. "Was the Qumran Settlement a mere Pottery Production Center? What Instrumental Neutron Activation revealed. in Holistic Qumran," in *Studies on the texts of the Dead Sea scrolls of Judah*, Gunneweg, Adriaens and Dik (eds). 39–66. (Leiden/Boston: Brill, 2010); see especially 46–47 & 54.
4. Gunneweg, J., and Balla, M. "Neutron Activation Analysis: Scroll jars and Common ware," see page 20 & Table 7.
5. Jacek Michniewicz and Miroslaw Krzysko. "The Provenance of Scroll Jars in the Light of Archaeometric Investigations," in *Qumran and 'Ain Feshkha, II: Studies of Anthropology, Physics and Chemistry*, Humbert, J.-B. and Gunneweg, J. (eds). 59–99. (Goettingen: Academic Press Fribourg, Vandenhoeck & Ruprecht, 2003), see page 62.

Addendum 3

Rock Breaking at the Cave of Letters

SHACHAR ARGOV
Sato and Associates, Honolulu, HI 96826, USA

PROJECT BACKGROUND

The Cave of Letters is considered to be one of the most important archeological sites discovered in the Middle East.[1] Found in the cave were treasure troves of artifacts—fragments from early copies of the Bible, incense shovels, bronze ritual objects, rare coins, textiles, household items, human skeletal remains, and written documents.[2]

In the time that passed between the time of inhabitance (2nd century AD) and the first discoveries in the 1960s, earthquakes[3] caused ceiling rock-falls to bury any remaining artifacts hidden by the cave's dwellers. The major obstacles encountered working in the cave was accessibility, and excavating without disturbing the sensitive cave environments.

After research in the cave in 1999 and 2000, the use of rock splitting equipment was added to the research design for the 2001 expedition. Archaeological probes, and various types of geophysical analysis, located anomalies within the roof fall that litters the cave's floor. During the 1999 and 2000 expeditions to the cave, difficulties were encountered with respect to removing some of the larger boulders that covered the floor, and thus overlaid artifacts buried beneath. A fiberscope, which uses a small camera mounted on a fiber-optic cable, was guided between the rocks in areas where geophysical analysis and archaeological probes indicated anomalies. In the vicinity of these anomalies, rock splitting equipment was used to break the boulders into manageable size pieces so that the Bar Kokhba-era floor beneath the boulders could be accessed.

EXPERIMENTATION

Dr. Richard Freund had requested Geotechnical Services Inc. (GSI) of Lincoln, Nebraska[4] to conduct experimentation with a goal of determining excavation methods, appropriate for the Cave of Letters. On February 27, 2001, GSI conducted a series of rock fracturing tests designed to determine the viability of lightweight, portable equipment and compounds for breaking boulders at Cave of Letters. The rock fracturing evaluation consisted of testing two hydro-expansive compounds and a hydraulic splitter. Rock drilling equipment and various drilling bits were also evaluated as part of the study.

The Rock Drill

GSI evaluated two electric Rock Drills, the Hilti TE 76 ATC and the Makita HR3851. Drilling time, ease of operation, and drill bit quality were evaluated. Both drills were roughly the same size and weight. Only carbide-tipped drill bits were utilized. Diamond coring bits were not evaluated for this study, as advised by a Hilti representative. Diamond percussion drills are not available at the hole-diameters that were tested.

Overall, the Hilti drill performed faster and showed to be more powerful than the Makita. An 18″ × 1–1/2″ hole required only 5 minutes with the Hilti drill, as compared to roughly 20 minutes when using the Makita drill. New versus used carbide bits also made a substantial difference in drill performance.

Darda Hydraulic Splitter

The Darda hydraulic splitter is a hand held device that is attached via hoses to a 10,000 P.S.I. hydraulic pump. The pump can be electric, gas engine, or air pneumatic powered. This tool contains a control valve and a piston that moves a wedge-like plug between two feathers. The plug and feather end is placed into a pre-drilled hole. The plug moves down between the two feathers forcing them against the wall of the hole. When the hydraulic expansive force exceeds the tensile strength of the rock material, a split in the rock occurs.

A #12 Darda hydraulic splitter with an air powered hydraulic pump was utilized for the February 27th experiment. The splitter was used to crack two 5-ton boulders. This splitter model requires the largest pilot bore hole of all the models available from Darda. Drilling the hole for the splitter took about 25 minutes, but the fracturing process took less than 30 seconds. After the splitter was retracted, about 3 to 5 cracks were visible around the boulders that were later manually prayed apart, with a crowbar and chisel. No further attempts were made to continue the fracturing process of the two boulders.

On March 2nd 2001, a #3 hydraulic splitter powered by an electric pump was evaluated (the #3 unit is equivalent to the #4 unit used later in the cave). This splitter, was used to fracture boulders that were not broken during the initial Betonamit splitting experiments. The smaller #3 splitter was more manageable than the #12 splitter due to its size and weight. Aligning the wedge with the desired crack direction can control the direction of the crack easily. Drilling the smaller holes for the smaller splitter took less than 5 minutes, which greatly expedited the process.

Betonamit

Betonamit is a non-explosive cracking agent that comes in powder form. The power is mixed with water and poured into predrilled holes. As the agent hardens, it expands exerting great pressure on the rock (up to 12,000 P.S.I). A type "R" Betonamit agent was tested for use in the Cave of Letters.

The Betonamit's chemical reaction requires that the ambient temperature is not below freezing. Therefore, a boulder that was to be tested was brought indoors four days before the experiment because of the cold temperatures in Lincoln, Nebraska at that time. A front-end loader was used for this task. A hole was drilled into the rock, and the rock's internal temperature was monitored by a thermometer, and controlled by increasing or lowering the room temperature (Figure A3.1).

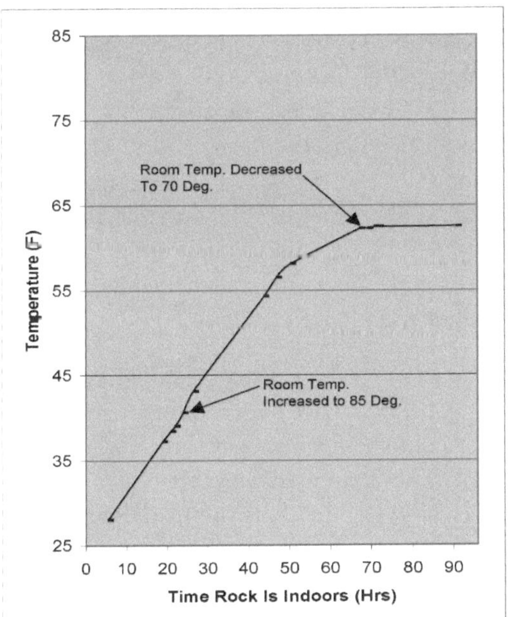

Figure A3.1: Graph depicting temperature changes associated with the rock that was moved indoors during the rock splitting experiment that was conducted in Lincoln, Nebraska.

On February 26th, a series of holes were drilled into the rock in a straight line on 35 centimeter (cm) centers. A 45 cm drill bit for this task. A single hole was also drilled on the opposite side of the boulder to evaluate the effectiveness of the Betonamit at higher loads. At 7:00 PM the holes were all filled with 5 Kilograms (kg) of the Betonamit slurry, and left to work overnight. The boulder was next examined on February 27th at 7:00 AM.

The results from the Betonamit experiment were better than expected. The entire boulder, including the area around the isolated hole showed extensive fracturing. Several additional fractures formed throughout the morning. The process appeared to have stopped around 11:00 AM due to the complete disintegration of the rock structure around the holes. After some manual work with a crowbar and chisel on the existing cracks, about 50% of the rock was broken into manageable size pieces (40 kg or lighter). It took two workers about one hour (not including hauling) to break the rest of the boulder into manageable pieces, using a Hilti drill and the #3 Darda Hydraulic Splitter. This time frame may have been significantly reduced if two rock drills were available for the task.

Experiment Recommendations

The tests showed that the drilling of the hole is perhaps the most important phase in the rock splitting process. The crew's work speed may depend primarily on the drilling output, especially if a Hydraulic Splitter is to be used. Moreover, the diameter of the drilled hole greatly affects the drilling speed. For example, it takes about 12 minutes less to drill a 3.8 cm hole, compared to a 4.5 cm hole.

GSI recommended using both Betonamit and a #2 or #3 Darda hydraulic splitter (or equivalent models) for the project. The clear advantage of the splitter over the Betonamit is speed (a few minutes vs. approximately 12 hours for Betonamit) and the control over the split alignment. The splitter's main disadvantages are its weight (17.7 or 21.8 kg depending on the model) and the potential for mechanical failures. Further recommendations include:

- Drill holes should not be bigger than required. A tight fit for the Hydraulic Splitter hole will also achieve better results.
- Spare drill bits and various size spare bits will be important. New carbide bits drill much faster than used worn bits.
- The Hilti TE 76 ATC drill was the recommended brand. Two or even three drills should be used simultaneously when possible to expedite the operation.
- A #2 or #3 splitter should be used with an electric powered hydraulic motor. The #9 splitter tested is too bulky for use in the cave.
- Spare parts for the splitter are essential, in particular, piston and feathers.

- Betonamit can be used in conjunction with the splitter, preferably at different locations in the cave or at different times, such as overnight.
- A generator, working in a similar fashion to the arrangement used in the 2000 season can power the drills and splitter. In 2000 the generator was placed on the flat plateau above the cave's entrances, and extension cords were run down the cliff face and into cave entrance #1. A 3% voltage drop is expected from the generator to the cave.

THE 2001 EXPEDITION TO THE CAVE OF LETTERS

The 2001 Expedition gathered at Metzokei Dragot in mid-July. The following rock breaking equipment was taken to the cave:

- Four 5 kg containers of Betonamit
- One Hilti TE 76 ATC Rock Drill with several drill bits
- Darda electric powerpack compressor-model EP, Portable (40 kg).
- #C 2 S Darda Hydraulic Splitter (17 kg).
- #C-4 Darda Hydraulic Splitter (20 kg). This model is equivalent to the #3 model used for the experiment in Lincoln
- Hydraulic hoses
- Electric power adaptors
- Miscellaneous tools, hydraulic fluids, and grease

Power Supply

The power supply for the project was a 12 KVA Diesel Generator (400 V @ 50 Hz). The cord length was around 200 M (five 6 mm Strands). In addition to the rock equipment, the generator supplied power to lighting, TV equipment, and other scientific instrumentation. The specifications for the electric motor for Hydraulic Splitter-model EP were 1.5 Horsepower, Single Phase, 60 Hz, 3450 RPM, and 220 Volts @ 10 Amps. For the Hilti drill (TE 76 ATC 230 (HLAL)) the specifications were 230 V, 1300 W, and 50–60 Hz.

The staging area for the expedition was located on the plateau above the cave near the ruins of the Roman encampment. From there, the expedition ascended every morning to the cave and returned before nightfall. The commute varied between 20 to 40 minutes (one way).

Day One, Sunday, July 22nd

Due to time constraints, the equipment was not available for testing before the trip to the site. Final equipment testing was performed in the morning before ascending to the cave on the first day. Hauling all of the equipment into the cave required four individual trips.

Mobilization and setup required most of the first day. Around 4:00 PM the equipment was set to start on the first task, which was a large boulder in the Hall B-C Passage. After two breaks using the #2 splitter, it was decided to attempt increasing productivity by operating the Hilty drill together with the Darda splitter. This attempt overloaded the generator, and tripped the breaker, shutting down the generator. It also tripped the breaker in the powerpack that distributed power to the various pieces of equipment in the cave.

Day Two, Monday, July 23rd

The powerpack was repaired during the morning of day two. Most of the remaining time was dedicated to the removal of the B-C Passage boulder that was started the day before. This massive 12-ton boulder had good potential for finds after the expedition identified fabric remains on both sides of the rock a year earlier. However, no significant finds were discovered. Before the end of day two, the equipment was relocated to the south end of Hall C. At this location, another boulder was identified for removal. Instead of using the splitter, a series of holes were drilled in the rock, which were then filled with Betonamit and left to work overnight.

Day Three, Tuesday, July 24th

The attempt to split the boulder at the south end of Hall C with Betonamit produced marginal results. The slurry had fractured most of the rock, but the use of the splitter was required to finish the task. The next task was to attempt to split a group of boulders along the west cave wall in Hall C. At the end of day three, some of the equipment was relocated to Hall A for the next day's work.

Day Four, Wednesday, July 25th

The morning of the last day was spent in Hall A, in the area located to the east of the passageway from Hall A to Hall B (the A-B Passage). The splitting and subsequent removal of that rock led to the discovery of a 2nd rebellion era coin. The coin was discovered using a metal detector. The remainder of day 4 was used to clean and pack the equipment.

CONCLUSION AND RECOMMENDATIONS FOR FUTURE WORK

The 2001 expedition lacked the time and resources to make a significant impact on large boulder removal in the cave. The rock-splitting team was composed of only two individuals, and we were limited to four work days in the cave. However, this first attempt proved promising, it indicated that it was indeed feasible to break up the rocks in the cave in terms of both logistics and the technology. A follow-up expedition with the goal of a comprehensive excavation in the cave is realistic in light of the success we had in breaking up some of the roof fall that covers the cave floor. This of course would help to address the question, what else may be hidden below the massive rock-fall in the Cave of Letters?

In terms of practical recommendations, with regard to cracking agents, the use of a non-explosive cracking agent, such as Betonamit, is not recommended for future use. The experiment in the cave showed that the method, although effective is too slow for a large-scale excavation. Some Betonamit may be brought along as a supplemental tool, but this method should not be relied on as the primary means of rock splitting.

With regard to hydraulic splitters, the spare #4 Darda Hydraulic Splitter was not used during the four-day expedition. Work started on day one with the #2 splitter, which proved to be more than adequate. The bigger and slower #4 splitter was therefore not utilized at all, and should not be included in the equipment list for any future work of this type. The #2 splitter performed better than expected in the cave environment. Not including drill time, it took only a few seconds to split the boulder. With very little practice, it was possible to direct the split and to break the rock in a desired manner. For natural rock excavations, especially within confined spaces, a small splitter is considerably easier to operate than a large one. Drill time for the smaller splitter is much shorter as well. The biggest disadvantage of the splitter is the potential for mechanical failure, as demonstrated in the cave. It is therefore highly recommended to carry spare parts.

Applicability for Future Projects

It is recommended that a future excavation team should include the following individuals with regard to the rock splitting component of the research:

1. Two to three drillers with Hilti TE 76 ATC (or equivalent)
2. One operator with a #2 Darda Hydraulic Rock Splitter. An additional rock splitter is not needed (one splitter will be able to keep up with three or more drillers)
3. Two more individuals with crowbars for manual removal of rubble

The electric power requirement should be designed to support all of the above tools, as well as lighting and other equipment. For a setup similar to the one used during the 2001 excavation, a 12 KVA Generator should suffice, however, with this setup the powerpack may not be operated simultaneously with the three drills unless the generator can produce 20 KVA or better. An ideal setup would include two generators and lighting and other essentials would be powered on a separate generator from the rock-breaking equipment. The power requirements should receive special attention in any future research at the Cave of Letters.

Additionally, a spare powerpack, drill, and leaf/wedge replacement parts are highly recommended. A spare #2 splitter be brought along as a replacement, but unlike the powerpack, the splitter is tough-built and is not as likely to break. The crew should also carry all the necessary tools, spare hoses, oil products, and additional extension cords.

For this type of work, two individuals can more than double the output of one individual. To minimize the potential for injuries, work should be performed in groups of two to three individuals, and all teams should work in close proximity to each other for support. The rock excavation team should be briefed and ready to work in a very challenging environment. The work to be performed is difficult, and is recommended for individuals with prior experience in such labor-intensive undertakings.

The technology used during the 2001 expedition is very suitable for projects with similar conditions and research designs. The rock splitting method works best on softer rocks, such as limestone, and it is not is not recommended for harder rocks such as basalt and granite. For projects where vibrations and/or noise are not an issue, Jackhammer tools should be considered. A large selection of electric and air jackhammers is widely available in the construction industry.

In conclusion, this pioneering work would not have been possible without the assistance of many individuals and private companies. I would especially like to acknowledge and that the following companies for their assistance:

Sato and Associates Inc., 2046 South King Street, Honolulu, HI 96826 (my current employer); Geotechnical Services Inc., 5701 North 58th Street, Lincoln, NE 68507–3256 (my employer at the time the research was completed); Darda GmbH Im Tal 1, D-78176 Blumberg, Germany (the company that made the hydraulic splitter); Elco International Inc., 111 Van Riper Avenue, Elmwood Park, NJ 07407 (the company that provided the hydraulic splitter for use in the cave); Peter Kiewit and Sons', Inc. 1000 Kiewit Plaza Omaha, NE 68131 (the company that provided the splitter during the evaluation phase in Nebraska); Martin Merietta Rock Quarry P.O. Box 448 Weeping Water, NE 68464 (the company that donated the limestone boulders for the rock splitting experiments in Nebraska); and Hilti 10260 Huntington Ave. Omaha, NE 68122 (the company that provided the drill used in this project).

NOTES

1. Yigael Yadin, *The Finds from the Bar Kokhba Period in the Cave of Letters.* (Judean Desert Studies; Jerusalem: Israel Exploration Society, 1963); Philip Reeder, et al., "Multidisciplinary Research at the Cave of Letters, Israel: A Melding of Physical and Social Sciences," in *Transkart 2004 Conference Proceedings* (Hanoi, Vietnam, 2004).
2. Yadin, *The Finds from the Bar Kokhba Period in the Cave of Letters.*
3. A Frumkin, "The Cave of Letters Sediments—Indication of an Early Phase of the Dead Sea Depression?" *The Journal of Geology* 109 (2001): 79–90.
4. Geotechnical Services Inc., 5701 North 58th Street, Lincoln, Nebraska 68507–3256.

Contributors

Shachar Argov is a civil engineer with over 25 years of experience in Engineering and Construction fields. For the past 10 years he has been a project manager (Frank Coluccio Construction), specializing in underground construction with a focus on tunneling methods projects.

Marta Balla, PhD is a geologist, archaeologist, and research associate at the Institute of Nuclear Techniques, Budapest University of Technology and Economics. Working in the Radiochemistry Laboratory, her main activity is the analysis of geological, biological and archaeological materials by the means of neutron activation analysis. Her research interest focuses on the provenance study of archaeological ceramics by chemical fingerprinting. Most of the projects concentrate on the investigation of pottery finds from Roman Pannonia, and from different archaeological sites in Israel.

Paul D. Bauman is the founder (1991) and President of Paul D. Bauman Geophysics, and the founder and Technical Director (since 1990) of the Geophysics group at Advisian WorleyParsons in Calgary, Alberta, Canada. Paul has a B.Sc.E. in Geological Engineering from Princeton University, and an M.Sc. in Earth Sciences from the University of Waterloo.

Walter C. Bouzard earned his MDiv and MTh degrees from Luther Theological Seminary in St. Paul, MN, and a PhD from Princeton Theological Seminary, Princeton, NJ. Besides numerous adult Bible studies, he is the author of "A Lutheran

Educator Reflects on Religion and Science." In *Together by Grace: Introducing the Lutherans*. Edited by Kathryn Kleinhans. Minneapolis: Augsburg Fortress, 2016.

Richard A. Freund is the Maurice Greenberg Professor of Jewish History and Director of the Maurice Greenberg Center for Judaic Studies at the University of Hartford. Most recent book: *Digging through History* (Rowman & Littlefield, 2012). PhD and MA, Jewish Theological Seminary in New York.

Amos Frumkin earned his PhD from The Hebrew University of Jerusalem, Jerusalem. He now serves as a professor at the Institute of Earth Sciences. His interests include karst and cave morphology, cave sediments as indicators of paleoclimate, paleohydrology, the development of karst aquifers, and geoarchaeology. The research is mostly associated with speleological features studied using earth-sciences methods, such as geomorphology, modern dating, and stable isotopes. He has authored or co-authored over 100-refereed articles and seven books. As an active speleologist, he founded and directs the Cave Research Center of the Hebrew University in Jerusalem.

Jan Gunneweg is Senior Lecturer Emeritus in Archaeology at The Hebrew University of Jerusalem, as well as the Head of Qumran Science Project Trans-Disciplinary Research.

Harry M. Jol earned his PhD from the University of Calgary (Alberta, Canada) and is presently a Professor at the University of Wisconsin-Eau Claire. He has published over 100 articles, is a Fellow of the Geological Society of America, and has worked at numerous archaeological sites in Israel.

Nili Liphschitz earned a PhD from Tel Aviv University. She is a fellow of the International Academy of Wood Science. She has published numerous books and articles.

Philip Reeder is currently the Dean of the Bayer School of Natural and Environmental Sciences at Duquesne University (DU) in Pittsburgh, Pennsylvania. Prior to that he was the Chair of the Environmental Science and Policy Division at the University of South Florida (USF) from 2006 to 2013. He received a doctoral degree in Geography with specializations in soils, hydrology and water resource management from the University of Wisconsin-Milwaukee in 1992. He has published scores of articles and delivered hundreds of invited lectures. He has conducted water resource management and/or geoarchaeologic research in the Central Kentucky Karst, west-central Florida, northeastern Wisconsin, the Driftless Area of Southwestern Wisconsin, the Huautla Plateau in Mexico, the island of Bohol in the Philippines, the Vaca Plateau and Belize River Valley in Belize, various locations in Israel, as well as in Rhodes Greece, Sobibor Poland, and Vilnius Lithuania.

Nicolae Roddy, PhD (University of Iowa), Professor of Hebrew Bible at Creighton University, directs and supervises the Bethsaida Archaeology Project and is a faculty associate for the Goren-Goldstein Center for Hebraic Studies, University of Bucharest. He has published numerous publications in archaeology and biblical studies.

Adolfo D. Roitman holds an MA in Anthropology from the University of Buenos Aires and in Comparative Religions from the Hebrew University of Jerusalem, as well as a PhD from The Hebrew University of Jerusalem. He holds a Doctorate Honoris Causa from the Rocky Mountain College, MN and a Gold Medal of Humanities from Anáhuac University, Mexico City. He is currently serving as the Lizbeth and George Krupp Curator of the Dead Sea Scrolls and Head of the Shrine of the Book at the Israel Museum, Jerusalem. He has published widely.

Carl E. Savage is Associate Professor of Biblical Archaeology in the Theological School of Drew University, as well as Assistant Director of Excavations at the Bethsaida Excavations Project. He received his A.B. (1975) in Sociology and Religion from Princeton University, his M.Div. (1978) and D.Min (1983) in History of Biblical Interpretation from the Wesley Theological School in Washington, D.C., and his M.Phil (2000) and Ph.D. (2007) in Origins of Christianity and Judaism from Drew University. His numerous publications include *Biblical Bethsaida: An Archaeological Study of the First Century CE* Lexington Books (2011) and "Political Propaganda in Gold" (2013), among others.

Orit Shamir is Curator of Organic Materials, Israel Antiquities Authority.

John (Jack) F. Shroder earned his PhD in 1967. He pursues research in the Middle East and South Asia for 45 years. He has written and edited over 46 books and over 150 professional papers. He retired from UNO in 2012 and now works for the Rocky Mountain Biological Laboratory in Colorado.

Fred Strickert is Professor Emeritus of Religion at Wartburg College. He received a Ph.D. from the University of Iowa. Select publications include *Bethsaida: Home of the Apostles, Philip's City*, and *The First Woman Portrayed on a Jewish Coin.*

Index

A

B

C

D

H

I

J

K

L

M

R

S

T

V

W

Y

Z

Crosscurrents

NEW STUDIES ON THE MIDDLE EAST

R. Kevin Lacey and Sari Nusseibeh
General Editors

This series will publish book-length manuscripts pertaining to the peoples of the Middle East. The Middle East is understood in the broadest sense associated with the term, and is reflective of widely shared socio-religious patterns, histories, and heritages.

For the purpose of this series, the Middle East will include what is more commonly referred to as the Near East (Lebanon, Syria, Jordan, Israel/Palestine); North Africa (Egypt, Libya, Tunisia, Algeria, Morocco, Mauritania, Mali, Chad, the Sudans, and Somalia); Turkey and Iran; Saudi Arabia, Yemen, and the countries of the Arab Gulf; and, finally, Afghanistan, Pakistan, and the Central Asian Republics. The series will be interdisciplinary and inclusive of diverse topics and methodologies. Representative fields will include art, art history, architecture, language and literature, history, politics, economics, and religion. Reinterpretations, as well as investigations of the hitherto uninvestigated, will be especially welcomed.

For additional information about this series or for the submission of manuscripts, please contact:

Peter Lang Publishing
Acquisitions Department
29 Broadway, 18th floor
New York, NY 10006

To order other books in this series, please contact our Customer Service Department:

800-770-LANG (within the U.S.)
212-647-7706 (outside the U.S.)
212-647-7707 FAX

Or browse online by series at:

www.peterlang.com